W0051414

Encyclopaedia of
Mathematical Sciences

Volume 27

Editor-in-Chief: R.V. Gamkrelidze

V.G. Maz'ya S.M. Nikol'skiĭ (Eds.)

Analysis IV

Linear and Boundary
Integral Equations

Springer-Verlag Berlin Heidelberg GmbH

Scientific Editors of the Series:
A.A. Agrachev, E.F. Mishchenko, N.M. Ostianu, L.S. Pontryagin
Editors: V.P. Sakharova, Z.A. Izmailova

Title of the Russian edition:
Itogi nauki i tekhniki, Sovremennye problemy matematiki,
Fundamental'nye napravleniya, Vol. 27, Analiz 4
Publisher VINITI, Moscow 1988

Mathematics Subject Classification (1980):
45-02, 45B05, 45C05, 45Exx, 45Fxx, 45P05, 47Axx, 47Bxx,
47G05, 58G15, 31Bxx

ISBN 978-3-642-63491-8

Library of Congress Cataloging-in-Publication Data
Analysis. (Encyclopaedia of mathematical sciences; v. 13)
Translation of: Analiz, issued as part of the serial: Itogi nauki i tekhniki.
Seriîa sovremennye problemy matematiki. Fundamental'nye napravleniîa.
Vol. 4—has subtitle: Linear and boundary integral equations
Vol. 4—edited by V. Maz'îa and S. Nikol'skiĭ. Includes bibliographical references.
1. Operational calculus. 2. Integral representations. 3. Integral transforms. 4. Asymptotic expansions.
I. Gamkrelidze, R.V. II. Maz'îa, V.G. III. Nikol'skiĭ, S.M. (Sergeĭ Mikhaĭlovich), 1905–
IV. Series: Encyclopaedia of mathematical sciences; v. 13, etc.
QA432.A6213 1991 515'.72 89-6163
ISBN 978-3-642-63491-8 ISBN 978-3-642-58175-5 (eBook)
DOI 10.1007/978-3-642-58175-5

This work is subject to copyright. All rights are reserved, whether the whole or part of the material in concerned,
specifically the rights of translation, reprinting, reuse of illustrations, recitation, broadcasting, reproduction on
microfilms or in other ways, and storage in data banks. Duplication of this publication or parts thereof is only
permitted under the provisions of the German Copyright Law of September 9, 1965, in its current version, and
a copyright fee must always be paid. Violations fall under the prosecution act of the German Copyright Law.
© Springer-Verlag Berlin Heidelberg 1991
Originally published by Springer-Verlag Berlin Heidelberg New York in 1991
Typesetting: Asco Trade Typesetting Ltd., Hong Kong
2141/3140-543210—Printed on acid-free paper

List of Editors, Contributors and Translators

Editor-in-Chief

R.V. Gamkrelidze, Academy of Sciences of the USSR, Steklov Mathematical
Institute, ul. Vavilova 42, 117966 Moscow, Institute for Scientific Information (VINITI), ul. Usievicha 20 a, 125219 Moscow, USSR

Consulting Editors

V.G. Maz'ya, Department of Mathematics, University of Linköping, S-58183
Linköping, Sweden
S.M. Nikol'skiĭ, Steklov Mathematical Institute, ul. Vavilova 42,
117333 Moscow, USSR

Contributors

V.G. Maz'ya, Department of Mathematics, University of Linköping, S-58183
Linköping, Sweden
S. Prössdorf, Karl-Weierstraß-Institut für Mathematik, Akademie der Wissen-
schaften, Mohrenstraße 39, O-1086 Berlin, FRG

Translators

A. Böttcher, Technische Hochschule Chemnitz, Reichenheiner Straße 70,
O-9000 Chemnitz, FRG
S. Prössdorf, Karl-Weierstraß-Institut für Mathematik, Akademie der Wissen-
schaften, Mohrenstraße 39, O-1086 Berlin, FRG

Contents

I. Linear Integral Equations

S. Prössdorf

Contents

Introduction

A *linear integral equation* is an equation of the form

$$\lambda a(x)\varphi(x) - \int_X k(x, y)\varphi(y)dv(y) = f(x), \qquad x \in X. \tag{1}$$

Here (X, v) is a measure space with σ-finite measure v, λ is a complex parameter, and a, k, f are given (complex-valued) functions, which are referred to as the *coefficient*, the *kernel*, and the *free term* (or the *right-hand side*) of equation (1), respectively. The problem consists in determining the parameter λ and the unknown function φ such that equation (1) is satisfied for almost all $x \in X$ (or even for all $x \in X$ if, for instance, the integral is understood in the sense of Riemann). In the case $f = 0$, the equation (1) is called *homogeneous*, otherwise it is called *inhomogeneous*. If a and k are matrix functions and, accordingly, φ and f are vector-valued functions, then (1) is referred to as a *system* of integral equations.

Integral equations of the form (1) arise in connection with many boundary value and eigenvalue problems of mathematical physics.

Three types of linear integral equations are distinguished: If $\lambda = 0$, then (1) is called an equation of the *first kind*; if $\lambda a(x) \neq 0$ for all $x \in X$, then (1) is termed an equation of the *second kind*; and finally, if a vanishes on some subset of X but $\lambda \neq 0$, then (1) is said to be of the *third kind*.

The analysis of certain specific integral equations dates back to the first half of the 19th century. We encounter such equations in the work of J. Fourier (1811), N.H. Abel (1826), or J. Liouville (1837), for example. Integral equations became the subject of special investigation after C. Neumann (1877) had succeeded in reducing the Dirichlet boundary value problem for the Laplace equation to an integral equation of the second kind and in solving the latter equation by means of the method of successive approximation. When dealing with the equation of a vibrating membrane, A. Poincaré (1896) was led to introducing the complex parameter λ in equation (1).

An important contribution to the development of the theory of linear integral equations was made by V. Volterra [1913] in 1896, who studied equations of the form

$$\varphi(x) - \mu \int_a^x k(x, y)\varphi(y)dy = f(x), \qquad a \leqq y \leqq x \leqq b. \tag{2}$$

Note that equation (2) can be viewed as a special case of equation (1): It results from (1) by choosing $X = [a, b]$, $dv(y) = dy$ (Riemann or Lebesgue measure) and by letting $k(x, y) = 0$ for $y > x$. Equation (2) is now usually called a *Volterra equation*. Volterra showed that if k and f are continuous functions, then (2) has exactly one solution φ for every choice of the complex parameter μ, and he constructed this solution by means of the method of successive approximation;

in this way the solution is obtained as a power series in μ (*Neumann's series*) and thus represents an entire function of μ.

In the case of a general linear integral equation (1) of the second kind, the Neumann series converges only if the parameter $\mu = 1/\lambda$ is sufficiently small. I. Fredholm's merit is that he studied the equation

$$\varphi(x) - \mu \int_a^b k(x, y)\varphi(y)dy = f(x), \qquad a \leq x \leq b, \tag{3}$$

(under the assumption that both the kernel and the right-hand side are continuous) for all possible values of μ (1900–1903; see Fredholm [1903]). He replaced the integral in (3) by its integral sums and approximated the equation (3) by the systems of linear algebraic equations

$$\varphi(x_i) - \mu \sum_{j=1}^n k(x_i, x_j)\varphi(x_j)\Delta x_j = f(x_i), \qquad i = 1, \ldots, n, \tag{4}$$

where $a \leq x_1 \leq \cdots \leq x_n \leq b$ is a partition of the interval $[a, b]$ and $\varphi(x_1), \ldots, \varphi(x_n)$ are the unknown variables. After solving the systems (4) and formally passing to the limit $n \to \infty$ (at the same time requiring that max $\Delta x_j \to 0$), Fredholm arrived at the formula

$$\varphi(x) = f(x) + \mu \int_a^b \Gamma(x, y; \mu)f(y)dy, \tag{5}$$

where $\Gamma(x, y; \mu) = D(x, y; \mu)/d(\mu)$ is the quotient of two entire functions of μ. The d is nothing but the limit of the determinants of the systems (4). The functions Γ and d are usually referred to as the *Fredholm resolvent kernel* and the *Fredholm determinant* of the equation (3), respectively. Formula (5) represents the (unique) solution of equation (3) whenever $d(\mu) \neq 0$. The zeros of the function d are called the *characteristic values* of equation (3); they have at most one accumulation point, namely, ∞. If μ is a characteristic value, then the corresponding homogeneous equation (i.e., equation (3) with $f = 0$) has nontrivial solutions, while the inhomogeneous equation (3) need not possess any solution. Fredholm also showed that the basic theorems of linear algebra extend to equations of the form (3) with a continuous kernel. He finally generalized his theory to systems of integral equations and to the case of kernels with a weak singularity. Fredholm's approach was extended to the case of square-integrable kernels by T. Carleman (1921).

D. Hilbert (1901–1904; also see Hilbert [1912]) erected a general theory of linear integral equations upon the basis of the theory of linear and bilinear forms with infinitely many variables and created a spectral theory of integral equations with Hermitian kernels (i.e. $k(x, y) = \overline{k(y, x)}$). The results of Hilbert were simplified and generalized by E. Schmidt (1905–1908). He proposed an easy method for solving Hermitian integral equations provided the characteristic values and the eigenfunctions are known. Furthermore, Schmidt found a straightforward and beautiful proof of the Fredholm theorems, which does not invoke determinants; it is based on the observation that every kernel can be decomposed into a sum of a degenerate kernel and a small kernel.

The theory of linear integral equations founded by Fredholm and elaborated by Hilbert, Schmidt, Carleman, and others originated the theory of linear operators on Hilbert and Banach spaces. Owing to the pioneering work of these mathematicians, it became clear that the basic properties of integral equations of the second kind with continuous kernels (for example, Fredholm's theorem which states that the index, i.e. the difference of the numbers of linearly independent solutions of two homogeneous equations transposed to each other, always equals zero) result from the compactness of the integral operators at hand and are not a consequence of their integral nature.

If the integral operator in (1) is compact (on an appropriate function space), then it is called a *Fredholm integral operator*, while equation (1) is referred to as a *Fredholm integral equation*. A detailed survey of the results on Fredholm integral equations assembled up to 1928 can be found in the remarkable monograph Hellinger, Toeplitz [1928].

All those results of Fredholm's theory which were not tied to determinants were extended by F. Riesz (1918) to what is now known as the general spectral theory of compact operators on Banach spaces. The theory of F. Riesz was subsequently completed by J. Schauder (1930) and A.F. Ruston (1954). The problem of extending the theory of Fredholm determinants to operators on abstract Banach spaces was solved independently by A. Grothendieck, T. Leżański, and A.F. Ruston in the early fifties. A new and straightforward approach to the construction of Fredholm determinants was proposed by A. Pietsch in 1963.

An important supplement to Fredholm's theory is the theorem on the square-summability of the sequence of the eigenvalues of an integral operator, established by I. Schur (1909) for continuous kernels and by T. Carleman (1921) for square-integrable kernels. This theorem along with several problems which arose in applications stimulated the search for sufficiently general and sharp estimates for the eigenvalues of integral operators in dependence on integrability and smoothness properties of the kernel. Many important results in this direction have been obtained during the last few decades.

At the present time the theories of large classes of integral equations tractable by the Fredholm theorems as well as of integral equations of the first and third kinds (see Chap. 2) have reached a rather advanced stage, and research in these directions is going on.

The most prominent representatives of non-Fredholmian linear integral equations are singular integral equations and Wiener-Hopf equations. Singular integral equations already appeared in Poincaré's investigation of the mathematical theory of tides. Equations which are now called Wiener-Hopf equations emerged from a problem of mathematical astrophysics and were first studied by N. Wiener and E. Hopf in their paper Wiener, Hopf [1931]. Since then both classes of equations have proved useful for many applications to problems of elasticity theory, hydromechanics, aeromechanics, prediction theory, antenna synthesis, and others. The crucial difference between these equations and Fredholm equations is that singular integrals and integrals of convolution type generate bounded operators which, however, are not compact (on appropriate spaces of functions). Another peculiarity of singular equations results from the decisive difference between the one- and higher-dimensional cases.

A *one-dimensional singular integral equation* (*with a Cauchy kernel*) has the form

$$A\varphi(t) := a(t)\varphi(t) + b(t)S_\Gamma\varphi(t) + T\varphi(t) = f(t), \qquad t \in \Gamma, \tag{6}$$

where a, b, f are given functions, T is a given Fredholm integral operator, Γ is a curve in the complex plane, and S_Γ is defined by

$$S_\Gamma\varphi(t) := \frac{1}{\pi i} \int_\Gamma \frac{\varphi(\tau)}{\tau - t} d\tau,$$

the integral understood in the Cauchy principal-value sense.

The investigation of equations of the form (6) was begun by Hilbert (1904–1905) and Poincaré (1910), almost simultaneously with the foundation of the theory of Fredholm equations. A special case of equation (6) had already been considered by Yu.V. Sokhotskiĭ (1873) in his doctoral dissertation. The first fundamental results of the theory of equations of the form (6) were obtained by F. Noether (1921) and T. Carleman (1922). Noether [1921] was the first to prove that (under certain conditions on the smoothness of the given and unknown functions in (6)) if Γ is a smooth closed curve and if the so-called *ellipticity condition* $a^2(t) - b^2(t) \neq 0$ ($t \in \Gamma$) is satisfied, then equation (6) is normally solvable (i.e. it has a solution whenever its right-hand side is orthogonal to all the solutions of the homogeneous transposed equation) and its index is equal to the increment of the function $(1/2\pi) \arg((a - b)/(a + b))$ along the curve Γ. Noether's approach was based on the idea of multiplying the operator A by an operator B of a form similar to (6) in order to obtain a representation $BA = I + T_1$, where I is the identity operator and T_1 is a Fredholm integral operator. This method, called the method of *left regularization*, had been earlier employed by Poincaré and Hilbert in various special situations. Carleman [1922] proposed a method for the explicit solution of certain special classes of equations of the form (6). This method consisted in reducing the *simplest singular equation* (6) (this is the equation with $T = 0$) to a so-called *Riemann-Hilbert boundary value problem* (see 2.4 of Chap. 3), or, what is the same, to the *factorization problem* for the function $(a - b)/(a + b)$ (see Sec. 4 of Chap. 4). On solving the boundary value problem, Carleman constructed the general solution of the simplest singular equation, and at the same time he solved the problem of regularizing the complete equation (6).

In the forties, the methods and results of Noether and Carleman were widely used and essentially generalized in the work of G. Giraud, F.D. Gakhov, N.I. Muskhelishvili, I.N. Vekua, N.P. Vekua, V.D. Kupradze, S.G. Mikhlin, B.V. Khvedelidze, and others. These authors mainly considered the case where Γ is a smooth curve and the given and unknown functions in (6) satisfy a Hölder condition on Γ (with the possible exception of a finite number of discontinuities) (see Gakhov [1977], Muskhelishvili [1968], Vekua [1970], Mikhlin [1948],

Khvedelidze [1956], [1975], Fichera [1958]). The work of Noether was also the beginning of the enormous development which resulted in the theory of normally solvable operators with a finite index on abstract spaces on the one hand and in the general index theory of operators on the other hand. Normally solvable operators with a finite index are now usually called *Noether operators* or *Fredholm operators*. The theory of these operators was worked out in the fifties (see Sec. 3 of Chap. 1).

The main questions arising in connection with equation (6) are as follows:

1) In which function spaces and under what assumptions on the integration contour Γ is the operator S_Γ bounded?

2) Under what assumptions on the coefficients a and b is the operator A Noetherian? What is its index in this case? Are there any methods for effectively solving the equation (6)?

In the case where Γ is a Lyapunov curve, the boundedness of S_Γ was proved by J. Plemelj (1908) and I.I. Privalov (1916) for the Hölder spaces $C^\alpha(\Gamma)$ $(0 < \alpha < 1)$, by M. Riesz (1928) for the spaces $L_p(\Gamma)$ $(1 < p < \infty)$, and by G.H. Hardy and J.E. Littlewood (1936) for spaces $L_p(\Gamma, \varrho)$ with a certain power weight ϱ.

A milestone in the development of the theory of singular operators was a paper published by B. Muckenhoupt in 1972. Therein he introduced a certain class (A_p) of weights which then turned out to be the key to a complete description of all the spaces $L_p(\Gamma, \varrho)$ on which S_Γ is bounded. The question whether S_Γ is bounded on $L_p(\Gamma)$ if Γ is a curve of the class C^1 remained open a long time. It was answered affirmatively by A.P. Calderón only in 1977.

The works cited above stimulated a large amount of further profound research into one- and higher-dimensional singular integral operators on non-smooth curves and manifolds, and even on spaces $L_p(\Gamma, \varrho)$ with fairly general weights (see the Chapters 3 and 4).

Since the late forties a large number of works concerning the problem of Noethericity and index computation have been published, and even at present a great deal of active research is being done on this topic (see the Sections 2–6 of Chapter 3 and the Sections 3–8 of Chapter 4). A remarkable feature of these works is the broad use of ideas and methods from functional analysis, notably from the theory of Banach algebras and from algebraic topology. In this connection, it should be mentioned that the interest does not primarily focus on the single operator defined by (6), but is directed towards the whole algebra generated by such operators; thus the above question 2) is usually answered for all operators in this algebra. A crucial role in the advanced theory of singular operators is played by the notion of the *symbol*, introduced by S.G. Mikhlin [1948] when he was studying equation (6) in the case of continuous coefficients. The symbol effects an isomorphism between an algebra of functions and an algebra of operators.

We finally remark that in recent years the theory of one-dimensional singular integral equations has been enriched by many important results in the following directions: non-elliptic equations, equations with shift, equations with fixed singularities, operators with a homogeneous kernel (in the study of the latter class of operators the Mellin transform plays the part that is traditionally performed by the Fourier transform; see Sections 3, 7, 8 of Chapter 3).

A *Wiener-Hopf equation* is of the form

$$\varphi(x) + \int_0^\infty k(x-y)\varphi(y)\,dy = f(x), \qquad 0 \leqq x < \infty, \qquad (7)$$

and it has a series of properties in common with singular integral equations. The first fundamental results on equations of the form (7) were obtained by N. Wiener and E. Hopf (1931), who proposed an effective method (now called the *method of Wiener-Hopf factorization*) for the solution of the homogeneous version of (7) provided the kernel $k(x)$ decreases exponentially as $|x| \to \infty$. Subsequently, the close relationship between Wiener-Hopf equations and boundary value problems for analytic functions was realized. For about 40 years enormous work has been done on Wiener-Hopf equations and a variety of results have been collected. We only remind of M.G. Krein (1958), who applied ideas from functional analysis to the study of equation (7) and its discrete analogue. Under the assumption that the kernel belongs to $L_1(-\infty, \infty)$, he developed an almost complete theory of Wiener-Hopf integral equations in various classes of Banach spaces, including $L_p(0, \infty)$ ($1 \leqq p \leqq \infty$). Krein established necessary and sufficient conditions for the operator generated by (7) to be Noetherian, and he also determined the asymptotic behavior of the solution of equation (7) for several special right-hand sides. Afterwards equation (7) was also studied in the case where k is not in $L_1(-\infty, \infty)$ but has an almost periodic or a piecewise continuous Fourier transform (see Gohberg, Feldman [1971], Douglas [1973], Duduchava [1979]).

The first significant results on *multidimensional singular integral equations* go back to F. Tricomi (1926–1928). A singular integral in the Euclidean space $\mathbb{R}^n$ has the form

$$Ku(x) := a(x)u(x) + \lim_{\varepsilon \to 0} \int_{|x-y|>\varepsilon} k(x, x-y)u(y)\,dy, \qquad (8)$$

where k satisfies

$$k(x, tz) = t^{-n}k(x, z) \quad \text{for } t > 0, \qquad \int_{|z|=1} k(x, z)\,dz = 0$$

and, in addition, is subject to certain integrability or smoothness conditions. Tricomi considered the case $n = 2$ and assumed that the kernel $k(x, z)$ is independent of x. He established a formula for the composition of two singular integrals, by means of which he then reduced the solution of the equation $Ku = f$ to the solution of a certain one-dimensional singular equation; however, note that he did not analyze the latter equation. The next important work on multidimensional singular integrals is due to G. Giraud (1934), who studied singular integrals on closed Lyapunov manifolds of arbitrary dimension. Assuming that the kernel satisfies some (quite strong) conditions, Giraud extended the Plemelj-

Privalov theorem on the boundedness of K on the space $C^\alpha(\Gamma)$ $(0 < \alpha < 1)$ to higher dimensions, and he also constructed a regularizer for K. In the work of both Tricomi and Giraud, there did not figure the so-called *ellipticity condition*, which is a necessary and sufficient condition for the singular operator to admit regularization. This condition first appeared in S.G. Mikhlin's papers [1936]. Mikhlin was also the first to introduce the concept of the *symbol* of the operator K (in the case $n = 2$). The symbol $\mathcal{K}$ of K was constructed by means of the expansion of the kernel k into the Fourier series with respect to spherical functions, and in terms of the symbol $\mathcal{K}$ the celebrated ellipticity condition assumes the simple form $\inf|\mathcal{K}| > 0$. It was A.P. Calderón and A. Zygmund (1952, 1956) who first applied Fourier transform techniques to singular integrals . Calderón and Zygmund as well as Mikhlin (1956) set up the extremely important formulas

$$\mathcal{K}(x, \xi) = a(x) + \hat{k}(x, \xi), \qquad K = F_{\xi \to x}^{-1} \mathcal{K}(x, \xi) F_{x \to \xi},$$

where F refers to the Fourier transform and $\hat{k}(x, \xi)$ denotes the Fourier transform of the kernel $k(x, z)$ with respect to the variable z. These formulas became the foundation of all the further work devoted to multidimensional singular integrals and integral equations. Moreover, the systematic use of the apparatus of the Fourier transform promoted a further synthesis of multidimensional singular integral equations and partial differential equations, which eventually culminated in the theory of pseudodifferential operators (J.J. Kohn, L. Nirenberg; 1965); in this connection, see also the monographs Eskin [1973], Shubin [1978], Treves [1982], Journé [1983].

S.G. Mikhlin (1937–1938, 1953) gave simple sufficient conditions for the boundedness of the operator (8) on the space L_2 in terms of the symbol and proved that if K is an elliptic operator then the equation $Ku = f$ can be reduced to an equivalent Fredholm equation; in contrast to the case of a single equation (scalar case), the index of a system of multidimensional singular equations may be nonzero. More general theorems on the boundedness of the singular operator (8) on the spaces $L_p(\mathbb{R}^n)$ $(1 < p < \infty)$ were established by Calderón and Zygmund (1952–1957, 1978). These fundamental results have led to the intensive development of research devoted to the problem of the boundedness of singular integral operators on function spaces (see the surveys Dyn'kin, Osilenker [1983], Journé [1983], Murai [1987] and the detailed exposition Garsia-Cuerva, Rubio de Francia [1985]).

The final solution of the index problem for a general elliptic operator (including as a special case the index problem for a singular integral operator) was given in the work of M. Atiyah, I. Singer, and R. Bott (1963–1964).

Many interesting results obtained during the last few years concern multi-dimensional singular integral equations on manifolds with boundary or with singularities, non-elliptic equations and operators with discontinuous symbols, multidimensional Wiener-Hopf equations, and bisingular equations. Notice that a major part of the results in the last three directions were obtained by resorting to so-called "local principles", the first of which was worked out by I.B. Simonenko (1964–1965). We note that these principles are, in a sense, analogues of the method of "freezing coefficients", known from the theory of differential equations.

In recent time interesting applications of the theory of multidimensional singular integral equations to boundary value problems for partial differential equations have been brought to light; the multidimensional singular equations of elasticity theory, of thermoelasticity, of moment elasticity, and of other mathematical models of solid deformable media and fluid flows have been studied in detail. Furthermore, the theory and practice of *approximation methods* for integral equations have now achieved a very advanced stage (see, for instance, the monographs Baker [1977], Ivanov [1968], Gohberg, Feldman [1971], Prössdorf, Silbermann [1977], Mikhlin, Prössdorf [1980], Gabdulkhaev [1980], Belotserkovskiĭ, Lifanov [1985], Fenyö, Stolle [1982–1984], Böttcher, Silbermann [1983], Heinig, Rost [1984], Prössdorf, Silbermann [1990]), and the same can also be said of the theory of *nonlinear integral equations*, which has its source in the work of A.M. Lyapunov (1906), E. Schmidt (1908), P.S. Uryson (1922), and A. Hammerstein (1930) (see, for example, the surveys Imanaliev, Khvedelidze, Gegelia, Babaev, Botashev [1982], Kosel, v. Wolfersdorf [1986] and v. Wolfersdorf [1974]). For lack of space we cannot embark on these topics here.

Let us conclude this introduction by some general remarks on the theory of integral equations, on its place in mathematics, and on the basic directions of its further development.

Over nearly a century, this theory has maintained its position among the central branches of mathematical analysis. On the one hand, its development has been stimulated by the needs of mechanics, physics, and numerous problems in engineering sciences; on the other hand, it has been strongly enlivened by its place at the intersection of several branches of pure mathematics, such as functional analysis, function theory, algebra, topology, or probability theory. It was the study of Volterra and Fredholm integral equations at the turn of the century that gave birth to functional analysis and laid the foundation for solving the classical problems of mathematical physics. In its early stage, the investigation of integral equations was based on the analogy between integral equations and algebraic systems and was so confined to compact operators. The next development stage had its origin in the study of singular integral operators. In that period such concepts as the index of an operator, the symbol, or the regularization of an equation were born. One realized the intimate connection between the boundary value problems of the theory of analytic functions and singular integral equations. Moreover, the notion of the symbol brought to light the close relationship between singular integral operators and the theory of Banach algebras, which subsequently led to the investigation of very general classes of integral operators. The problem of computing the index of an integral operator finally required working with algebraic topology. Since the classical tools were no longer available for answering the basic questions on the boundedness of the operators on appropriate function spaces, new techniques of "hard analysis" were elaborated. In the end, the calculus of singular integral operators led to the creation of pseudodifferential and Fourier integral operators, which, in turn, have left their imprint on the face of the modern theory of partial differential equations and boundary value problems.

A major part of the topics mentioned above continue to be a subject of active research, and to elucidate some of the most recent developments and trends in the theory of integral equations is the main purpose of this survey.

Whenever possible, our exposition will focus on the operator theory point of view and will try to use uniform and consistent terminology. For this reason, some of the most important facts from abstract operator theory and the theory of Banach algebras are recorded in the first chapter. An overall view of what is dealt with in this survey may be obtained by a glance at the table of contents and at the introductory remarks preceding each chapter.

It is a pleasure to express my sincere appreciation to A. Böttcher, V.G. Maz'ya, E. Meister, B. Silbermann, and F.-O. Speck, who read the manuscript of this survey and made a series of valuable remarks.

Chapter 1
Some Facts from Abstract Operator Theory

As already pointed out in the introduction, the theory of integral equations with continuous kernels founded by Fredholm at the beginning of this century was the starting-point for the theory of linear operators on Hilbert or Banach spaces. Today it has been realized that many results on integral equations represent nothing but special cases of more general results from functional analysis. For instance, the problem of finding the eigenvalues and eigenfunctions of an integral operator is now naturally placed into the spectral theory of linear operators. All results of the Fredholm theory which can be stated without resorting to determinants were extended by F. Riesz to results of the general theory of compact operators. The application of concepts and methods of functional analysis simplifies and generalizes the theory of integral equations to a high degree, and so a modern account of this theory which pays no due attention to the functional analytic background is impossible.

We first give a summary of the basic facts from the theory of Banach algebras and shall then consider Noether and Fredholm operators, Riesz' operator theory, the Grothendieck-Pietsch theory of Fredholm determinants of nuclear operators on Banach spaces, and the spectral theory of compact operators on Hilbert space.

§ 1. Banach Algebras

The overwhelming majority of classes of integral operators which will be studied in the following chapters are actually Banach algebras. Therefore the problem of deciding whether an operator is invertible or Noetherian can be solved by using I.M. Gelfand's theory of maximal ideals. For the reader's convenience, we shall first summarize the fundamentals of the Gelfand theory (for more about this topic see, e.g., Naimark [1968]).

1.1. Definitions and Examples

1.1.1. A complex Banach space $\mathfrak{A}$ with a norm $\|\cdot\|$ is said to be a *Banach algebra* if an operation on $\mathfrak{A}$ is defined that sends the elements $A, B \in \mathfrak{A}$ to their so-called product $AB \in \mathfrak{A}$ and that has the following properties: it is associative, it is distributive with respect to addition, it commutes with multiplication by scalars, and it satisfies the inequality $\|AB\| \leq \|A\| \|B\|$.

A linear subspace $\mathfrak{B}$ of a Banach algebra $\mathfrak{A}$ is called a *subalgebra* if AB belongs to $\mathfrak{B}$ whenever A and B are in $\mathfrak{B}$.

A Banach algebra $\mathfrak{A}$ is said to be *commutative* if $AB = BA$ for all $A, B \in \mathfrak{A}$. If $\mathfrak{A}$ contains an element I such that $AI = IA = A$ for all $A \in \mathfrak{A}$, then $\mathfrak{A}$ is called a *Banach algebra with identity I*.

If a Banach algebra $\mathfrak{A}$ has no identity, it can be supplied with an identity in such a way that one obtains a larger Banach algebra which contains $\mathfrak{A}$ as a closed subalgebra of codimension 1.

If (A_n) and (B_n) are two convergent sequences in a Banach algebra whose limits are A and B, respectively, then, obviously, $A_n B_n$ converges to AB.

1.1.2. Examples of Commutative Banach Algebras with Identity

$1°$. The algebra $C(X)$ of all complex-valued functions f that are continuous and bounded on a metric space X; the norm is given by $\|f\|_{C(X)} := \sup_{t \in X} |f(t)|$.

$2°$. The *Hölder algebra (space)* $C^\alpha(X)$ $(0 < \alpha \leq 1)$ of all functions $f \in C(X)$ such that

$$\|f\|_\alpha := \|f\|_{C(X)} + \sup_{x,y \in X} \frac{|f(x) - f(y)|}{d^\alpha(x, y)} < \infty,$$

where d denotes the metric on X. The functions in $C^\alpha(X)$ are said to satisfy a *Hölder condition* with exponent α.

$3°$. The *Wiener algebra* W of all complex-valued functions A on the unit circle of the complex plane $\mathbb{C}$ with absolutely convergent Fourier series:

$$A(z) = \sum_{n=-\infty}^{\infty} a_n z^n \quad (|z| = 1), \qquad \|A\|_W := \sum_{n=-\infty}^{\infty} |a_n| < \infty.$$

$4°$. The algebra $W(\mathbb{R})$ of all functions of the form

$$F(x) = c + \int_{-\infty}^{\infty} e^{ixt} f(t)\, dt \qquad (c \in \mathbb{C}, x \in \mathbb{R}),$$

where $f \in L_1(\mathbb{R})$; the norm in this algebra is defined by $\|F\| := |c| + \int_{-\infty}^{\infty} |f(t)| dt$.

1.2. Regular Elements. Resolvent. Spectrum. Let $\mathfrak{A}$ be a Banach algebra with identity.

1.2.1. An element $A \in \mathfrak{A}$ is said to be *regular* or *invertible* if there exists an element $B \in \mathfrak{A}$ such that $AB = BA = I$. The element B defined in this way is uniquely determined (if it exists); it is called the *inverse* of A and is denoted by A^{-1}. A non-invertible element is said to be *singular*.

1.2.2. The set $G\mathfrak{A}$ of all regular elements of $\mathfrak{A}$ is a group. If $\|A\| < 1$, then $I - A \in G\mathfrak{A}$ and

$$(I - A)^{-1} = \sum_{n=0}^{\infty} A^n \qquad (A^0 = I), \tag{1.1}$$

the series (which is referred to as the *Neumann series* of A) being convergent in the norm of $\mathfrak{A}$. As an easy consequence of this, one sees that $G\mathfrak{A}$ is an open subset of $\mathfrak{A}$. Furthermore, the mapping $A \to A^{-1}$ defined on $G\mathfrak{A}$ is continuous.

1.2.3. The set $\varrho(A)$ of all $\lambda \in \mathbb{C}$ such that $A - \lambda I$ is invertible is called the *resolvent set* of the element $A \in \mathfrak{A}$. Every number $\lambda \in \varrho(A)$ is termed a *regular point* and the element $R_\lambda = R_\lambda(A) := (A - \lambda I)^{-1}$ is referred to as the *resolvent* of the element $A \in \mathfrak{A}$. The complement $\sigma(A) := \mathbb{C} \setminus \varrho(A)$ is called the *spectrum* of the element A.

Any two points $\lambda, \mu \in \varrho(A)$ satisfy the so-called *resolvent equality* or *Hilbert identity* $R_\lambda - R_\mu = (\lambda - \mu) R_\lambda R_\mu$.

1.2.4. If $\lambda_0 \in \varrho(A)$ and $|\lambda - \lambda_0| \|R_{\lambda_0}\| < 1$ $(\lambda \in \mathbb{C})$ then, by (1.1), λ belongs to $\varrho(A)$ and we have

$$R_\lambda = \sum_{n=0}^{\infty} (\lambda - \lambda_0)^n R_{\lambda_0}^{n+1}. \tag{1.2}$$

If follows that $\varrho(A)$ is an open set containing, in particular, all λ with $|\lambda| > \|A\|$ and that $\sigma(A)$ is a closed and bounded set.

1.2.5. The number $r_A := \max_{\lambda \in \sigma(A)} |\lambda|$ is called the *spectral radius* of the element A. I.M. Gelfand showed that

$$r_A = \lim_{n \to \infty} \|A^n\|^{1/n} \leqq \|A^m\|^{1/m} \qquad (m = 1, 2, \ldots),$$

the existence of the limit being part of the conclusion. For any λ with $|\lambda| > r_A$, the resolvent can be represented in the form $R_\lambda = - \sum_{n=0}^{\infty} \frac{1}{\lambda^{n+1}} A^n$.

1.2.6. From (1.2) we infer that the resolvent $R_\lambda \colon \varrho(A) \to \mathfrak{A}$ is an analytic function. The convergence radius ϱ_0 of the series (1.2) is no smaller than $\|R_{\lambda_0}\|^{-1}$. Moreover, we have

$$\varrho_0 = \left(\limsup_{n \to \infty} \|R_{\lambda_0}^{n+1}\|^{1/n} \right)^{-1}.$$

As the resolvent is bounded at infinity, we conclude from Liouville's theorem that it cannot be analytic throughout the complex plane. Hence the spectrum of an arbitrary element of a Banach algebra is never empty.

This is turn immediately implies that a Banach field is isometrically isomorphic to the field $\mathbb{C}$; indeed, if $\mathfrak{A}$ is a field and $A \in \mathfrak{A}$, then there is exactly one point $\lambda \in \mathbb{C}$ such that $A - \lambda I$ is singular and in that case we have $A - \lambda I = 0$, i.e. $A = \lambda I$.

1.3. Poles of the Resolvent. Again let $\mathfrak{A}$ be a Banach algebra with identity.

1.3.1. A function $f \colon D \to \mathfrak{A}$ defined on an open subset D of $\mathbb{C}$ is said to be *analytic* or *holomorphic* in D if every point $\lambda_0 \in D$ has a neighborhood U such that

$$f(\lambda) = \sum_{n=0}^{\infty} (\lambda - \lambda_0)^n A_n$$

with certain $A_n \in \mathfrak{A}$ for all $\lambda \in U$.

1.3.2. A point $\lambda_0 \in D$ is called a *pole* of the analytic function $f \colon D \backslash \{\lambda_0\} \to \mathfrak{A}$ if there exist a natural number p and elements $A_{-n} \in \mathfrak{A}$, $n = 1, \dots, p$, $A_{-p} \neq 0$, such that the function

$$f_0(\lambda) = f(\lambda) - \sum_{n=1}^{p} (\lambda - \lambda_0)^{-n} A_{-n} \qquad (1.3)$$

is analytic in a neighborhood of λ_0 (including the point λ_0). It is easily seen that the representation (1.3) is unique. The number p is referred to as the *order* of the pole and A_{-1} is termed its *residue*. The sum in (1.3) is called the *principal part* and f_0 is called the *regular part* of the *Laurent series expansion* of f at the pole λ_0.

1.3.3. Let $A \in \mathfrak{A}$ and let λ_0 be a pole of the order p of the resolvent $R_\lambda = R_\lambda(A)$ of A. Then $\lambda_0 \in \sigma(A)$. The residue A_{-1} possesses the following properties: $A_{-1}^2 = A_{-1}$, $A_{-1} A = A A_{-1}$,

$$(A - \lambda_0 I)^n A_{-1} \neq 0, \quad n = 0, 1, \dots, p - 1; \qquad (A - \lambda_0 I)^p A_{-1} = 0.$$

For $\lambda \in \varrho(A)$, we have a decomposition $R_\lambda = S_\lambda + T_\lambda$ where $S_\lambda = \sum_{n=1}^{p} (\lambda - \lambda_0)^{-n} (A - \lambda_0 I)^{n-1} A_{-1}$ and T_λ is a function which is analytic in $\varrho(A) \cup \{\lambda_0\}$. If B_{-1} is the residue of another pole of the resolvent R_λ, then $A_{-1} B_{-1} = B_{-1} A_{-1} = 0$.

1.4. Maximal Ideals. The Gelfand Homomorphism

1.4.1. A subset $\mathfrak{J}$ of a Banach algebra $\mathfrak{A}$ with identity is said to be an *ideal* if $\mathfrak{J}$ is a linear subspace of $\mathfrak{A}$ and if $AB \in \mathfrak{J}$ and $BA \in \mathfrak{J}$ whenever $A \in \mathfrak{A}$ and $B \in \mathfrak{J}$. In what follows ideal will always mean *proper* ideal, i.e. $\mathfrak{J} \neq \varnothing$ and $\mathfrak{J} \neq \mathfrak{A}$. The closure of an arbitrary ideal is again an ideal.

1.4.2. If $\mathfrak{J}$ is a closed ideal of $\mathfrak{A}$, then the quotient space $\mathfrak{A}/\mathfrak{J}$ is a Banach algebra with identity element; the norm in $\mathfrak{A}/\mathfrak{J}$ is defined as

$$\|\tilde{A}\| := \inf_{A \in \tilde{A}} \|A\| \qquad (\tilde{A} \in \mathfrak{A}/\mathfrak{J})$$

and the multiplication is defined by $\tilde{A}\tilde{B} := \widetilde{AB}$. The algebra $\mathfrak{A}/\mathfrak{J}$ is called the *quotient algebra*.

1.4.3. Every ideal consists of singular elements only. An element $A \in \mathfrak{A}$ is regular if and only if there is no ideal containing A.

1.4.4. Each ideal is contained in some *maximal ideal*, i.e. in an ideal which is not a proper subset of any larger ideal. Maximal ideals are always closed.

1.4.5. In the remainder of Sec. 1.4 we suppose that $\mathfrak{A}$ is a *commutative* Banach algebra with identity. In that case an element $A \in \mathfrak{A}$ is regular if and only if there is no maximal ideal containing A.

The only maximal ideals of the Wiener algebra W are the ideals constituted by all functions in W which vanish at some point of the unit circle. Thus, what was said in the preceding paragraph immediately yields the following *Theorem of Wiener*: If $f \in W$ and $f(z) \neq 0$ for all z with $|z| = 1$, then $1/f \in W$.

1.4.6. If M is a maximal ideal, then the quotient algebra $\mathfrak{A}/M$ does not contain any (proper) ideal and is therefore a Banach field. So, by 1.2.6, there is an isometrical isomorphism $\varphi_M \colon \mathfrak{A}/M \to \mathbb{C}$. This isomorphism induces a continuous multiplicative linear functional f_M on $\mathfrak{A}$ (that is, a functional $f_M \in \mathfrak{A}'$ such that $f_M(AB) = f_M(A)f_M(B)$ for all A, $B \in \mathfrak{A}$), namely, the functional f_M defined by $f_M(A) = \varphi_M(\tilde{A})$. Note that $\ker f_M = M$.

Conversely, every multiplicative linear functional is continuous, has norm 1, and its kernel is a maximal ideal.

1.4.7. Because the norm of a multiplicative linear functional equals 1, all these functionals are located on the unit sphere of the dual space $\mathfrak{A}'$ and they form a set $\mathfrak{M}$ which, provided with the weak *-topology of the dual space, is a compact Hausdorff space. The space $\mathfrak{M}$ is called the *maximal ideal space* of the algebra $\mathfrak{A}$.

1.4.8. For $A \in \mathfrak{A}$ and $M \in \mathfrak{M}$, put $\hat{A}(M) = f_M(A)$. The mapping $\gamma_{\mathfrak{A}} \colon \mathfrak{A} \to C(\mathfrak{M})$ defined by $\gamma_{\mathfrak{A}}(A) = \hat{A}$ is referred to as the *Gelfand homomorphism*. This homomorphism has the following important properties:
  (i) $\|\hat{A}\|_{C(\mathfrak{M})} = \lim_{n\to\infty} \|A^n\|^{1/n} \leq \|A\|$;
  (ii) im $\hat{A} = \sigma(A)$;
  (iii) $A \in \mathfrak{A}$ is regular $\Leftrightarrow A(M) \neq 0 \ \forall M \in \mathfrak{M}$;
  (iv) if $M_1 \neq M_2$, then there is an $A \in \mathfrak{A}$ such that $\hat{A}(M_1) \neq \hat{A}(M_2)$.

1.4.9. The kernel of the Gelfand homomorphism is called the *radical* of the algebra $\mathfrak{A}$. Notice that, obviously, the radical coincides with the intersection of all maximal ideals. Furthermore, the radical consists of all $A \in \mathfrak{A}$ for which $\|A^n\|^{1/n} \to 0$ as $n \to \infty$.

1.4.10. An algebra $\mathfrak{A}$ is said to be *semisimple* or to be a *function algebra* if its radical consists only of the zero element. If $\mathfrak{B}$ is a semisimple commutative Banach algebra with identity, then every algebra homomorphism $\psi \colon \mathfrak{A} \to \mathfrak{B}$ is continuous.

1.5. Algebras with an Involution. C*-Algebras

1.5.1. A Banach algebra $\mathfrak{A}$ is called an *algebra with an involution* if a mapping $A \to A^*$ of $\mathfrak{A}$ into itself is defined such that $(A^*)^* = A, (\lambda A + \mu B)^* = \bar{\lambda}A^* + \bar{\mu}B^*$, $(AB)^* = B^*A^*$ for all $A, B \in \mathfrak{A}$ and $\lambda, \mu \in \mathbb{C}$. If, in addition, $\|A^*A\| = \|A\|^2$ for all $A \in \mathfrak{A}$, then $\mathfrak{A}$ is said to be a *C*-algebra*. The element A^* is referred to as the *adjoint* of the element A.

1.5.2. An important example of a C*-algebra is the algebra $\mathscr{L}(H)$ of all bounded linear operators on a Hilbert space H; in this case the involution is passage to the adjoint operator (see Sec. 2). Moreover, it turns out that actually every C*-algebra is isometrically isomorphic to some subalgebra of $\mathscr{L}(H)$ (for appropriately chosen H); this is the *Gelfand-Naimark theorem*.

Also notice that $C(X)$ is a C*-algebra with the isometric involution $f^*(x) := \overline{f(x)}$ $(x \in X, f \in C(X))$.

1.5.3. Let $\mathfrak{A}$ and $\mathfrak{B}$ be two Banach algebras with involutions $*$ and $+$, respectively.

A homomorphism $\varphi: \mathfrak{A} \to \mathfrak{B}$ such that $\varphi(A^*) = \varphi(A)^+$ for all $A \in \mathfrak{A}$ is called a **-homomorphism*.

Theorem (I.M. Gelfand and M.A. Naimark). *If $\mathfrak{A}$ is a commutative C*-algebra with identity, then the Gelfand transform $\mathfrak{A} \to C(\mathfrak{M})$ is an isometrical *-isomorphism of $\mathfrak{A}$ onto $C(\mathfrak{M})$.*

We remark that the famous Stone-Weierstrass theorem is a particular case of this theorem.

1.5.4. An element A of an algebra with an involution is called *self-adjoint* if $A^* = A$ and is termed *normal* if $A^*A = AA^*$.

The following useful properties of a C*-algebra $\mathfrak{A}$ can be easily verified:
1) An element $A \in \mathfrak{A}$ is regular if and only if A^* is so; in this case $(A^{-1})^* = (A^*)^{-1}$.
2) $\sigma(A^*) = \{\lambda : \bar{\lambda} \in \sigma(A)\}$, and $R_\lambda(A^*) = [R_{\bar{\lambda}}(A)]^*$ for all $\lambda \in \varrho(A^*)$.
3) If A is normal, then the resolvent $R_\lambda(A)$ is also normal. If A is self-adjoint and $\lambda \in R$, then $R_\lambda(A)$ is also self-adjoint.
4) If A is normal, then $\|A^*\| = \|A\|$ and $r_A = \|A^n\|^{1/n}$ for $n = 1, 2, \ldots$.
5) The order of a pole of a normal element is equal to 1 and its residue is again a normal element.

1.6. The Symbol. Let $\mathfrak{A}$ be a Banach algebra with identity, let $\mathfrak{J}$ be a closed ideal of $\mathfrak{A}$, and suppose that the quotient algebra $\tilde{\mathfrak{A}} := \mathfrak{A}/\mathfrak{J}$ is commutative. Denote the maximal ideal space of $\tilde{\mathfrak{A}}$ by $\mathfrak{M}$ and let $\pi_{\tilde{\mathfrak{A}}} : \mathfrak{A} \to \tilde{\mathfrak{A}}$ denote the canonical projection. The mapping

$$\sigma_{(\mathfrak{A}, \mathfrak{J})} := \gamma_{\tilde{\mathfrak{A}}} \circ \pi_{\tilde{\mathfrak{A}}} : \mathfrak{A} \to C(\mathfrak{M})$$

is called a *symbol of the algebra* $\mathfrak{A}$. For $A \in \mathfrak{A}$, the function $\sigma_{(\mathfrak{A}, \mathfrak{J})}(A) \in C(\mathfrak{M})$ is called the *symbol of the element* A (with respect to the ideal $\mathfrak{J}$).

It is clear that the symbol $\sigma_{(\mathfrak{A},\mathfrak{J})}$ is a homomorphism and that $\sigma_{(\mathfrak{A},\mathfrak{J})}(A) = \sigma_{(\mathfrak{A},\mathfrak{J})}(B)$ whenever $A - B \in \mathfrak{J}$. Furthermore, from the definition of the symbol we immediately infer that

$$\|\sigma_{(\mathfrak{A},\mathfrak{J})}(A)\|_{C(\mathfrak{M})} \leqq \|A\| \qquad \forall A \in \mathfrak{A},$$

i.e. that a symbol is a continuous mapping of $\mathfrak{A}$ into $C(\mathfrak{M})$. N.Ya. Krupnik [1984] constructed a matrix analogue of the Gelfand transform and matrix symbols for some interesting non-commutative Banach algebras (in this connection, see also Sec. 3 of Chap. 3).

§2. Algebras of Linear Operators

2.1. Basic Definitions

2.1.1. Let E and F be complex normed spaces and let D be a linear subset of E. A mapping $A: D \to F$ is called a *linear operator* if $A(\alpha x + \beta y) = \alpha A(x) + \beta A(y)$ for all $x, y \in D$ and $\alpha, \beta \in \mathbb{C}$.

We shall usually abbreviate $A(x)$ to Ax. The set D is the *domain* of the operator A and it is denoted by $D(A)$. The linear sets $N(A) = \ker A := \{x \in D(A) : Ax = 0\}$ and $R(A) = \operatorname{im} A := \{Ax : x \in D(A)\}$ are referred to as the *kernel* and the *range* (*image space*) of the operator A, respectively.

The quotient space $F/\overline{\operatorname{im} A}$ is called the *cokernel* of the operator A and is denoted by coker A. If at least one of the integers $\alpha(A) := \dim \ker A$, $\beta(A) := \dim$ coker A is finite, then their difference, $\alpha(A) - \beta(A)$, is termed the *index* of the operator A and is designated by Ind A.

2.1.2. A linear operator A is said to be *continuous* at a point $x_0 \in D(A)$ if $x_n \to x_0$ ($x_n \in D(A)$) implies that $Ax_n \to Ax_0$. If $D(A) = E$ and A is continuous at any point (e.g. at 0), then A is continuous at all points of E, i.e. is *continuous on* the whole space E.

2.1.3. A linear operator A whose domain coincides with E is said to be *bounded* if $\|Ax\|_F \leqq C\|x\|_E \; \forall x \in E$, where C is some constant independent of $x \in E$. The least of the numbers C for which this inequality holds is called the *norm* of the operator A and is denoted by $\|A\|_{E \to F}$. In the case $F = E$ we shall simply write $\|A\|$ in place of $\|A\|_{E \to F}$. From the definition of the norm we also obtain that

$$\|A\|_{E \to F} = \sup_{x \in E} \frac{\|Ax\|_F}{\|x\|_E} = \sup_{\|x\|_E = 1} \|Ax\|_F.$$

For a linear operator to be continuous it is necessary and sufficient that it be bounded.

Theorem (on the extension by continuity). *Suppose D is a dense subset of E and F is a Banach space. Then every continuous linear operator A_0 from D into F admits a unique extension to a continuous linear operator A of E into F; moreover, one has $\|A\|_{E \to F} = \|A_0\|_{D \to F}$.*

Thus, in order to define a bounded linear operator of E into F it suffices to define the action of the operator on a dense subset D of E.

2.1.4. The following theorem is one of the fundamental results of the theory of linear operators.

Open Mapping Theorem. *If A is a continuous linear mapping of a Banach space E onto a Banach space F, then $A(U)$ is open whenever U is open.*

2.1.5. A continuous linear operator P of E into itself is called a *projection* if $P^2 = P$. If P is a projection, then the operator $Q = I - P$ is clearly also a projection (I is the identity operator); it is referred to as the *complementary projection* of P. Since im $P = \ker Q$, the range of a projection is always closed. Also notice that, obviously, P is the complementary projection of Q and that $PQ = QP = 0$. The norm of a non-zero projection is never less than 1.

2.1.6. Let M and N be closed subspaces of E such that $E = M + N$ and $M \cap N = \{0\}$. In this case E is said to be the *direct sum* of its subspaces M and N, and this is written in the form $E = M \dotplus N$. If $E = M \dotplus N$, then every element $x \in E$ has a unique decomposition into a sum $x = x_M + x_N$ with $x_M \in M$ and $x_N \in N$. The projection P defined by $Px := x_M$ is called the *projection of E onto M along N*; note that

$$M = \operatorname{im} P, \qquad N = \ker P, \tag{2.1}$$

and that M is isomorphic to E/N.

Vice versa, if P is any projection, then the space E decomposes into the direct sum of the two spaces given by (2.1). In this context, M and N are called *complementary subspaces*, and either of them is referred to as a *direct complement* of the other one.

If M is a finite-dimensional subspace of E or if M is a closed subspace of E whose *codimension*, codim $M := \dim(E/M)$, is finite, then M possesses a direct complement. Furthermore, every closed subspace M of a Hilbert space H is complementable: for example, the orthogonal complement $N = H \ominus M$ is a direct complement of M. A projection of a Hilbert space onto a closed subspace along the orthogonal complement of that subspace is said to be an *orthogonal projection*; note that its norm equals 1.

2.2. The Space of Linear Operators. Let E and F be complex normed spaces.

2.2.1. The collection of all continuous linear operators from E to F can be made into a normed space, $\mathscr{L}(E, F)$, by defining $(\alpha A + \beta B)x$ as $\alpha Ax + \beta Bx$ ($\alpha, \beta \in \mathbb{C}$, $x \in E$) and the norm of an operator A as $\|A\|_{E \to F}$. If F is a Banach space, then $\mathscr{L}(E, F)$ is also a Banach space.

2.2.2. In the case $E = F$ we shall abbreviate $\mathscr{L}(E, E)$ to $\mathscr{L}(E)$. For $A, B \in \mathscr{L}(E)$, define $AB \in \mathscr{L}(E)$ by $(AB)(x) = A(Bx)$ ($x \in E$). It is easy to see that $\|AB\| \leqq \|A\| \, \|B\|$.

In particular, if E is a Banach space, then $\mathscr{L}(E)$ is a Banach algebra with identity. The identity is nothing else than the identity operator I. Note that $\mathscr{L}(E)$ is not commutative unless E is one-dimensional.

The operator $[A, B] := AB - BA$ is called the *commutator* of the operators A and B. If $AB = BA$, then A and B are said to *commute* with each other.

2.3. Dual Spaces and Adjoint Operators

2.3.1. Given a normed space E, we let E' denote the so-called *dual* (or *adjoint*) *space* of E, i.e. the space $\mathscr{L}(E, \mathbb{C})$. The elements of E' are called *continuous linear functionals* or simply *functionals*. We remark that the dual space E' is always complete, even if E is not so.

2.3.2. Sometimes it is more convenient to introduce the algebraic operations on $\mathscr{L}(E, \mathbb{C})$ not as in 2.2.1, but by defining $(\alpha f + \beta g)(x) = \bar{\alpha} f(x) + \bar{\beta} g(x)$ ($\alpha, \beta \in \mathbb{C}$, $f, g \in E'$, $x \in E$). In this way $\mathscr{L}(E, \mathbb{C})$ becomes also a Banach space; this space is likewise referred to as the *dual space* of E, but it is denoted by E^*.

If H is a Hilbert space, then, by a well-known theorem of F. Riesz, H^* is isometrically isomorphic to H itself, in view of which H^* is often identified with H.

For the convenience's sake (and in analogy to the notation for the scalar product in Hilbert space), we shall henceforth write $\langle x, f \rangle$ in place of $f(x)$ ($x \in E, f \in E^*$).

For example, we have $[L_p(0, 1)]^* = L_{p'}(0, 1)$, where $p' = p/(p - 1)$ for $1 < p < \infty$ and $p' = \infty$ for $p = 1$. This identification of $[L_p(0, 1)]^*$ and $L_{p'}(0, 1)$ amounts to the statement that $f \in [L_p(0, 1)]^*$ if and only if there is a $y \in L_{p'}(0, 1)$ for which $\langle x, f \rangle = \int_0^1 x(t)\overline{y(t)}\, dt$ for all $x \in L_p(0, 1)$; in this case we have $\|f\| = \|y\|_{L_{p'}}$ (also see Sec. 1.2 of Chap. 2).

2.3.3. Let $A \in \mathscr{L}(E, F)$. The *adjoint operator* of A is the (uniquely determined) operator $A^* \in \mathscr{L}(F^*, E^*)$ satisfying

$$\langle Ax, f \rangle = \langle x, A^*f \rangle \qquad \forall x \in E, \qquad \forall f \in F^*.$$

We have $\|A^*\| = \|A\|$ and $(\alpha A + \beta B)^* = \bar{\alpha} A^* + \bar{\beta} B^*$ for $A, B \in \mathscr{L}(E, F)$ and α, $\beta \in \mathbb{C}$. If $A, B \in \mathscr{L}(E)$, then $(AB^*) = B^*A^*$.

If H is a Hilbert space, then $\mathscr{L}(H)$ is a C^*-algebra, the involution being passage to the adjoint operator.

2.4. One-Sided Invertible Operators. Again let E and F be Banach spaces. An operator $A \in \mathscr{L}(E, F)$ is said to be *left-* (resp. *right-*) *invertible* if there exists an operator $A_l^{-1} \in \mathscr{L}(F, E)$ (resp. $A_r^{-1} \in \mathscr{L}(F, E)$) such that $A_l^{-1}A = I_E$ (resp. $AA_r^{-1} = I_F$), where I_E is the identity operator on E. The operator A_l^{-1} (resp. A_r^{-1}) is called a *left* (resp. *right*) *inverse* of A. If A is both left and right invertible, then all left and right inverses are equal to one another; in that case A is said to be *invertible* and the unique operator A^{-1} satisfying $A^{-1}A = AA^{-1} = I$ is referred to as the *inverse* of A.

The open mapping theorem gives the following invertibility criteria:

$1°$. An operator $A \in \mathcal{L}(E, F)$ is invertible if and only if it is one-to-one and onto, i.e. if and only if $\ker A = \{0\}$ and $\operatorname{im} A = F$. This is *Banach's theorem*. Notice that one-to-one and onto operators are frequently called *bijective* operators.

$2°$. A is left-invertible if and only if $\ker A = \{0\}$ and there exists a projection of F onto $\operatorname{im} A$.

$3°$. A is right-invertible if and only if $\operatorname{im} A = F$ and there is a projection of E onto $\ker A$.

Finally, from the results of 1.2.2 we deduce that the set of all invertible (left- or right-invertible) operators is an open subset of $\mathcal{L}(E, F)$.

2.5. Transposed Operators. Let E and F be normed spaces.

2.5.1. We shall say that E and F form a *dual system* $\langle E, F \rangle$ with respect to a mapping $\langle \cdot, \cdot \rangle \colon E \times F \to \mathbb{C}$ if this mapping has the following properties:

1. $\langle \cdot, \cdot \rangle$ is a bounded bilinear form (i.e., the mapping is linear with respect to either of the variables and one has $|\langle x, y \rangle| \leq \gamma \|x\| \|y\|$ $\forall x \in E$, $\forall y \in F$ with some $\gamma > 0$ independent of x and y).

2. If $\langle x, y_0 \rangle = 0$ for all $x \in E$ and some $y_0 \in F$, then $y_0 = 0$.

3. If $\langle x_0, y \rangle = 0$ for all $y \in F$ and some $x_0 \in E$, then $x_0 = 0$.

In case E and F form a dual system, we define mappings $\char`^$ of F into E^* (resp. E into F^*) by $\hat{y}(x) = \langle x, y \rangle$ (resp. $\hat{x}(y) = \langle x, y \rangle$). It is easy to see that in this way natural embeddings $F \to E^*$ and $E \to F^*$ are given.

2.5.2. Let $\langle E, F \rangle$ be a dual system. Two operators $A \in \mathcal{L}(E)$ and $A^T \in \mathcal{L}(F)$ are said to be *transposed* to each other if

$$\langle Ax, y \rangle - \langle x, A^T y \rangle \qquad \forall x \in E, \qquad \forall y \in F.$$

Note that not every operator $A \in \mathcal{L}(E)$ (or $A^T \in \mathcal{L}(F)$) does have a transposed operator; however, if the transposed operator exists, it is, obviously, uniquely determined.

2.5.3. Once more assume $\langle E, F \rangle$ is a dual system. The collection of all operators $A \in \mathcal{L}(E)$ which possess a transposed operator $A^T \in \mathcal{L}(F)$ forms a linear space. This space will be denoted by $\mathcal{A}(E, F)$. It contains the identity operator I_E, and if $A, B \in \mathcal{A}(E, F)$ then AB also belongs to $\mathcal{A}(E, F)$; furthermore, one has $I_E^T = I_F$, $(\alpha A + \beta B)^T = \alpha A^T + \beta B^T$, $(AB)^T = B^T A^T$.

We define a norm on $\mathcal{A}(E, F)$ by $|A| := \max\{\|A\|, \|A^T\|\}$. It is readily seen that $|AB| \leq |A| |B|$ and $|I_E| = 1$. If both E and F are Banach spaces, then $\mathcal{A}(E, F)$ is a Banach algebra with identity.

2.5.4. Examples

$1°$. If E is any normed space, then E and its dual space E' form the so-called *natural dual system* $\langle E, E' \rangle$. The corresponding bilinear form is given by $\langle x, f \rangle = f(x)$ $(x \in E, f \in E')$. We have $\mathcal{A}(E, E') = \mathcal{L}(E)$ and $|A| = \|A\| = \|A^T\|$ for all $A \in \mathcal{L}(E)$. The operator $A^T \in \mathcal{L}(E')$ is determined by the same equality as the operator $A*$ (see 2.3.3); it is often referred to as the *dual operator* of A and denoted by A'.

2°. $\langle C[0, 1], C[0, 1]\rangle$ is a dual system with respect to the bilinear form $\langle x, y\rangle = \int_0^1 x(t)y(t)\varrho(t)\,dt$, where $\varrho \in C[0, 1]$ is an arbitrary positive function. Every integral operator K of the form

$$(Kx)(t) = \int_0^1 k(t, s)\varrho(s)x(s)\,ds$$

whose kernel k is continuous belongs to $\mathscr{A}(C[0, 1], C[0, 1])$ and its transposed operator is an integral operator of the same type; the kernel of the transposed operator is given by $k^T(t, s) = k(s, t)$.

3°. If $I \leq p \leq \infty, 1 \leq q \leq \infty$, and $1/p + 1/q = 1$, then both $\langle L_p(0, 1), L_q(0, 1)\rangle$ and $\langle C[0, 1], L_1(0, 1)\rangle$ are dual system with respect to the form $\langle x, y\rangle = \int_0^1 x(t)y(t)\,dt$.

2.6. Compact Operators. Let E, F, G be Banach spaces.

2.6.1. A linear operator $A: E \to F$ is said to be *compact* or to be *completely continuous* if the closure $\overline{A(M)}$ of $A(M)$ is a compact subset of F whenever M is a bounded subset of E. The collection of all compact operators from E into F will be denoted by $\mathscr{K}(E, F)$ $(\mathscr{K}(E) := \mathscr{K}(E, E))$.

It is clear that $\mathscr{K}(E, F) \subseteq \mathscr{L}(E, F)$. The identity operator I_E is compact if and only if E is a finite-dimensional space (*F. Riesz' theorem*).

2.6.2. An operator $A \in \mathscr{L}(E, F)$ is compact if and only if each sequence (x_n) in E with $\|x_n\| \leq 1$ contains a subsequence (x_{n_k}) such that (Ax_{n_k}) is a convergent sequence in F.

2.6.3. J. Schauder showed that A belongs to $\mathscr{K}(E, F)$ if and only if $A*$ is in $\mathscr{K}(F*, E*)$.

2.6.4. Let $A \in \mathscr{K}(E, F)$. Then the range im A is separable. Furthermore, if (x_n) is any weakly convergent sequence in E, then (Ax_n) is a norm convergent sequence in F.

2.6.5. $\mathscr{K}(E, F)$ is a closed subspace of $\mathscr{L}(E, F)$. If $A \in \mathscr{K}(E, F)$ or $B \in \mathscr{K}(F, G)$, then $BA \in \mathscr{K}(E, G)$.

In particular, $\mathscr{K}(E)$ is a closed ideal of the Banach algebra $\mathscr{L}(E)$. The quotient algebra $\mathscr{L}(E)/\mathscr{K}(E)$ is usually referred to as the *Calkin algebra* (in recognition of J.W. Calkin). For $A \in \mathscr{L}(E, F)$, the norm

$$\|A\| := \inf\{\|A - T\| : T \in \mathscr{K}(E, F)\}$$

is called the *essential norm* of A.

2.6.6. Finite rank operators provide the simplest examples of compact operators.

An operator $K \in \mathscr{L}(E, F)$ is said to be a *finite rank operator* if its range is finite-dimensional. A finite rank operator K can be represented in the form

$K = \sum_{j=1}^{m} f_j \otimes y_j$ with certain $y_j \in F$ and $f_j \in E'$ ($j = 1, \ldots, m$); here $f_j \otimes y_j$ is defined by $(f_j \otimes y_j)(x) = f_j(x)y_j$.

If the range of a compact operator is closed, then the operator is of finite rank.

The collection of all finite rank operators $K \in \mathcal{L}(E, F)$ is denoted by $\mathcal{F}(E, F)$. It is obvious that the set $\mathcal{F}(E) := \mathcal{F}(E, E)$ is an ideal of the algebra $\mathcal{L}(E)$. It turns out that every ideal $\mathfrak{J}$ of the algebra $\mathcal{L}(E)$ contains $\mathcal{F}(E)$ and is contained in $\mathcal{K}(E)$, i.e., we have $\mathcal{F}(E) \subseteqq \mathfrak{J} \subseteqq \mathcal{K}(E)$ (*Calkin's theorem*).

2.6.7. An important subset of $\mathcal{K}(E, F)$ is the class of nuclear operators. An operator $A \in \mathcal{L}(E, F)$ is said to be *nuclear* if it is of the form $A = \sum_{j=1}^{\infty} f_j \otimes y_j$, where

$$y_j \in F, f_j \in E', \quad \text{and} \quad \sum_{j=1}^{\infty} \|f_j\|_{E'} \|y_j\|_F < \infty.$$

It is clear that nuclear operators can be represented in the form $A = \sum_{j=1}^{\infty} \sigma_j g_j \otimes x_j$, where $\sigma_j \geq 0$, $x_j \in F$, $g_j \in E'$, $\|x_j\| = \|g_j\| = 1$, $\sum_{j=1}^{\infty} \sigma_j < \infty$.

We remark that every nuclear operator can be approximated in the norm of $\mathcal{L}(E, F)$ by finite rank operators as closely as desired.

2.6.8. Every compact operator on a separable Hilbert space is the uniform limit of a sequence of finite rank operator (see also Sec. 7.1).

There exist separable Banach spaces E with the property that not every compact operator on E is the uniform limit of finite rank operators. The first example of such a space was constructed by P. Enflo in 1972, who so at the same time solved two fundamental problems of functional analysis: the basis problem posed by S. Banach in 1932 and the approximation problem raised by A. Grothendieck in 1955.

For more about the topics touched upon in this section see, for instance, Jörgens [1970] or Pietsch [1978].

§3. Noether and Fredholm Operators

The linear integral operators that will be considered in the following chapters possess an extremely important property: when thought of as acting on appropriate Banach spaces, they are normally solvable and have a finite index. Such operators will be called Noether operators. The present section concentrates upon the basic properties of Noether operators and upon the connection between the Noether property and a-priori estimates on the one hand and the solvability theory of the corresponding operator equations on the other hand.

Throughout this chapter, we shall assume that all spaces occuring are complex Banach spaces unless the contrary is stated explicitly.

3.1. Let $A \in \mathcal{L}(E, F)$ and $y \in F$. If the equation

$$Ax = y \tag{3.1}$$

has a solution $x \in E$, then, obviously,

$$\langle y, f \rangle = 0 \qquad \text{for all } f \in \ker A^*. \tag{3.2}$$

If the converse of this is true, that is, if condition (3.2) implies that equation (3.1) is solvable, then the operator A is said to be *normally solvable* (in the sense of Hausdorff). The so-called *Hausdorff lemma* says that for A to be normally solvable it is necessary and sufficient that the image im A be a closed subset of F. Furthermore, the following statements are equivalent (*Banach-Hausdorff theorem*):

(i) A is normally solvable;
(ii) im A^* is a closed subset of E^*;
(iii) im $A^* = \{ f \in E^* : \langle x, f \rangle = 0 \ \forall x \in \ker A \}$.

3.2. An important subset of the collection of all normally solvable operators is constituted by the generalized invertible operators. An operator $A \in \mathscr{L}(E, F)$ is said to be *generalized invertible* if there exists an operator $B \in \mathscr{L}(F, E)$ such that $ABA = A$.

It is not difficult to see that an operator A is generalized invertible if and only if it is normally solvable and both the spaces $\ker A \subseteq E$ and $\operatorname{im} A \subseteq F$ are complementable. Moreover, if A is generalized invertible then there exists even an operator $B \in \mathscr{L}(F, E)$ such that

$$ABA = A \qquad \text{and} \qquad BAB = B. \tag{3.3}$$

Any operator B satisfying (3.3) is called a *generalized inverse* of A and is denoted by $A^{(-1)}$. It can be easily verified that (3.3) is equivalent to

$$BA = I_E - P, \qquad AB = I_F - Q, \qquad PB = BQ = 0, \tag{3.4}$$

where P is a projection of E onto $\ker A$ and $I_F - Q$ is a projection of F onto im A.

It is clear that every left inverse and every right inverse of an operator is also a generalized inverse. If equation (3.1) is solvable and if a generalized inverse is known, then the set of all solutions of (3.1) admits the following description: $x_0 = A^{(-1)}y$ is one of the solutions, and the general solution is given by $x = x_0 + u - A^{(-1)}Au$, where u is an arbitrary element of E. The equality $AA^{(-1)}y = y$ is necessary and sufficient for equation (3.1) to be solvable.

3.3. A normally solvable operator $A \in \mathscr{L}(E, F)$ is called a *Noether operator* or a *Φ-operator* (or is said to be *Noetherian*) if the integers $\alpha(A) := \dim \ker A$ and $\beta(A) := \alpha(A^*)$ are both finite, and it is called a *semi-Noether* operator if at least one of the integers $\alpha(A)$ and $\beta(A)$ is finite. A semi-Noether operator A for which $\alpha(A) < \infty$ is termed a *Φ_+-operator* while semi-Noether operators with $\beta(A) < \infty$ are referred to as *Φ_--operators*.

The collection of all $\Phi(\Phi_\pm)$-operators $A \in \mathscr{L}(E, F)$ will be denoted by $\Phi(E, F)$ $(\Phi_\pm(E, F))$. By what was said above, $A \in \Phi_\pm(E, F) \Leftrightarrow A^* \in \Phi_\mp(F^*, E^*)$. If $A \in \Phi(E, F)$, then Ind $A = \alpha(A) - \alpha(A^*)$ and Ind $A = -$ Ind A^*.

Remark. The term "Φ-operator" was introduced by Gohberg, Krein [1957]. Frequently, and in English literature almost consistently, Φ-operators are called

"Fredholm" operators. However, in the author's opinion, the epithet "Noether" seems to be more correct from the historical point of view: normally solvable operators with a nonvanishing index first appeared in the fundamental paper of Noether [1921], in which the foundation of the general theory of one-dimensional singular integral equations with Hilbert kernels was laid. Note that, moreover, Noether was the first to establish an explicit index formula for the corresponding integral operators. In this way he also repaired some inaccuracies admitted by D. Hilbert (1904) when studying the singular integral equation to which the Riemann-Hilbert boundary value problem had been reduced by him (see also Hilbert [1912], p. 219, Theorem 43, and Mikhlin, Prössdorf [1980], p. 189).

3.4. The Noether property and the essential norm of an operator are intimately connected with a certain a-priori estimate (see Kohn, Nirenberg [1965], Mikhlin, Prössdorf [1980]). We shall say that a seminorm $|\cdot|$ defined on E is compact with respect to a norm $\|\cdot\|$ on E if every sequence of elements of E which is bounded in the norm $\|\cdot\|$ contains a subsequence which is a Cauchy sequence in the seminorm $|\cdot|$.

Theorem 1. *Let $A \in \mathscr{L}(E, F)$. Then $A \in \Phi_+(E, F)$ if and only if there exist a seminorm $|\cdot|$ compact with respect to $\|\cdot\|_E$ and a number $C > 0$ such that*

$$\|x\|_E \leqq C\|Ax\|_F + |x| \qquad \forall x \in E. \tag{3.5}$$

Theorem 2. *Let H be a Hilbert space and let $A \in \mathscr{L}(H)$ be an operator with the following property: for some constant $K > 0$ and for every $\varepsilon > 0$ there exists a seminorm $|\cdot|$ which is compact with respect to the norm $\|\cdot\|$ (and is allowed to depend on ε) such that*

$$\|Ax\| \leqq (K + \varepsilon)\|x\| + |x| \qquad \forall x \in H.$$

Then $\|\!|\!|A|\!|\!| \leqq K$.

3.5. From the results listed above we obtain the following basic properties of Noether operators.

(i) An operator $A \in \mathscr{L}(E, F)$ is a Φ-operator if and only if there are operators $B, C \in \mathscr{L}(F, E)$, $T_E \in \mathscr{K}(E)$, and $T_F \in \mathscr{K}(F)$ such that

$$BA = I_E + T_E, \qquad AC = I_F + T_F. \tag{3.6}$$

Moreover, if $A \in \Phi(E, F)$ then there exists a generalized inverse $A^{(-1)} \in \mathscr{L}(F, E)$ such that both $A^{(-1)}A - I_E$ and $AA^{(-1)} - I_F$ are finite rank operators.

In particular, we see that an operator $A \in \mathscr{L}(E)$ is Noetherian if and only if the coset $\tilde{A}$ of the Calkin algebra $\mathscr{L}(E)/\mathscr{K}(E)$ containing A is invertible.

(ii) If $A \in \Phi_\pm(E, F)$ and $B \in \Phi_\pm(F, G)$, then $BA \in \Phi_\pm(E, G)$ and Ind $BA =$ Ind $A +$ Ind B.

(iii) If $A \in \Phi_\pm(E, F)$ and $T \in \mathscr{K}(E, F)$, then $A + T \in \Phi_\pm(E, F)$ and Ind$(A + T) =$ Ind A.

An important special case of this is the *Theorem of F. Riesz*: If $T \in \mathscr{K}(E)$, then $I + T \in \Phi(E)$ and Ind$(I + T) = 0$.

(iv) Given an operator $A \in \Phi_{\pm}(E, F)$, there exists a number $\varrho > 0$ such that $A + C \in \Phi_{\pm}(E, F)$, $\mathrm{Ind}(A + C) = \mathrm{Ind}\, A$, $\alpha(A + C) \leqq \alpha(A)$, $\beta(A + C) \leqq \beta(A)$ whenever $C \in \mathscr{L}(E, F)$ and $\|C\| < \varrho$.

Thus, the sets $\Phi_{\pm}(E, F)$ are open subsets of $\mathscr{L}(E, F)$ and the indices of two homotopic operators $A_0, A_1 \in \Phi(E)$ are equal to each other.

The first part of Theorem (ii) can be reversed. Let $A \in \mathscr{L}(E, F)$ and $B \in \mathscr{L}(F, G)$.

(v) If $BA \in \Phi_+(E, G)$, then $A \in \Phi_+(E, F)$; if $BA \in \Phi_-(E, G)$, then $B \in \Phi_-(F, G)$.

(vi) If $BA \in \Phi(E, G)$, then either both A and B are Φ-operators or none of them is Noetherian.

3.6. An operator $A \in \Phi(E, F)$ is said to be *Fredholm* if $\mathrm{Ind}\, A = 0$. For an operator $A \in \mathscr{L}(E, F)$ to be Fredholm it is necessary and sufficient that it be representable in the form $A = D + T$ where D is an invertible operator in $\mathscr{L}(E, F)$ and T belongs to $\mathscr{K}(E, F)$ (*Theorem of S.M. Nikol'skiĭ*).

The *equation* $Ax = y$ is said to be *Noetherian* or *Fredholm* if the operator A has the corresponding property. The F. Riesz theorem provides the classical example of a Fredholm equation: $x + Tx = y$ ($T \in \mathscr{K}(E)$). Equations of this form are frequently called *Riesz-Schauder equations* (or *equations of the second kind*).

From what has been said above we infer the following fundamental properties of Noether and Fredholm equations. If $A \in \Phi(E, F)$, then the *Noether theorems* hold:

(i) Either the equation (3.1) has at least one solution for every right-hand side $y \in F$ or the equation $A^*f = 0$ has a nontrivial solution $f \in F^*$.

(ii) Each of the equations $Ax = 0$ and $A^*f = 0$ has at most finitely many linearly independent solutions.

(iii) The equation (3.1) has a solution if and only if condition (3.2) is satisfied.

In case $A \in \Phi(E, F)$ is even a Fredholm operator, the *Fredholm theorems* are valid:

(a) (*Fredholm alternative*) Either the equation (3.1) has a unique solution for every right-hand side or the equation $Ax = 0$ has a nontrivial solution.

(b) The equations $Ax = 0$ and $A^*f = 0$ have the same number of linearly independent solutions.

(c) For equation (3.1) to have a solution it is necessary and sufficient that (3.2) be fulfilled.

3.7. Recall the definitions of a dual system $\langle E, F \rangle$ and of the Banach algebra $\mathscr{A}(E, F)$ (Sec. 2.5). The reason for giving these definitions is that F may be viewed as a certain substitute for the dual space E^* and that the transposed operator A^T may take the place of the adjoint operator A^*, which is especially useful if E^* or A^* are of a too complicated nature or are even unknown (see the examples of 2.5.4).

Theorem. *Let $A \in \mathscr{A}(E, F)$ and assume $A \in \Phi(E)$, $A^T \in \Phi(F)$, and $\mathrm{Ind}\, A = -\mathrm{Ind}\, A^T$. Then $\ker A^T = \ker A^*$, $\ker A = \ker(A^T)^*$, and consequently, $\alpha(A) = \beta(A^T)$, $\alpha(A^T) = \beta(A)$.*

Thus, under the hypothesis of this theorem, the Noether and Fredholm theorems continue to hold if the equation $A^*f = 0$ is replaced by the equation $A^T f = 0$.

3.8. Every Noether equation can be reduced to a Riesz-Schauder equation by means of two linear transformations. Before demonstrating this, it is convenient to introduce the following notions.

Any operators B and C satifying (3.6) with $T_E \in \mathcal{K}(E)$ and $T_F \in \mathcal{K}(F)$ are referred to as *left* and *right regularizers* of $A \in \mathcal{L}(E, F)$, respectively. The construction of regularizers is one of the most powerful methods of proving that a given equation is Noetherian. Note also that any two regularizers of a Φ-operator differ by a compact operator only.

If we are given a left regularizer B and a right regularizer C, equation (3.1) can be transformed into the equation

$$x + T_E x = By, \tag{3.7}$$

or, by means of the substitution $x = Cu$, into the equation

$$u + T_F u = y.$$

A left regularizer B is said to be an *equivalent* one if $\ker B = \{0\}$, and a right regularizer C is called *equivalent* if $\operatorname{im} C = E$. An operator possesses a left (right) equivalent regularizer if and only if it is Noetherian and its index is non-negative (non-positive).

If B (resp. C) is an unbounded operator acting from F into E such that (3.6) is satisfied, then it is called an *unbounded* left (resp. right) *regularizer*.

There is a class of equations which, on the one hand, frequently occur in several applied problems and, on the other hand, can be easily regularized. This is the class of equations

$$x - Sx = y \tag{3.8}$$

where $S \in \mathcal{L}(E)$ is a so-called *quasicompact* operator, which means that there exist an operator $T \in \mathcal{K}(E)$ and a positive integer m such that $\|S^m - T\| < 1$. In that case the operator $A := I - S^m + T$ is clearly invertible and so, with $D := I + S + \cdots + S^{m-1}$, we have

$$A^{-1}D(I - S) = I - A^{-1}T, \qquad (I - S)DA^{-1} = I - TA^{-1}.$$

Thus, $A^{-1}D$ is simultaneously a left and right regularizer of equation (3.8).

If, in particular, $S^m = T \in \mathcal{K}(E)$, then an equivalent left and right regularizer of equation (3.8) is given by $B := (\varepsilon_1 I - S) \ldots (\varepsilon_{n-1} I - S)$ where $\varepsilon_k = e^{ik\pi/n}$ and $n \geq m$ is chosen so that all the operators $\varepsilon_k I - S$ $(k = 1, \ldots, n - 1)$ are invertible. Note that actually $B(I - S) = (I - S)B = I - S^n$.

3.9. All results of the present section (except for those whose formulation makes sense only in the Banach space case) remain valid in the setting of locally convex linear topological spaces (H.H. Schaefer, 1956). The only modification is that the normal solvability of an operator, $A \in \mathcal{L}(E, F)$ say, must be everywhere supplemented by the requirement that A be a topological homomorphism, i.e. that, for every open subset U of E, the image $A(U)$ be an open subset of im A (for the topology induced by F) (see Schaefer [1971], Chap. III; also see 2.1.4 and 2.4).

For more about the topics dealt with in this section see, for example, Gohberg, Krein [1957], Gohberg, Krupnik [1973], Jörgens [1970], Prössdorf [1974], Mikhlin, Prössdorf [1980].

§4. Classification of the Points of the Spectrum of a Linear Operator

We are going to combine the general spectral theory in Banach algebras with the theory of Noether operators. First notice that a number $\lambda \in \mathbb{C}$ belongs to the spectrum $\sigma(A)$ of an operator $A \in \mathscr{L}(E)$ if and only if at least one of the conditions $\alpha(A - \lambda I) = 0$, $\beta(A - \lambda I) = 0$, $A - \lambda I \in \Phi(E)$ is violated.

4.1. Let $A \in \Phi(E)$. The set $\varphi(A) := \{\lambda \in \mathbb{C} : A - \lambda I \in \Phi(E)\}$ will be called the *Noether resolvent set* of A. It is clear that $\varrho(A)$ is a subset of $\varphi(A)$. From Theorem 3.5(iv) we deduce that $\varphi(A)$ is an open set and that the index of $A - \lambda I$ is constant on the connected components of $\varphi(A)$. If such a component contains a point $\lambda_0 \in \varrho(A)$, then $A - \lambda I$ is Fredholm for all λ belonging to that component. Components of this type may only contain isolated points of the spectrum of A, which, moreover, are poles of the resolvent $R_\lambda(A)$, and all the operators occuring in the principal part of the Laurent expansion of $R_\lambda(A)$ at these points have finite rank.

4.2. The points $\lambda \in \mathbb{C}$ for which $|\lambda| > r_A$ belong all to the resolvent set $\varrho(A)$. Hence there is a connected component of $\varphi(A)$ which entirely contains the exterior of the spectral disk, and on this component the operator $A - \lambda I$ is Fredholm.

4.3. The set $\sigma_e(A) := \mathbb{C} \backslash \varphi(A)$ $(\subseteq \sigma(A))$ is called the *essential spectrum* of the operator A. If dim $E = \infty$, then $\sigma_e(A)$ coincides with the spectrum of the corresponding coset $\tilde{A}$ of the Calkin algebra $\mathscr{L}(E)/\mathscr{K}(E)$ and consequently, $\sigma_e(A)$ cannot be empty. If in particular $A \in \mathscr{K}(E)$, then, by the F. Riesz theorem, $\sigma_e(A) = \{0\}$.

It may happen that $\sigma_e(A) = \sigma(A) = \{0\}$, in which case clearly $r_A = 0$. A classical example of an operator A with this property is the (compact) *Volterra integral operator*

$$(Ax)(t) = \int_0^t k(t, s)x(s)\, ds \qquad \text{(see Sec. 1 of Chap. 2),}$$

considered as acting on $L_p(0, 1)$ $(1 \le p \le \infty)$ or on $C[0, 1]$; here the kernel k is supposed to be continuous on the triangle $0 \le s \le t \le 1$. On the other hand, there exist integral operators (for instance, singular ones; see Chapters 3 and 4) whose essential spectrum consists of line segments and circular arcs in the complex plane.

4.4. The following classification of the points of the spectrum of an operator $A \in \mathscr{L}(E)$ is in common use: if $\alpha(A - \lambda I) \neq 0$, then λ is said to belong to the *point spectrum* $\sigma_p(A)$; if $\alpha(A - \lambda I) = 0$ but $\beta(A - \lambda I) \neq 0$, then λ is referred to as a point of the *residual spectrum* $\sigma_r(A)$; and if $\alpha(A - \lambda I) = \beta(A - \lambda I) = 0$ but $A - \lambda I \notin \Phi(E)$ (implying that A is not normally solvable), then λ is included into the *continuous spectrum* $\sigma_c(A)$. Note that obviously $\sigma_c(A) \subseteq \sigma_e(A)$.

Thus, the whole complex plane is divided into the union of four pairwise disjoint sets:

$$\mathbb{C} = \varrho(A) \cup \sigma_p(A) \cup \sigma_r(A) \cup \sigma_c(A).$$

If $\alpha(A - \lambda I) \neq 0$, then the number λ is called an *eigenvalue* of the operator A and the integer $\gamma(\lambda) := \alpha(A - \lambda I)$ is referred to as the *order* or the *geometric multiplicity* of the eigenvalue λ. The subspace $\ker(A - \lambda I)$ is called the *eigensubspace*, and each nonzero element $u \in \ker(A - \lambda I)$, i.e. each nonzero solution of the equation $Au = \lambda u$, is referred to as an *eigenvector* corresponding to the eigenvalue λ.

If $\beta(A - \lambda I) \neq 0$, then λ is said to be a *deficiency value* of order $\beta(A - \lambda I)$, and the space $[\mathrm{im}(A - \lambda I)]^{\perp}$ of all functionals $f \in E^*$ which annihilate $\mathrm{im}(A - \lambda I)$ is called the *deficiency space* of λ.

It is clear that $\bar{\lambda}$ is an eigenvalue of the adjoint operator A^* if and only if λ is a deficiency value of A; in that case their orders are equal to each other. From the results recorded in Sec. 3 one easily obtains the following relations betweeen the spectra σ_p, σ_r, σ_c of the operators A and A^* (for $M \subseteq \mathbb{C}$, we let $M^* := \{\bar{\lambda} : \lambda \in M\}$):

$$[\sigma_c(A)]^* \subseteq \sigma_c(A^*) \cup \sigma_r(A^*), \qquad [\sigma_r(A)]^* \subseteq \sigma_p(A^*),$$

$$[\sigma_p(A)]^* \subseteq \sigma_p(A^*) \cup \sigma_r(A^*)$$

If E is a reflexive space, then $[\sigma_c(A)]^* = \sigma_c(A^*)$ and thus $[\sigma_r(A^*)]^* \subseteq \sigma_p(A)$.

4.5. From 1.3.3 we infer that if λ_0 is a pole of the order p of the resolvent $R_\lambda(A)$, then λ_0 is an eigenvalue of the operator A. Moreover, if A_{-1} is the residue of the pole λ_0 then (recall that $N(A) := \ker A$, $R(A) := \mathrm{im}\, A$)

$$\{0\} \subseteq \cdots \subseteq N((A - \lambda_0 I)^{p-1}) \subseteq N((A - \lambda_0 I)^p),$$

$$E \supseteq \cdots \supseteq R((A - \lambda_0 I)^{p-1}) \supseteq R((A - \lambda_0 I)^p),$$

$$R(A_{-1}) = N((A - \lambda_0 I)^n), \qquad N(A_{-1}) = R((A - \lambda_0 I)^n)$$

for all $n \geq p$, and all these subspaces of E are invariant subspaces of A. Finally notice that E decomposes into the direct sum

$$E = N((A - \lambda_0 I)^n \dotplus R((A - \lambda_0 I)^n) \qquad (n \geq p).$$

4.6. Let λ be a nonzero eigenvalue of A. The number $\mu = 1/\lambda$ is said to be a *characteristic value* of A and the number

$$\varrho_A := \sup\{r : I - \mu A \text{ is Fredholm for } |\mu| < r\}$$

is called the *Fredholm radius* of A. We clearly have $\varrho_A \geq 1/|||A|||$, where $|||A|||$ is the essential norm of A. If A is compact then $\varrho_A = \infty$.

4.7. It is frequently more convenient to work with the operator

$$A(\mu) := A(I - \mu A)^{-1} \qquad (\mu^{-1} \in \varrho(A) \text{ or } \mu = 0)$$

instead of the resolvent $R_\lambda(A)$; the operator $A(\mu)$ is called the *Fredholm resolvent* of A and, as a rule, it has better properties than $R_\lambda(A)$. The Fredholm resolvent is especially useful in studying integral equations (see Chap. 2). Notice that if $\mu \neq 0$ and $\lambda = 1/\mu$, then $A(\mu) = -\lambda A R_\lambda(A)$.

A simple computation gives that

$$(I - \mu A)^{-1} = I + \mu A(\mu) \qquad (\mu^{-1} \in \varrho(A) \text{ or } \mu = 0).$$

The function $A(\cdot)$ shares the analyticity property with the resolvent $R_\lambda(A)$: it is analytic on the set $\{\mu \in \mathbb{C} : \mu^{-1} \in \varrho(A)\} \cup \{0\}$. From what was said in 4.5, we conclude that each pole of the function $A(\cdot)$ is a characteristic value of the operator A.

For more about the subject matter of the present section we refer to Jörgens [1970], Riesz, Sz.-Nagy [1979], Pietsch [1978] for example.

§5. Theory of Riesz Operators

F. Riesz studied operators of the form $I - \lambda S$, where S is a compact operator. In 1954, A.F. Ruston introduced the notion of a Riesz operator and extended the basic results of Riesz to this more general class of operators.

5.1. An operator $S \in \mathscr{L}(E)$ is called a *Riesz operator* if $I - \lambda S \in \Phi(E)$ for all $\lambda \in \mathbb{C}$.

From what was said at end of Sec. 3.8 we infer that S is a Riesz operator if $S^m \in \mathscr{K}(E)$ for some natural number m. Other examples of Riesz operators are the strongly singular and strongly co-singular operators. An operator $S \in \mathscr{L}(E)$ is said to be *strongly singular* if there is no infinite-dimensional subspace M of E such that the restriction $S|M$ of S to M is one-to-one (injective), and an operator $S \in \mathscr{L}(E)$ is called *strongly co-singular* if there is no subspace N of E with infinite codimension such that $\pi_N S$ is onto (surjective), π_N being the canonical projection of E onto E/N.

If S is a Riesz operator, then, obviously, zS is a Riesz operator for all $z \in \mathbb{C}$. Also note that by virtue of the equality $r_{\tilde{S}} = \lim_{n \to \infty} |||S^n|||^{1/n} = 0$, where $\tilde{S}$ refers to the coset of the Calkin algebra containing S, every Riesz operator is quasi-compact. This simple observation is a key ingredient of the proof of the following basic result on the structure of Riesz operators.

5.2. Theorem. *If S is a Riesz operator, then*

(i) *the spectrum $\sigma(S)$ is an at most countable set the only possible accumulation point of which is zero,*

(ii) *the nonzero points λ_k ($k = 1, 2, \ldots$) of the spectrum are poles, of the order p_k, of the resolvent $R_\lambda(S)$ and are thus eigenvalues of the operator S.*

As already mentioned, the space E decomposes into the direct sum

$$E = N((S - \lambda_k I)^{p_k}) \dotplus R((S - \lambda_k I)^{p_k}), \tag{5.1}$$

and either of these two spaces is an invariant subspace of S.

5.3. The finite-dimensional space $N_k := N((S - \lambda_k I)^{p_k})$ is referred to as the *generalized eigenspace* for the eigenvalue λ_k, and its dimension $\alpha(\lambda_k) := \dim N_k$ is called the *algebraic multiplicity* of the eigenvalue λ_k. The geometric multiplicity $\gamma(\lambda_k) = \dim N((S - \lambda_k I))$ does clearly not exceed the algebraic multiplicity $\alpha(\lambda_k)$, and, since $N_k \neq N((S - \lambda_k I)^{p_k-1})$, both multiplicities coincide if and only if $p_k = 1$. Also notice that $p_k \leqq \alpha(\lambda_k)$.

5.4. The decomposition (5.1) induces a representation of the Riesz operator S as a sum

$$S = S_k' + S_k'', \tag{5.2}$$

where $S_k' = SP_k$, $S_k'' = S(I - P_k)$, and P_k is the projection of E onto N_k along $R((S - \lambda_k I)^{p_k})$. Since the spaces to the right of (5.1) are invariant under the operator S, we obtain that

(1) λ_k is the only point of the spectrum of the operator S_k,

(2) $\sigma(S_k'') = \sigma(S)\backslash\{\lambda_k\}$,

(3) $S_k' S_k'' = S_k'' S_k' = 0$.

5.5. Because $P_k P_j = 0$ for $k \neq j$ (see 1.3.3), the space E decomposes into the direct sum $E = N_0 \dotplus N_1 \dotplus \cdots \dotplus N_n$, where $N_0 = P_0 E$, $P_0 = I - \sum_{k=1}^n P_k$, and n is an arbitrary natural number which does not exceed the number of nonzero eigenvalues of S. The operator S maps each of the subspaces N_j ($j = 0, \ldots, n$) into itself, and its restriction S_j' to N_j has the spectrum $\sigma(S_j) = \{\lambda_j\}$ if $j \geqq 1$, while $\sigma(S_0') = \{0, \lambda_{n+1}, \lambda_{n+2}, \ldots\}$. This decomposition produces a representation of the resolvent $R_\lambda(S)$ as the sum of the principal parts corresponding to the poles λ_1, $\lambda_2, \ldots, \lambda_n$ and some remainder which is analytic in $\mathbb{C}\backslash\sigma(S_0')$.

5.6. The following theorem shows that each generalized eigenspace N_k of a Riesz operator has a basis with respect to which the matrix representation of the operator $S_k' = S|N_k$ is in Jordan canonical form.

Theorem. *Let λ_0 be a nonzero point of the spectrum of a Riesz operator S, let p be the order of λ_0, and let P denote the projection of E onto the generalized eigenspace $N = N((S - \lambda_0 I)^p)$ along $R((S - \lambda_0 I)^p)$. Then there exists a basis $\{e_{j,k}\}$ ($k = 1, \ldots, n; j = 1, \ldots, m := \gamma(\lambda_0)$) in N such that*

$$(S - \lambda_0 I)e_{j,k} = \begin{cases} 0 & \text{for } k = 1, \\ e_{j,k-1} & \text{for } k = 2, \ldots, n_j; \end{cases} \tag{5.3}$$

one has $\max\{n_j : j = 1, \ldots, m\} = p$ *and* $\sum_{j=1}^m n_j = \alpha(\lambda_0)$.

There is also a basis $\{f_{j,k}\}$ $(k = 1, \ldots, n_j; j = 1, \ldots, m)$ *in the generalized eigenspace* $N((S^* - \bar{\lambda}_0 I)^p)$ *such that*

$$(S^* - \bar{\lambda}_0 I)f_{j,k} = \begin{cases} 0 & \text{for } k = 1, \\ f_{j,k-1} & \text{for } k = 2, \ldots, m \end{cases}$$

and, moreover, the biorthogonality condition

$$\langle e_{j,k}, f_{l,h} \rangle = \delta_{j,l}\delta_{k,n_l-h+1}$$

is satisfied.

The collection of elements $\{e_{j,k}\}$ and functionals $\{f_{j,k}\}$ appearing in the preceding theorem is called the *canonical basis* of the operator S corresponding to the eigenvalue λ_0. In terms of the canonical basis, the projection P can be written in the form

$$Pu = \sum_{j=1}^m \sum_{k=1}^{n_j} \langle u, f_{j,n_j-k+1} \rangle e_{j,k}, \tag{5.4}$$

so that the operator SP (see (5.2)) assumes the form

$$SPu = \lambda_p Pu + \sum_{j=1}^m \sum_{k=1}^{n_j-1} \langle u, f_{j,n_j-k} \rangle e_{j,k}. \tag{5.5}$$

In a sufficiently small neighborhood of the point λ_0 the resolvent $R_\lambda(S)$ has the expansion

$$R_\lambda(S) = \sum_{k=1}^p (\lambda - \lambda_0)^{-k} S_{-k} + \sum_{n=0}^\infty (\lambda - \lambda_0)^n S_n, \tag{5.6}$$

where $S_{-k} = -(S - \lambda_0 I)^{k-1} P$ $(k = 1, \ldots, p)$ and $S_n = S_0^{n+1}$ $(n \geq 1)$, S_0 being some operator in $\mathcal{L}(E)$. Note that the series (5.6) converges in the norm of $\mathcal{L}(E)$.

For the results of this section we refer to Jörgens [1970], Riesz, Sz.-Nagy [1979], Pietsch [1978].

§ 6. The Grothendieck-Pietsch Theory of Fredholm Determinants

Determinants of integral operators were introduced by Fredholm in 1903, who described the exact solution of integral equations of the second kind with continuous kernels in terms of these determinants (see also Sec. 3 of Chap. 2). Proceeding in analogy to the theory of linear algebraic equations, Fredholm studied this type of equations almost completely.

The abstract theory of determinants for nuclear operators on Banach spaces was developed by Grothendiek, Ruston, and Leżański in the early fifties. Here we give an account of the approach of A. Pietsch (1963; also see the monograph Pietsch [1978]), which is based on a straightforward construction of the determinants and Fredholm divisors. This approach has its sources in the construction of the characteristic polynomial of a matrix, in the definition of infinite determinants going back to H. von Koch (1900), and in Fredholm's theory of integral equations. The first step is the construction of determinants and Fredholm divisors for nuclear operators on the space l_1, and using the concept of related operators, after this the results are extended to some more general classes of operators, including nuclear operators acting on arbitrary Banach spaces.

6.1. We begin by stating a characterization of Riesz operators which relies on the fact that their Fredholm resolvent is a meromorphic function.

Theorem. *An operator* $S \in \mathcal{L}(E)$ *is a Riesz operator if and only if for each* $\lambda_0 \in \mathbb{C}$ *there exists a number* $\varrho > 0$ *such that*

$$S(\lambda) = \sum_{k=1}^{p} (\lambda - \lambda_0)^{-k} S_{-k}(\lambda_0) + \sum_{n=0}^{\infty} (\lambda - \lambda_0)^n S_n(\lambda_0)$$

for $0 < |\lambda - \lambda_0| < \varrho$, *where* $S_{-p}(\lambda_0), \dots, S_{-1}(\lambda_0)$ *are any finite rank operators and* $S_0(\lambda_0), S_1(\lambda_0), \dots$ *are any bounded operators of E.*

Moreover, the Fredholm resolvent $S(\lambda)$ *has a pole at each point* λ_0 *which is a characteristic value of S.*

This theorem can be derived from the properties of Riesz operators recorded in the previous section.

6.2. Because the set of characteristic values of a Riesz operator S does not have finite accumulation points (Theorem 5.2), a well-known theorem of Weierstrass can be applied to deduce that there exists an entire function d on the complex plane whose set of zeros coincides with the set of the characteristic values of S. Furthermore, this function d can be chosen so that the order of each zero λ_0 is equal to the algebraic multiplicity $\alpha(1/\lambda_0)$ of λ_0. Each such function is called a *Fredholm divisor* of the operator S.

6.3. Fredholm divisors possess the following important property.

Theorem. *Let* $S \in \mathcal{L}(E)$ *be a Riesz operator and let* $d(\lambda) = \sum_0^\infty \delta_n \lambda^n$ $(\lambda \in \mathbb{C})$ *be a Fredholm divisor of S. Then there exists an entire* $\mathcal{L}(E)$-*valued function* $D(\lambda) = \sum_0^\infty D_n \lambda^n$ $(\lambda \in \mathbb{C})$ *such that* $S(\lambda) = D(\lambda)/d(\lambda)$ *for all* $\lambda \in \mathbb{C}$ *with* $1/\lambda \in \varrho(S)$. *Moreover, one has the following recursion formulas:*

$$D_0 = \delta_0 S, \qquad D_n = \delta_n S + D_{n-1} S \qquad (n = 1, 2, \dots). \tag{6.1}$$

To prove this theorem, observe first that the function $D(\lambda) := d(\lambda)S(\lambda)$ is analytic on $\{\lambda \in \mathbb{C} : 1/\lambda \in \varrho(S)\}$ and that it admits analytic continuation onto the entire complex plane. By comparing the coefficients of the expansions of both sides of the equality $D(\lambda)(I - \lambda S) = d(\lambda)S$, one then arrives at (6.1).

6.4. The constructions that will follow below heavily rely upon the following result on the representation of nuclear operators on l_1.

Theorem. *An operator $S \in \mathscr{L}(l_1)$ is nuclear if and only if there exists an infinite matrix $(\sigma_{i,k})$ such that*

$$\mathscr{N}(S) := \sum_{i=1}^{\infty} \sup_k |\sigma_{i,k}| < \infty$$

and S acts on l_1 by the rule

$$S : (\xi_k)_{k=1}^{\infty} \rightarrow \left(\sum_{k=1}^{\infty} \sigma_{i,k}\, \xi_k \right)_{i=1}^{\infty}.$$

The number $\operatorname{tr}(S) := \sum_{j=1}^{\infty} \sigma_{j,j}$ is referred to as the *trace* of the nuclear operator $S \in \mathscr{L}(l_1)$.

In what follows we shall freely identify operators with their matrix representation.

6.5. Let $(\sigma_{i,k})$ be an arbitrary infinite matrix generating a nuclear operator on l_1. Put

$$\sigma \begin{pmatrix} i_1, \ldots, i_n \\ k_1, \ldots, k_n \end{pmatrix} := \det(\sigma_{i_p, k_q})_{p,q=1}^{n},$$

$$\sigma_0 := 1, \qquad \delta_n := \frac{(-1)^n}{n!} \sum_{j_1, \ldots, j_n = 1}^{\infty} \sigma \begin{pmatrix} j_1, \ldots, j_n \\ j_1, \ldots, j_n \end{pmatrix},$$

and define $d(\lambda) := \sum_0^{\infty} \delta_n \lambda^n$ ($\lambda \in \mathbb{C}$). The function d is called the *Fredholm determinant* and the numbers δ_n are referred to as the *Fredholm coefficients* of $(\sigma_{i,k})$.

The Fredholm determinant is an entire function. This follows immediately from the estimates $|\delta_n| \leqslant n^{n/2}/n!\; \mathscr{N}(S)^n$ ($n = 0, 1, \ldots$), which, in turn, are an easy consequence of the well-known *Hadamard inequality*:

$$|\det(\sigma_{p,q})_{p,q=1}^{n}| \leq \prod_{p=1}^{n} \left(\sum_{q=1}^{n} |\sigma_{p,q}|^2 \right)^{1/2}.$$

6.6. In an analogous fashion we now define the first Fredholm minor. Given an infinite matrix $(\sigma_{i,k})$ corresponding to a nuclear operator on l_1, set

$$\Delta_{i,k}^0 := \sigma_{i,k} \quad \text{and} \quad \Delta_{i,k}^n := \frac{(-1)^n}{n!} \sum_{j_1, \ldots j_n = 1}^{\infty} \sigma \begin{pmatrix} i, j_1, \ldots, j_n \\ k, j_1, \ldots, j_n \end{pmatrix}.$$

The operator induced by the matrix $D_n := (\Delta_{i,k}^n)$ can be shown to be nuclear on l_1. Application of Hadamard's inequality gives the estimates

$$\mathscr{N}(D_n) \leqq \frac{(n+1)^{(n+1)/2}}{n!}\, \mathscr{N}(S)^{n+1} \qquad \text{for } n = 0, 1, \ldots .$$

Hence the function $D : \mathbb{C} \rightarrow \mathscr{L}(l_1)$ defined by $D(\lambda) := \sum_0^{\infty} D_n \lambda^n$ ($\lambda \in \mathbb{C}$) is an entire function whose values are nuclear operators on l_1. The function D is called the *first Fredholm minor* of $(\sigma_{i,k})$.

6.7. By expanding $\Delta_{i,k}^n$ with respect to the first row one easily gets the recursion formulas (6.1), from which one may consecutively derive the formulas $\delta_n =$

$-1/n \sum_{k=0}^{n-1} \delta_k \mathrm{tr}(S^{n-k})$ $(n = 1, 2, \ldots)$. The latter formulas show that the Fredholm determinant is uniquely determined by the numbers $\mathrm{tr}(S^k)$ $(k = 1, 2, \ldots)$.

6.8. Here are the basic properties of Fredholm determinants and minors.

Theorem. (i) *If* $d(\lambda) \neq 0$, *then* $1/\lambda \in \varrho(S)$,

$$S(\lambda) = \frac{D(\lambda)}{d(\lambda)} \qquad \text{and} \qquad \mathrm{tr}(S(\lambda)) = -\frac{d'(\lambda)}{d(\lambda)}.$$

(ii) *If* $d(\lambda_0) = 0$, *then* λ_0 *is a characteristic value of* S.

Indeed, (6.1) implies that $D(\lambda) = d(\lambda)S + \lambda D(\lambda)S = d(\lambda)S + \lambda S D(\lambda)$, whence

$$\left(I + \lambda \frac{D(\lambda)}{d(\lambda)} \right)(I + \lambda S) = (I - \lambda S)\left(I + \lambda \frac{D(\lambda)}{d(\lambda)} \right) = I.$$

This shows that if $d(\lambda) \neq 0$, then $1/\lambda \in \varrho(S)$. Furthermore,

$$\mathrm{tr}(S(\lambda)) = -\frac{1}{d(\lambda)} \sum_{1}^{\infty} n\delta_n \lambda^{n-1} = -\frac{d'(\lambda)}{d(\lambda)}.$$

In case $d(\lambda_0) = 0$, λ_0 is a pole of the Fredholm resolvent and, consequently, a characteristic value of S.

6.9. Using Theorem 6.8 and taking into account the decomposition (5.2) one can show that the Fredholm determinant of a nuclear operator on l_1 is a Fredholm divisor of that operator.

6.10. In order to extend the results of Subsections 6.4.–6.9 to nuclear operators on an arbitrary Banach space E we make use of the fact that every such operator is related to some nuclear operator on l_1. That a nuclear operator $S \in \mathscr{L}(F)$ is *related to a nuclear operator* $T \in \mathscr{L}(l_1)$ means, by definition, that there exist operators $A \in \mathscr{L}(E, l_1)$ and $B \in \mathscr{L}(l_1, E)$ such that $S = BA$ and $T = AB$.

The operators A and B may be constructed as follows. Let $S = \sum_{j=1}^{\infty} \sigma_j f_j \otimes x_j$ be the so-called nuclear representation of S; herein $x_j \in E, f_j \in E', \|x_j\| = \|f_j\| = 1$, and $(\sigma_j) \in l_1$. Put

$$A x := (\sigma_j \langle x, f_j \rangle), \qquad B(\xi_j) := \sum_{1}^{\infty} \xi_j x_j.$$

It is easily seen that A and B are bounded, that $S = BA$ and $T = AB$, and that T is nuclear on l_1, its matrix being $(\sigma_j \langle x_k, f_j \rangle)$.

6.11. In an analogous fashion one may define what it means that $S \in \mathscr{L}(E)$ and $T \in \mathscr{L}(E)$ are related to each other.

Related operators have many properties in common: 1) their resolvent sets coincide; 2) two related operators are either simultaneously Riesz operators or not. Moreover, two related operators have the same set of characteristic values, the algebraic multiplicities of corresponding characteristic values being equal to each other.

6.12. The latter property immediately implies that related Riesz operators have the same collection of Fredholm divisors. Thus, by virtue of 6.10, all results of 6.4–6.9 extend to nuclear operators on arbitrary Banach spaces:

Theorem. *Let $S \in \mathscr{L}(E)$ be a nuclear operator. Given any nuclear representation $S = \sum_1^\infty \sigma_j f_j \otimes x_j$, put $\delta_0 := 1$ and*

$$\delta_n := \frac{(-1)^n}{n!} \sum_{j_1,\ldots,j_n = 1}^\infty \sigma_{j_1,\ldots,j_n} \det(\langle x_{j_p}, f_{j_r} \rangle).$$

Then $d(\lambda) := \sum_0^\infty \delta_n \lambda^n \ (\lambda \in \mathbb{C})$ is a Fredholm divisor of S.

6.13. It is unknown whether every nuclear operator $S \in \mathscr{L}(E)$ is related to some nuclear integral operator $T \in \mathscr{L}(C[0, 1])$ with continuous kernel. An affirmative answer to this question would allow us to derive the determinant theory of nuclear operators from Fredholm's classical theory.

§7. Compact Operators on Hilbert Space

Throughout this section H always denotes an infinite-dimensional complex separable Hilbert space with scalar product $(\cdot, \cdot)$.

Operators on Hilbert space have a series of important properties which are not shared by operators on arbitrary Banach spaces. Above all this concerns operators which are normal, self-adjoint, or Hilbert-Schmidt. Results on such operators have important applications to integral equations in the space L_2 (see Chap. 2). For the material of the present section we refer, e.g., to Gohberg, Krein [1965], Jörgens [1970].

7.1. Two General Properties

7.1.1. If $T \in \mathscr{K}(H)$ then, by the Riesz theory, the space H admits the following orthogonal decompositions:

$$H = N(I - T) \oplus R(I - T^*) = R(I - T) \oplus N(I - T^*).$$

7.1.2. Let (e_j) be a complete orthogonal system in H and let P_n denote the orthogonal projection defined by $P_n x = \sum_{j=1}^n (x, e_j)e_j \ (n = 1, 2, \ldots)$. Then if $T \in \mathscr{K}(H)$, the "truncated" operators $P_n T P_n =: T_n$ converge in the norm of $\mathscr{L}(H)$ to T as $n \to \infty$.

7.2. Compact Normal Operators

7.2.1. Recall that an operator $A \in \mathscr{L}(H)$ is said to be *normal* (resp. *self-adjoint*) if $A^*A = AA^*$ (resp. $A^* = A$).

If A is a normal operator then $\|Ax\| = \|A^*x\| \ (x \in H)$, $N(A - \lambda I) = N(A^* - \bar{\lambda}I) \ (\lambda \in \mathbb{C})$, and $N(A - \lambda I) \perp N(A - \mu I) \ (\lambda \neq \mu)$.

7.2.2. The following result describes the structure of the spectrum of a normal operator.

Theorem. *Let $A \in \mathscr{L}(H)$ be a normal operator. Then $\alpha(A - \lambda I) = \beta(A - \lambda I)$ for all $\lambda \in \mathbb{C}$. The residual spectrum $\sigma_r(A)$ is empty. The "unessential" spectrum, $\sigma(A) \cap \varphi(A)$, coincides with the at most countable set of all eigenvalues $\lambda_1, \lambda_2, \dots$, and the only possible accumulation points of this set belong to the essential spectrum. Each eigenvalue λ_0 is a pole of the first order of the resolvent $R_\lambda(A)$ and the residue P is given by*

$$Pu = \sum_{j=1}^{n} (u, e_j)e_j$$

where $e_1, \dots, e_n$ is any orthonormal basis in the eigensubspace $N(\lambda_0 I - A)$. Moreover, the generalized eigenspace for λ_0 coincides with the corresponding eigensubspace and, thus, the algebraic multiplicity of λ_0 is equal to the geometric one.

7.2.3. The existence of eigenvalues for nontrivial compact normal operators is ensured by the following result.

Theorem (Hilbert). *Every nonzero compact normal operator K possesses an eigenvalue λ_0 such that $|\lambda_0| = \|K\|$.*

Indeed, K is a normal element of the C^*-algebra $\mathscr{L}(H)$ such that $r_K = \|K\| > 0$, and so there must exist a $\lambda_0 \in \sigma(K)$ for which $|\lambda_0| = \|K\|$. From Theorem 5.2 (ii) we deduce that λ_0 is even an eigenvalue.

7.2.4. From the Riesz theory we know that the nonzero part of the spectrum of a compact operator $K \in \mathscr{L}(H)$ consists of its eigenvalues $\lambda_1, \lambda_2, \dots$ and that zero is the only possible accumulation point. Each eigenvalue has a finite algebraic multiplicity. In the following, when speaking of the ordered sequence of eigenvalues of a compact operator K we have in mind any sequence (λ_j) constituted by its eigenvalues and meeting the following requirements: each eigenvalue is repeated according to its algebraic multiplicity and $|\lambda_{j+1}| \leq |\lambda_j|$ for all j; in case K has only n eigenvalues (counted up to algebraic multiplicity), we put $\lambda_j = 0$ for $j > n$. In either case, $\lambda_j \to 0$ as $j \to \infty$.

The above results of this section give the following theorem.

Hilbert's Spectral Theorem. *Let $K \neq 0$ be a compact normal operator on H and let (λ_j) be its ordered sequence of eigenvalues. Then there exists an orthonormal sequence (e_j) in H such that e_j is an eigenvector for λ_j. Every element $u \in H$ has an expansion*

$$u = u_0 + \sum_j (u, e_j)e_j \tag{7.1}$$

with some element $u_0 \in \ker K$, and K acts on H by the rule

$$Ku = \sum_j \lambda_j(u, e_j)e_j. \tag{7.2}$$

The series (7.2) converges in the norm of H.

7.2.5. Now let $K \neq 0$ be a compact self-adjoint operator on H. Its eigenvalues are obviously real. Consequently, its ordered sequence of eigenvalues can be divided into two monotone sequences: the sequences (λ_j^+) and (λ_j^-) built up by the positive and negative eigenvalues, respectively. The corresponding ortho-normal sequences of eigenvectors, whose existence is guaranteed by the theorem in 7.2.4, will be denoted by (e_j^+) and (e_j^-). Formula (7.2) then takes the form

$$Ku = \sum_j \lambda_j^+ (u, e_j^+)e_j^+ + \sum_j \lambda_j^- (u, e_j^-)e_j^-. \tag{7.3}$$

From (7.3) one can derive the following important theorem, which is referred to as *Courant's variational principle* (and was established by R. Courant in 1920). To simplify notation, we put $\lambda_j^+ = 0$ (resp. $\lambda_j^- = 0$) for $j > n$ if K has only n positive (resp. negative) eigenvalues.

Theorem. *Let $K \in \mathscr{K}(H)$ be a self-adjoint operator. Then*

$$\lambda_1^+ = \sup_x \{(Kx, x) : \|x\| \leq 1\}, \lambda_1^- = \inf_x \{(Kx, x) : \|x\| \leq 1\};$$

$$\lambda_n^+ = \inf_{x_1, \dots, x_{n-1} \in H} \sup_x \{(Kx, x) : \|x\| \leq 1, x \perp x_j, j = 1, \dots, n - 1\},$$

$$\lambda_n^- = \sup_{x_1, \dots, x_{n-1} \in H} \inf_x \{(Kx, x) : \|x\| \leq 1, x \perp x_j, j = 1, \dots, n - 1\}.$$

7.3. Singular Values and Schatten-von Neumann Classes. There is some reason for expecting that there ought to be a connection between growth estimates for the eigenvalues of a compact operator and its "degree" of compactness (see also Sec. 2 of Chap. 2). First of all the question arises how the compactness of an operator can be "measured". It turns out that such a "measure" is given by the so-called s-numbers, which were introduced by E. Schmidt in 1907, when he was studying integral equations with non-symmetric kernels.

7.3.1. Let $K \in \mathscr{K}(H)$. Then the operator K^*K is clearly self-adjoint, and since $(K^*Kx, x) = \|Kx\|^2 \geq 0$ for all $x \in H$, it is even non-negative. Hence, if λ is a nonzero eigenvalue of K^*K, then $\lambda > 0$. The (positive) root $s = \sqrt{\lambda}$ is referred to as a *singular value* or an *s-number* or a *Schmidt number* of K. If $K \neq 0$ then $K^*K \neq 0$ and so, by Theorem 7.2.3, K has at least one s-number. We denote by $s_n = s_n(K)$, $n = 1, 2, \dots$, all s-numbers of the operator K listed so that $s_1 \geq s_2 \geq \dots \geq 0$, each s_j being repeated according to the algebraic multiplicity of the corresponding eigenvalue $\lambda_j = s_j^2$.

First notice that $s_1(K) = \|K\|$. From Courant's variational principle one can easily deduce the following approximation property of the s-numbers: $s_n(K) = \inf\{\|K - T\| : T \in \mathscr{F}_{n-1}\}$, where $\mathscr{F}_{n-1}$ is the set of all operators whose range has dimension at most $n - 1$. This implies that $|s_n(K) - s_n(L)| \leq \|K - L\|$ for all K, $L \in \mathscr{K}(H)$. Here are some more important properties of the s-numbers:
  1) $s_n(K) = s_n(K^*)$;

2) $s_n(AK) \leq \|A\| s_n(K)$, $s_n(KA) \leq \|A\| s_n(K)$ for $A \in \mathcal{L}(H)$;

3) if $K, L \in \mathcal{K}(H)$, then $s_{m+n-1}(K + L) \leq s_m(K) + s_n(L)$ and $s_{m+n-1}(KL) \leq s_m(K)s_n(L)$ $(m, n = 1, 2, \ldots)$.

7.3.2. Let e_j be any eigenvector of the operator K^*K for the eigenvalue s_j^2 and put $f_j := s_j^{-1} K e_j$. From Hilbert's spectral theorem (7.2.4) we now obtain the following result.

Theorem. *Let $K \in \mathcal{K}(H)$ and let (s_j) be its sequence of s-numbers. Then there exist orthonormal sequences (e_j) and (f_j) in H such that $Ke_j = s_j f_j$, $K^*f_j = s_j e_j$ $(j = 1, 2, \ldots)$, and*

$$Ku = \sum_j s_j(u, e_j)f_j \qquad (u \in H). \tag{7.4}$$

7.3.3. The expansion (7.4), which is usually called the *Schmidt expansion* of K, can even be taken as a characterization for compact operators.

Theorem. *An operator $K \in \mathcal{L}(H)$ is compact if and only if there exist two orthonormal sequences (e_j) and (f_j) in H and a sequence of positive numbers $s_1 \geq s_2 \geq \cdots \geq 0$ which is either finite or converges to zero such that (7.4) holds.*

The "only if" part of the theorem is a consequence of Theorem 7.3.2. Its "if" portion follows from the compactness of the finite rank operators K_n given by $K_n u := \sum_{j=1}^n s_j(u, e_j)f_j$ together with the estimate

$$\|(K - K_n)u\|^2 = \sum_{j=n+1}^{\infty} s_j^2 |(u, e_j)|^2 \leq s_{n+1}^2 \|u\|^2.$$

7.3.4. The following important theorem was established by H. Weyl in 1949.

Theorem. *Let K be a compact operator on H and let (λ_j) and (s_j) be its ordered sequences of eigenvalues and s-numbers, respectively. Then, for $n = 1, 2, \ldots$,*

$$|\lambda_1 \lambda_2 \ldots \lambda_n| \leq s_1 s_2 \ldots s_n \qquad (Weyl's\ inequality)$$

and if $p > 0$,

$$\sum_{j=1}^n |\lambda_j|^p \leq \sum_{j=1}^n s_j^p. \tag{7.5}$$

We remark that if $K \in \mathcal{K}(H)$ is a normal operator, then the last inequality becomes an equality, because $\sigma_j = |\lambda_j|$ in this case. Also note that in the case $p = 2$ the inequality (7.5) yields *Schur's inequality*:

$$\sum_j |\lambda_j|^2 \leq \sum_j s_j^2. \tag{7.6}$$

The inequality (7.6) holds for every $K \in \mathcal{K}(H)$; it was first proved by I. Schur (1909) for integral operators on L_2 with continuous kernels and then extended by T. Carleman (1921) to the case of square-summable kernels (also see Sections 1–3 of Chapter 2).

7.3.5. For $p > 0$, we denote by $\mathscr{S}_p = \mathscr{S}_p(H)$ the collection of all operators $K \in \mathscr{K}(H)$ for which

$$N_p(K) := \left[\sum_j s_j^p(K) \right]^{1/p} < \infty; \qquad (7.7)$$

the sets $\mathscr{S}_p(K)$ are called *Schatten-von Neumann classes* (after R. Schatten and J.v. Neumann).

If $p < q$ then $\mathscr{S}_p \subset \mathscr{S}_q$, because $[N_q(K)]^q \leq s_1^{q-p}[N_p(K)]^p$ in that case. Note that there exist operators $K \in \mathscr{K}(H)$ which do not belong to any of the classes $\mathscr{S}_p(H)$ $(p > 0)$; for example, such an operator may be given by (7.4) with $s_j = 1/\log(1 + j)$ and arbitrary orthonormal sequences (e_j) and (f_j).

Since an operator and its adjoint have the same s-numbers, we have $N_p(K) = N_p(K^*)$. From the properties of the s-numbers listed above we also infer that, for $1 \leq p < \infty$, $N_p(K)$ is a symmetric norm on $\mathscr{S}_p$, i.e., a norm with the additional property that

$$N_p(AKB) \leq \|A\| N_p(K) \|B\| \qquad (A, B \in \mathscr{L}(H); K \in \mathscr{K}(H)).$$

It is natural to define $\mathscr{S}_\infty(H) = \mathscr{K}(H)$ and $N_\infty(K) := \sup_j s_j(K)$. Note that the norm $N_\infty(K)$ is clearly equivalent to the operator norm $\|K\|$ on $\mathscr{S}_\infty(H)$.

From what has been said above we obtain the following result.

Theorem. $\mathscr{S}_p$ $(1 \leq p \leq \infty)$ *is a separable ideal of* $\mathscr{L}(H)$ *with a symmetric norm,* $N_p(K)$. *The set of finite rank operators is dense in* $\mathscr{S}_p$ *and*

$$\inf_{T \in \mathscr{F}_n} \|K - T\| = \left(\sum_{j=n+1}^{\infty} s_j^p(K) \right)^{1/p} \qquad (n = 1, 2, \ldots)$$

for all $K \in \mathscr{S}_p$.

It is not difficult to see that $\mathscr{S}_1(H)$ actually coincides with the ideal of all *nuclear operators* on H. If $K \in \mathscr{S}_1(H)$ and (e_j) is any orthonormal basis in H, then the number $\operatorname{tr} K := \sum_{j=1}^{\infty} (Ke_j, e_j)$ does not depend on the particular choice of the basis (e_j); this number is referred to as the *trace* of K. A fundamental result of B.V. Lidskiĭ (1958) states that $\operatorname{tr} K = \sum_{j=1}^{\infty} \lambda_j$.

The operators of the ideal $\mathscr{S}_2(H)$ are called *Hilbert-Schmidt operators*. The perhaps most important examples of such operators are the *Hilbert-Schmidt integral operators*, i.e., the integral operators with square-summable kernels on L_2. Note that every operator $K \in \mathscr{S}_2(H)$ is unitarily equivalent to some Hilbert-Schmidt integral operator on $L_2(0, 1)$.

7.3.6. Combining (7.5) and (7.4) we arrive at the following estimates for the eigenvalues and singular values of an operator $K \in \mathscr{S}_p(H)$: $\sum_j |\lambda_j|^p \leq [N_p(K)]^p$ and $|\lambda_n| \leq N_p(K)n^{-1/p}$, $s_n \leq N_p(K)n^{-1/p}$ $(n = 1, 2, \ldots)$.

7.3.7. From Theorem 7.3.2 a simple characterization of Hilbert-Schmidt operators can be derived:

Theorem. *For an operator $K \in \mathcal{L}(H)$ the following are equivalent:*
(i) *K is Hilbert-Schmidt;*
(ii) *$\sum_{j=1}^{\infty} \|Kx_j\|^2 < \infty$ for some orthonormal basis (x_j) in H;*
(iii) *$\sum_{j=1}^{\infty} \|Kx_j\|^2 < \infty$ for every orthonormal system (x_j) in H.*

7.3.8. The following result is a consequence of Theorems 7.3.2 and 7.3.7.

Theorem. *An operator $K \in \mathcal{L}(H)$ is nuclear if and only if it is the product of two Hilbert-Schmidt operators.*

Chapter 2
Fredholm Integral Equations

In this chapter we apply the general results of the preceding chapter (bounded-ness and compactness criteria, Fredholm determinants and minors, structure of the generalized eigenspace, estimates for the eigenvalues and singular values) to integral operators on the spaces C and L_p. In this more concrete setting, those general results can be made more precise, and, on the other hand, they can be enriched and enlivened by plenty of remarkable insights into the nature of integral operators. We finally also consider Fredholm integral equations of the first, second, and third kinds.

§1. Linear Integral Operators

These are operators of the form

$$(Kf)(x) = \int_X k(x, y)f(y)d\mu(y). \tag{1.1}$$

Here (X, μ) is a measure space with σ-finite measure μ and k is a $(\mu \times \mu)$-measurable function on $X \times X$, which is referred to as the *kernel* of the integral operator K.

Dictated by the needs of several applications, the solution of the integral equation $Kf = g$ is usually sought in the space of continuous functions or in the space $L_p(X, \mu)$. Therefore it is useful to have conditions ensuring that the given integral operator is bounded or even compact on these spaces. In this section we shall establish such boundedness and compactness criteria.

1.1. Integral Operators on the Space $C(X)$. Throughout Section 1.1 let X be a subset of $\mathbb{R}^n$ carrying a σ-finite positive measure μ.

1.1.1. Boundedness and Compactness Conditions

1. In the case where $X = [a, b]$ is a finite interval, J. Radon (1919) found the following descriptions of continuous and compact operators on $C(X)$.

Theorem 1. *For T to be a linear continuous operator mapping the space C[a, b] into itself it is necessary and sufficient that it be representable in the form*

$$(Tf)(x) = \int_a^b f(y)\, d\tau_x(y),$$

where $\tau_x(y)$ is a function which is defined on the square $a \leq x \leq b, a \leq y \leq b$ and has the following properties: $\tau_x(a) = 0$, the total variation of $\tau_x(y)$ with respect to y does not exceed some constant independent of x, and the expressions $\tau_x(b)$ and $\int_a^\xi \tau_x(y)\, dy$ are continuous functions of x for each fixed $\xi \in [a, b]$.

Theorem 2. *An operator $K \in \mathscr{L}(C[a, b])$ is compact if and only if it can be represented in the form*

$$(Kf)(x) = \int_a^b k(x, y)f(y)\, d\sigma(y),$$

where σ is some non-decreasing function, and the function $k(x, \cdot)$ is summable with respect to σ for each fixed x and satisfies

$$\int_a^b |k(x, y) - k(z, y)|\, d\sigma(y) \to 0 \qquad as\ z \to x.$$

2. Now let X be an arbitrary measurable subset of $\mathbb{R}^n$ and let μ be a non-negative measure on X.

Theorem 1. *Let k be a complex-valued and continuous function on $X \times X$ which has the following properties:*
1) *for each $x \in X$, the function $k(x, \cdot)$ is summable with respect to μ and*

$$m_k := \sup_{x \in X} \int_X |k(x, y)|\, d\mu(y) < \infty; \tag{1.2}$$

2) *for each $x \in X$ and each $\varepsilon > 0$ there exists a number $\delta(x) > 0$ such that*

$$\int_X |k(x, y) - k(x', y)|\, d\mu(y) < \varepsilon \tag{1.3}$$

whenever $x' \in X$ and $|x - x'| < \delta(x)$.
Then the operator K given by (1.1) belongs to $\mathscr{L}(C(X))$ and $\|K\| = m_K$.

If X is compact, then every integral operator with continuous kernel is compact on $C(X)$; this is an immediate consequence of the well-known Arzela-Ascoli theorem.

However, if X is not compact, then such an operator need not be compact on $C(X)$ unless an additional condition is fulfilled:

Theorem 2. *Let the function k have the two properties listed in Theorem 1 and suppose that k also satisfies the following condition:*
3) *for each $\varepsilon > 0$ there exists a compact subset M of X such that $\int_X |k(x, y)|\, d\mu(y) < \varepsilon$ for all $x \in X \setminus M$.*
Then $K \in \mathscr{L}(C(X))$.

As the kernel $k(x, y) = a(x)b(y)$ shows, condition 3) is not necessary for K to be compact.

1.1.2. The Transposed Integral Operator. If $\mu(X) < \infty$, then through

$$\langle f, g \rangle = \int_X f(x)g(x)\, d\mu(x) \tag{1.4}$$

a dual system $\langle C(X), C(X) \rangle$ is given (see 2.5 of Chap. 1). If $\mu(X) = \infty$, then the right-hand side of (1.4) is not defined for every pair $f, g \in C(X)$, but it makes sense in the case where at least one of the two functions $f, g \in C(X)$ is summable. This motivates the introduction of the space

$$C_1(X, \mu) := \{f \in C(X) : \|f\|_1 < \infty\}, \qquad \|f\|_1 := \int_X |f(x)|\, d\mu(x).$$

Notice that $C_1(X, u)$ is a Banach space with respect to the norm $\|f\| = \max\{\|f\|_{C(X)}, \|f\|_1\}$. The bilinear form (1.4) obviously generates two dual systems: $\langle C(X), C_1(X, \mu) \rangle$ and $\langle C_1(X, \mu), C(X) \rangle$.

The following theorem gives an answer to the important question about conditions ensuring that the integral operator with the transposed kernel $k^T(x, y) := k(y, x)$,

$$K^T f(x) := \int_X k(y, x)f(y)\, d\mu(y), \tag{1.5}$$

is the transposed operator of K with respect to the dual system $\langle C(X), C_1(X, \mu) \rangle$.

Theorem. *Let the kernels k and k^T satisfy the conditions of Theorem 1 of 1.1.2. Then K and K^T are transposed to each other with respect to the dual systems $\langle C(X), C_1(X, \mu) \rangle$ and $\langle C_1(X, \mu), C(X) \rangle$, that is, $\langle Kf, g \rangle = \langle f, K^T g \rangle$ for all $f \in C(X), g \in C_1(X, \mu)$ and all $f \in C_1(X, \mu), g \in C(X)$.*

Now suppose that X is a compact set. We denote by $\mathscr{A}(X, \mu)$ the algebra of all operators $A \in \mathscr{L}(C(X))$ that possess a transposed operator A^T with respect to the form (1.4) (recall 2.5 of Chap. 1). Furthermore, we let $\mathscr{F}(X, \mu)$ denote the collection of all (finite rank) integral operators of the form (1.1) with a *degenerate kernel*, i.e., a kernel of the form $k(x, y) = \sum_{j=1}^m a_j(x)b_j(y)$, where $a_j, b_j \in C(X)$. We finally let $\mathscr{K}(X, \mu)$ refer to the closure of $\mathscr{F}(X, \mu)$ in the norm of $\mathscr{A}(X, \mu)$ (see 2.5.3 of Chap. 1). It turns out that $\mathscr{K}(X, \mu)$ coincides with the set of all operators $A \in \mathscr{A}(X, \mu)$ for which A^T is also compact.

1.1.3. Integral Operators with a Diagonal Kernel. These operators constitute an important subclass of $\mathscr{K}(X, \mu)$, where X is a compact subset of $\mathbb{R}^n$. A kernel k is said to be *diagonal* if it is continuous on $X_A := \{(x, y) \in X \times X : x \neq y\}$ and if for each $x \in X$ the limits

$$h(x) := \lim_{\varepsilon \to 0} \int_X |k(x, y)| \eta_\varepsilon(|x - y|)\, d\mu(y)$$

and

$$h^T(x) := \lim_{\varepsilon \to 0} \int_X |k(y, x)| \eta_\varepsilon(|x - y|) \, d\mu(y)$$

exist uniformly with respect to x; here η_ε ($\varepsilon > 0$) is the cut-off function given by

$$\eta_\varepsilon(t) := \begin{cases} 0 & \text{for } 0 \le t \le \varepsilon/2, \\[2mm] \dfrac{2}{\varepsilon} t - 1 & \text{for } \varepsilon/2 \le t \le \varepsilon < 1, \\[2mm] 1 & \text{for } \varepsilon \le t < \infty. \end{cases}$$

A moment's thought reveals that the functions h and h^T are continuous and non-negative. As the above definition is symmetric with respect to the variables x and y, the transposed kernel of a diagonal kernel must also be diagonal.

We denote by K_ε and K_ε^T the integral operators with the kernels $k_\varepsilon(x,y) :=$ $k(x, y)\eta_\varepsilon(|x - y|)$ and $k_\varepsilon^T(x, y) := k_\varepsilon(y, x)$, respectively. The following theorem can be proved without difficulty.

Theorem. *If k is a diagonal kernel, then there exists an operator $K \in \mathcal{K}(X, \mu)$ such that*

$$\|K - K_\varepsilon\| \to 0 \quad \text{and} \quad \|K^T - K_\varepsilon^T\| \to 0 \quad \text{as } \varepsilon \to 0.$$

Moreover, $\|K\| = \max_{x \in X} h(x)$, $\|K^T\| = \max_{x \in X} h^T(x)$, and the kernel $k: X_\Delta \to \mathbb{C}$ is uniquely determined by the operator K.

This theorem justifies the notation

$$(Kf)(x) = \int_X k(x, y)f(y) \, d\mu(y) := \lim_{\varepsilon \to 0} \int_X k_\varepsilon(x, y)f(y) \, d\mu(y).$$

There are measures μ such that not every continuous kernel is diagonal. For instance, it is easy to see that the kernel $k(x, y) \equiv 1$ is diagonal if and only if

$$\lim_{\varepsilon \to 0} \int_X [1 - \eta_\varepsilon(|x - y|)] \, d\mu(y) = 0, \tag{1.6}$$

uniformly with respect to $x \in X$. It is clear that Lebesgue measure satisfies (1.6).

If (1.6) is satisfied, then every bounded and continuous kernel $k: X_\Delta \to \mathbb{C}$ is diagonal. Furthermore, in that case the product $M = LK$ of two integral operators L, $K \in \mathcal{K}(X, \mu)$ with diagonal kernels l and k, respectively, is again an operator with diagonal kernel; the latter kernel is given by

$$m(x, y) = \lim_{\varepsilon, \varepsilon' \to 0} \int_X l_\varepsilon(x, z)k_{\varepsilon'}(z, y) \, d\mu(z), \tag{1.7}$$

the limit in (1.7) being uniform on the set $\{(x, y) \in X \times X : |x - y| \ge \delta\}$ for each $\delta > 0$.

Here are some important examples of diagonal kernels, which frequently arise in several applications:

1°. *Volterra kernels.* Let $X = [a, b] \subset \mathbb{R}$ and assume $k(x, y)$ is continuous for $a \le y \le x \le b$ and vanishes for $a \le x \le y \le b$. We let μ be Riemann measure and define K by

$$(Kf)(x) = \int_a^x k(x, y)f(y)\, dy \qquad (a \le x \le b, f \in C[a, b]).$$

2°. *Abel kernels.* In case $X = [a, b] \subset \mathbb{R}$ and

$$k(x, y) = \begin{cases} h(x, y)/|x - y|^\alpha & \text{for } a \le y \le x \le b, \\ 0 & \text{for } a \le x \le y \le b, \end{cases}$$

where h is some continuous function and $0 < \alpha < 1$, we have an Abel kernel.

3°. *Kernels with a weak singularity* (or *kernels of potential type*). In this case X is a measurable subset of $\mathbb{R}^n$ and $k(x, y) = h(x, y)/|x - y|^\alpha$ on X_Δ, where $0 < \alpha < n$ and h is some function which is bounded on $X \times X$ and continuous on X_Δ. The measure μ is again Riemann measure.

If k and l are kernels with a weak singularity such that

$$|k(x, y)| \le C_1|x - y|^{-\alpha}, \qquad |l(x, y)| \le C_2|x - y|^{-\beta}$$

with certain constants C_1 and C_2, then the kernel of the product LK equals

$$m(x, y) = \int_X l(x, z)k(z, y)\, dz$$

and admits the following estimate:

$$|m(x, y)| \le \begin{cases} C|x - y|^{-\alpha - \beta + n} & \text{for } \alpha + \beta > n, \\ C|\log(C'|x - y|)| & \text{for } \alpha + \beta = n, \\ C & \text{for } \alpha + \beta < n. \end{cases}$$

In particular, the so-called *m*th *iterated kernel* defined by

$$k_m(x, y) := \int_X k_{m-1}(x, z)k(z, y)\, dz, \qquad k_1(x, y) := k(x, y),$$

and being nothing else than the kernel of the operator K^m, is bounded if only $m > n/(n - \alpha)$.

1.1.4. Positive Operators. Let E and F be certain Banach spaces of functions. A linear operator K of E into F is said to be *positive* if it sends non-negative functions of E into non-negative functions of F, and it is called *regular* if there exists a positive operator K_0 of E into F such that $|Ku| \le K_0|u|$ for all $u \in E$. Regular operators acting from L_q into L_p (or C) are continuous.

R. Jentzsch (1912) established the following result, which provides conditions for a positive operator to possess nonzero eigenvalues.

Theorem. *Suppose an operator $K \in \mathcal{K}(X, \mu)$ has the following two properties:
a) K is a positive operator of $C(X)$ into $C(X)$ and b) there is a function $f_0 \in C(X)$
such that $f_0 \geq 0, f_0 \neq 0$, and $K f_0 \geq \gamma f_0$ with some $\gamma > 0$.*

*Then $\lambda_0 = r_K(\geq \gamma)$ is a eigenvalue of K and there exists a non-negative eigen-
function for K.*

1.1.5. The Case of a Non-Compact Set X. We now assume that $X \subseteq \mathbb{R}^n$ is a
non-compact set. Both the definition of a diagonal kernel and the basic theorem
of 1.1.3 can be literally extended to this setting if one only requires that the kernels
k_ε satisfy the hypotheses of Theorem 1 of 1.1.1.2 for all $\varepsilon > 0$.

Let K be an operator with a diagonal kernel k. For such operators we have a
simple sufficient compactness condition: If, for each $\varepsilon > 0$, there is a compact
subset X_ε of X such that $h(x) < \varepsilon$ and $h^T(x) < \varepsilon$ for all $x \in X \backslash X_\varepsilon$, then both K
and K^T are in $\mathcal{K}(C(X))$.

In the case at hand, the Banach algebra $\mathcal{A}(X, \mu)$ consists of all operators
$A \in \mathcal{L}(C(X)) \cap \mathcal{L}(C_1(X, \mu))$ possessing a transposed operator $A^T \in \mathcal{L}(C(X)) \cap
\mathcal{L}(C_1(X, \mu))$ such that $\langle Af, g \rangle = \langle f, A^T g \rangle$ for all $f \in C(X), g \in C_1(X, \mu)$ or for
all $f \in C_1(X, \mu), g \in C(X)$.

It turns out that if $K \in \mathcal{A}(X, \mu)$ and its transposed operator K^T are compact
on $C(X)$, then K is also compact on $C_1(X, \mu)$; moreover, the spectrum and the
generalized eigenspace of K are the same regardless whether it is considered on
$C(X)$ or on $C_1(X, \mu)$. More about the matter of this subsection can be found in
Jörgens [1970], Riesz and Sz.-Nagy [1979], Zabreĭko et. al. [1968], Fenyö and
Stolle [1982–84].

1.1.6. Fredholm Determinants and Minors. We now apply the results of Sec. 6
of Chap. 1 to integral operators of the form

$$(Su)(x) = \int_\Omega k(x, t)u(t)\, dt,$$

where Ω is a compact subset of $\mathbb{R}^m$ and k belongs to $C(\Omega \times \Omega)$.

In this case the Fredholm resolvent $S(\lambda)$ is an integral operator,

$$(S(\lambda)u)(x) = \int_\Omega \Gamma(x, t, \lambda)u(t)\, dt,$$

whose kernel is

$$\Gamma(x, t, \lambda) = \frac{D(x, t, \lambda)}{d(\lambda)}.$$

Fredholm provided the following series expansion for the functions d and D,
which are now called the *Fredholm series*:

$$d(\lambda) = \sum_0^\infty d_n \lambda^n, \qquad D(x, t, \lambda) = \sum_0^\infty D_n(x, t)\lambda^n,$$

where

$$d_0 := 1, \qquad D_0(x, t) := K(x, t),$$

$$d_n := \frac{(-1)^n}{n!} \int_\Omega D_{n-1}(x, x)\, dx, \qquad n \geq 1,$$

$$D_n(x, t) := \frac{(-1)^n}{n!} \int_\Omega \cdots \int_\Omega k\binom{x, t_1, \ldots, t_n}{t, t_1, \ldots, t_n}\, dt_1 \ldots dt_n,$$

and

$$k\binom{x_1, \ldots, x_n}{t_1, \ldots, t_n} := \det(k(x_p, t_q))^n_{p,q=1}.$$

Invoking Hadamard's inequality, Fredholm proved that the series for d and D in fact represent entire functions of λ.

The recursion formulas (6.1) of Chapter 1 now assume the form

$$D_n(x, t) = d_n k(x, t) + \int_\Omega k(x, y) D_{n-1}(y, t)\, dy,$$

and the trace formula of Theorem 6.8 of Chapter 1 can now be written as follows:

$$\operatorname{tr} S(\lambda) = -\frac{d'(\lambda)}{d(\lambda)} = \sum_1^\infty A_n \lambda^n,$$

where the A_n's are the so-called *traces of the kernel* $k(x, t)$,

$$A_n = \int_\Omega k_n(x, x)\, dx,$$

with the nth iterated kernel $k_n(x, t)$ defined as

$$k_n(x, t) := \int_\Omega \cdots \int_\Omega k(x, y_1) k(y_1, y_2) \ldots k(y_{n-1}, t)\, dy_1 dy_2 \ldots dy_n.$$

In the case under consideration the Fredholm minors of order p can be given by the formula

$$D\binom{x_1, \ldots, x_p}{t_1, \ldots, t_p}; \lambda = k\binom{x_1, \ldots, x_p}{t_1, \ldots, t_p}$$

$$+ \sum_{n=1}^\infty \frac{(-1)^n}{n!} \lambda^n \int_\Omega \cdots \int_\Omega k\binom{x_1, \ldots, x_p, y_1, \ldots, y_n}{t_1, \ldots, t_p, y_1, \ldots, y_n}\, dy_1 \ldots dy_n.$$

For any p and any $x_k, t_k \in \Omega$, the latter series defines an entire function of λ. Note that, obviously, $D(x, t, \lambda) = D(^x_t; \lambda)$. In Jörgens [1970] and Zabreĭko et al. [1968] it is shown how the eigenfunctions of S can be expressed in terms of the Fredholm minors.

We conclude by remarking that the functions d and D played the central part in Fredholm's original theory of integral equations with continuous kernels, which included in particular the theorem on the square-summability of the sequence of the eigenvalues (see also 7.3 of Chap. 1).

1.2. Integral Operators on the Spaces $L_p(X, \mu)$. Let (X, μ) and (Y, ν) be any measure spaces with σ-finite measures μ and ν, respectively. We let $L_p(X, \mu)$ $(1 \leq p \leq \infty)$ denote the Banach space of μ-measurable complex-valued functions f on X such that

$$\|f\|_p := \left[\int_X |f(x)|^p \, d\mu(x) \right]^{1/p} < \infty, \qquad 1 \leq p < \infty,$$

$$\|f\|_\infty := \inf\{C : |f(x)| \leq C \ \mu\text{-a.e. on } X\}.$$

In case $d\mu(x) = dx$ is Lebesgue measure, we shall write $L_p(X)$ in place of $L_p(X, \mu)$. As usual, we put $p' := p/(p-1)$ if $1 < p < \infty$ and set $p' = 1$ if $p = \infty$ and $p' = \infty$ if $p = 1$.

For each $p \in [1, \infty]$, we can define a dual system $\langle L_p, L_{p'} \rangle$ via

$$\langle f, g \rangle := \int_X f(x)g(x) \, d\mu(x) \qquad (f \in L_p, g \in L_{p'}).$$

If $p \in [1, \infty)$, then $L_{p'}(X, \mu) = [L_p(X, \mu)]'$ (recall Sec. 2.3 of Chap. 1).

The concern of this section is conditions ensuring that the integral operator

$$(Kf)(x) = \int_Y k(x, y)f(y) \, d\nu(y) \tag{1.8}$$

be bounded or compact, thought of as acting from $L_q(Y, \nu)$ into either of the spaces $L_p(X, \mu)$ or $C(X)$.

1.2.1. General Boundedness and Compactness Conditions

1. We first consider the case where $X = Y$ is a compact subset of $\mathbb{R}^n$ and where $\mu = \nu$ is Lebesgue measure. We also suppose that X has positive (Lebesgue) measure. In that case the kernels of all integral operators of the form (1.8) mapping $L_p(X)$ into $C(X)$ are completely described by the following two theorems, which, in essence, were established by Radon.

Theorem 1. *An integral operator K given by (1.8) maps $L_q(X)$ $(1 \leq q \leq \infty)$ into $C(X)$ if and only if*
   a) *$k(x, \cdot) \in L_{q'}(X)$ and $h(x) := \|k(x, \cdot)\|_{q'} \leq M < \infty$ for all $x \in X$;*
   b) *for every measurable subset $D \subseteq X$ and every $x_0 \in X$ the equality*

$$\lim_{x \to x_0} \int_D k(x, y) \, dy = \int_D k(x_0, y) \, dy$$

   holds.
Moreover, if conditions a) and b) are satisfied then $\|K\|_{L_q \to C} = \|h\|_\infty$.

Theorem 2. *Assume K maps $L_q(X)$ $(1 \leq q \leq \infty)$ into the space $C(X)$. Then for K to be compact it is necessary and sufficient that, for each $x_0 \in X$,*

$$\lim_{x \to x_0} \|k(x, \cdot) - k(x_0, \cdot)\|_{q'} = 0.$$

Note that if K is a mapping of $L_q(X)$ $(1 \leq q \leq \infty)$ into $C(X)$, then the adjoint operator $K^*: L_1(X) \to L_{q'}(X)$ is an integral operator whose kernel is $k^*(x, y) = \overline{k(y, x)}$.

2. Unfortunately, no general criterion for k to generate a bounded or compact integral operator $K: L_q \to L_p$ is known at the present moment. Below we shall state some sufficient conditions for this to hold, and in a certain important special case we shall even provide conditions that are necessary and sufficient ones. For other sufficient conditions see also Zabreĭko et al. [1968], Kantorovich and Akilov [1977], Okikivlu [1971].

We let P_M refer to the operator of multiplication by the characteristic function χ_M of a set $M \subseteq X$:

$$(P_M u)(x) = \chi_M(x) u(x).$$

The Lebesgue measure of a measurable set X will be denoted by $|X|$. Recall that throughout this section X always stands for a compact subset of $\mathbb{R}^n$ with positive Lebesgue measure.

Theorem. *Let k be a measurable function on $X \times X$ such that the integral operator K generated by the kernel k maps $L_q(X)$ into $L_p(X)$ and suppose at least one of the following conditions if fulfilled:*
  1) $1 < q \leq \infty, 1 \leq p < \infty, q > p$, K *is regular;*
  2) $1 < q \leq \infty, 1 \leq p < \infty, q \leq p$, K *is regular and*

$$\lim_{|M|+|N| \to 0} \|P_M K P_N\|_{L_q \to L_p} = 0;$$

3) $q > 1, p < \infty$, K *satisfies*

$$\lim_{|M| \to 0} \|P_M K\|_{L_q \to L_p} = \lim_{|N| \to 0} \|K P_N\|_{L_q \to L_p} = 0.$$

Then K is a compact operator of $L_q(X)$ into $L_p(X)$.

1.2.2. Hille-Tamarkin Operators. Given any measure spaces (X, μ) and (Y, ν) and any $p, q \in [1, \infty]$, we denote by $\mathcal{H}_{pq}(X, Y)$ the collection of all integral operators of the form (1.8) which map $L_q(Y, \nu)$ into $L_p(X, \mu)$ and have a $(\mu \times \nu)$-measurable kernel k such that $h(x) := \|k(x, \cdot)\|_{q'} < \infty$ for μ-almost all $x \in X$ and h is in $L_p(X, \mu)$. We also put $|K|_{pq} = \|h\|_p$.

The operators K in $\mathcal{H}_{pq}(X, Y)$ are called *Hille-Tamarkin operators* (in recognition of E. Hille and J.D. Tamarkin, who, about 1930, made essential contributions to the theory of integral equations). Notice that $\mathcal{H}_{pp'}(X, Y)$ equals $L_p(X \times Y, \mu \times \nu)$; in particular, $\mathcal{H}_{22}(X, Y)$ is nothing but the Hilbert space of all *Hilbert-Schmidt integral operators* of $L_2(Y)$ into $L_2(X)$.

The transposed operator K^T of an operator $K \in \mathcal{H}_{pq}(X, Y)$ is defined as the integral operator whose kernel is $k^T(x, y) = k(y, x)$. By Fubini's theorem, $\langle Kf, g \rangle = \langle f, K^T g \rangle$ for all $f \in L_q(Y, \nu)$ and $g \in L_{p'}(X, \mu)$. If $p \in (1, \infty)$, then K^T may be identified with the dual operator K' in the sense of 2.5 of Chap. 1; if $p = \infty$, then $L_1(X, \mu)$ is isomorphic to some closed subspace of $[L_\infty(X, \mu)]'$ and K^T is the restriction of K' to $L_1(X, \mu)$.

$\mathcal{H}_{pq}(X, Y)$ is a Banach space with respect to the norm $|\cdot|_{pq}$, and we have $\|K\| \leq |K|_{pq}$ (resp. $\|K\| \leq |K^T|_{q'p'}$) for $K \in \mathcal{H}_{pq}(X, Y)$ (resp. $K \in \mathcal{H}_{q'p'}(Y, X)$). If $1 \leq p < \infty$ and $1 < q \leq \infty$, then every operator $K \in \mathcal{H}_{pq}(X, Y)$ is compact from $L_q(Y, v)$ into $L_p(X, \mu)$; moreover, if $q' \leq p$, then K can be approximated in the norm $|\cdot|_{pq}$ by finite rank operators as closely as desired, and if $p \leq q'$ then K^T is the limit in the norm of $\mathcal{H}_{q'p'}(Y, X)$ of finite rank operators (see Jörgens [1970], Okikivlu [1971]).

1.2.3. Convolution Integral Operators. Recently H. Hansson (1982) found necessary and sufficient conditions for a convolution operator $K: L_q(\mathbb{R}^n) \to L_p(\mathbb{R}^n, \mu)$ $(1 < q \leq p < \infty)$,

$$(Kf)(x) = (K * f)(x) := \int_{\mathbb{R}^n} K(x - y)f(y) \, dy,$$

to be continuous or compact; here $K(x) := k(|x|)$ $(x \in \mathbb{R}^n)$, where k is a positive and nondecreasing function on $(0, \infty)$. It is also supposed that k is subject to the conditions

$$\int_0^1 k(r)r^{n-1}dr < \infty \qquad \text{and} \qquad \int_1^\infty [k(r)]^{q'}r^{n-1}dr < \infty \qquad (1.9)$$

$(q' = q/(q - 1))$. We here confine ourselves to the especially interesting case where $K \notin L^{q'}(\mathbb{R}^n)$.

Theorem 1. *The operator K is continuous if and only if $\sup_{E \subset \mathbb{R}^n} \mu(E)/c(E)^{p/q} < \infty$, and it is compact if and only if it is continuous and the measure μ satisfies the following two conditions:*
a) $\lim_{\delta \to 0} \sup_{\text{diam}(E) \leq \delta} \mu(E)/c(E)^{p/q} = 0$,
b) $\lim_{R \to \infty} \sup_{E \subset \{x:|x| > R\}} \mu(E)/c(E)^{p/q} = 0$.
Here $c(E)$ refers to the L_q-capacity of a subset E of $\mathbb{R}^n$, defined by $c(E) := \inf \int g^q \, dx$, the infimum over all non-negative functions g such that $K * g \geq 1$ on E.

Theorem 2. *Suppose that k is continuous and $r^\delta k(r)$ is an increasing function on $0 < r < \infty$ for some $\delta > 0$. Then the operator $K: L_q(\mathbb{R}^n) \to L_p(\mathbb{R}^n, \mu)$ is continuous for arbitrary $q < p$, $1 < q < n/(n - \delta)$ if and only if $\sup_{B_R} \mu(B_R)/h(R) < \infty$, where $B_R := \{x \in \mathbb{R}^n : |x| \leq R\}$ and $h(R) := R^{-pn/q'}k(R)^{-p}$. The operator $K: L_q(\mathbb{R}^n) \to L_p(\mathbb{R}^n, \mu)$ is compact for arbitrary $q < p$, $1 < q < n/(n - \delta)$ if and only if*

$$\alpha) \lim_{\varepsilon \to 0} \sup_{R \leq \varepsilon} \mu(B_R)/h(R) = 0$$

and

$$\beta) \lim_{\varrho \to \infty} \sup_{B_R \subset \{|x| > \varrho\}} \mu(B_R)/h(R) = 0.$$

Theorem 1 extends a criterion of V.G. Maz'ya (1962; see Maz'ya [1985]) to a large class of integral operators with difference kernels, while Theorem 2 generalizes a result of D.R. Adams (1973) pertaining to kernels of potential type: $k(r) = r^{-n-\alpha}$ with $0 < \alpha q < n$; in the latter case $h(r) = r^m$, where $m = (m - \alpha q)p/q$.

In this connection mention must also be made of the extensive investigations devoted to translation invariant operators (e.g., by L. Hörmander (1960)) and to the multiplier problem (see, e.g., Larsen [1971] and Maz'ya and Shaposhnikova [1985]).

1.2.4. Generalized Eigenspaces of Integral Operators. Let K be the integral operator defined by (1.1). We suppose that $X \subseteq \mathbb{R}^n$ has positive measure $\mu(X)$, and we assume that K is compact from $L_p(X, \mu)$ $(1 \leq p \leq \infty)$ into itself or into $C(X)$. Let λ_0 be any nonzero eigenvalue of K.

In the case at hand, the canonical basis of K for λ_0 (recall Sec. 5 of Chap. 1) consists of functions $e_{j,k} \in L_p(X, \mu)$ (or $C(X)$) such that

$$Ke_{j,k} - \lambda_0 e_{j,k} = \begin{cases} 0 & \text{for } k = 1, \\ e_{j,k-1} & \text{for } k = 2, \dots, n_j \end{cases}$$

and of functions $f_{j,k} \in L_{p'}(X, \mu)$ (or $L_1(X, \mu)$), $k = 1, \dots, n_j, j = 1, \dots, m$, satisfying

$$\int_X k(y, x)f_{j,k}(y) \, d\mu(y) - \lambda_0 f_{j,k}(x) = \begin{cases} 0 & \text{for } k = 1, \\ f_{j,k-1}(x) & \text{for } k = 2, \dots, n_j \end{cases}$$

and

$$\int_X e_{j,k}(x)f_{l,h}(x) \, d\mu(x) = \delta_{jl}\delta_{k,n_l-h+1}.$$

The operators P, KP, and K_{-l} (see (5.4)–(5.6) of Chap. 1) are integral operators of the form (1.1) with the degenerate kernels

$$p(x, y) = \sum_{j=1}^{m} \sum_{k=1}^{n_j} e_{j,k}(x)f_{j,n_j-k+1}(y),$$

$$k_0(x, y) = \lambda_0 p(x, y) + \sum_{j=1}^{m} \sum_{k=1}^{n_j-1} e_{j,k}(x)f_{j,n_j-k}(y),$$

$$k_{-l}(x, y) = -\sum_{j=1}^{m} \sum_{k=1}^{n_j} e_{j,k-l+1}(x)f_{j,n_j-k+1}(y) \quad (l = 1, 2, \dots, p),$$

respectively.

§2. Estimates for the Eigenvalues and Singular Values of Integral Operators

The classical problem of determining the behavior of the Fourier coefficients of a periodic function in dependence on its smoothness, which was first considered by B. Riemann (1854), is a special case of the problem of finding asymptotic estimates for the eigenvalues of integral operators. Indeed, every periodic function f generates a convolution integral operator,

$$(Ku)(x) = \int_0^1 f(x - y)u(y)\, dy,$$

whose eigenvalues are exactly the Fourier coefficients of f.

The pioneering work on the problem of estimating the eigenvalues and s-numbers of integral operators in terms of integrability and smoothness properties of the kernel was done by Fredholm, Schur, Carleman, and Weyl, to mention only some of the principal figures. The most significant contributions to the subject made in the subsequent period are due to E. Hille, J.D. Tamarkin, F. Smithies, M.G. Krein, and A.O. Gelfond. In recent time, the needs of several applications have attracted considerable interest in sufficiently general and sharp estimates for the λ- and s-numbers. The supply of classes of integral operators for which these problems were settled is now fairly good (it includes kernels which are not Hilbert-Schmidt, kernels with singularities on the diagonal, unbounded integration domains, general measures in place of Lebesgue measure etc.). Many important results in these directions have been obtained during the last 15–20 years. The most general and complete results on this topic can be found in the surveys by M.Sh. Birman and M.Z. Solomyak [1977] and A. Pietsch [1980–83, 1987]. In this section we attempt to outline some of the significant pieces of the theory developed in the works cited above.

2.1. Triebel Classes

2.1.1. Let H_1 and H_2 be separable Hilbert spaces. In 1967, H. Triebel introduced certain classes $\mathscr{S}_{p,q}(0 < p, q \leq \infty)$, which are generalizations of the Schatten-von Neumann classes $\mathscr{S}_p$ (see 7.3 of Chap. 1) and represent analogues of the Lorentz spaces $l_{p,q}$: the space $\mathscr{S}_{p,q}$ is, by definition, the collection of all operators $K \in \mathscr{K}(H_1, H_2)$ such that $(s_n(K)) \in l_{p,q}$, i.e. such that

$$N_{p,q}(K) := \begin{cases} \left(\sum_n [n^{1/p - 1/q} s_n(K)]^q \right)^{1/q} & (0 < q < \infty), \\[2ex] \sup_n [n^{1/p} s_n(K)] & (q = \infty) \end{cases}$$

is finite.

It is clear that $\mathscr{S}_{p,p} = \mathscr{S}_p$. The classes $\mathscr{S}_{p,\infty}$ $(1 < p < \infty)$ consist of all operators whose s-numbers admit "individual" estimates of the form $s_n(K) = O(n^{-1/p})$. Note that, for $p, q \in (0, \infty]$, the class $\mathscr{S}_{p,q}$ is a quasinormed space under the quasinorm $N(\cdot) = N_{p,q}(\cdot)$; by a quasinorm we mean a functional which satisfies the usual axioms of a norm except for the triangle inequality, which is now replaced by the inequality $N(K_1 + K_2) \leq c[N(K_1) + N(K_2)]$, the constant $c = c(p, q) \geq 1$ being of course independent of K_1 and K_2. If $1 < p \leq \infty$, $1 < q \leq \infty$ and also if $p = q = 1$, then the spaces $\mathscr{S}_{p,q}$ are actually normed ones. If $p < 1$ or $q < 1$ and equally if $p = 1$, $1 < q \leq \infty$, then the spaces $\mathscr{S}_{p,q}$ are metrizable.

We have the following inclusions:

$$\mathcal{S}_{p_1,q_1} \subset \mathcal{S}_{p_2,q_2} \qquad (p_1 < p_2; q_1, q_2 \text{ arbitrary}),$$

$$\mathcal{S}_{p,q_1} \subset \mathcal{S}_{p,q_2} \qquad (0 < p \leqq \infty; 0 < q_1 < q_2 \leqq \infty),$$

$$\mathcal{S}_{p,q} \subset \overset{\circ}{\mathcal{S}}_{p,\infty} \qquad (0 < p < \infty; 0 < q < \infty),$$

where $\overset{\circ}{\mathcal{S}}_{p,\infty}$ refers to the separable subspace of $\mathcal{S}_{p,q}$ ($0 < p < \infty$) constituted by all operators K for which $s_n(K) = o(n^{-1/p})$.

2.1.2. By applying Weyl's inequality (see Sec. 7.3 of Chap. 1), we get the following inequalities between the eigenvalues and singular values of operators $K \in \mathcal{S}_{p,q}$:

$$\sum_n n^{q/p-1}|\lambda_n(K)|^q \leqq c \, N^q_{p,q}(K) \qquad \text{for } 0 < p \leqq \infty, 0 < q \leqq \infty,$$

$$\sup_n n^{1/p}|\lambda_n(K)| \leqq c \, N_{p,\infty}(K) \qquad \text{for } 0 < p \leqq \infty, q = \infty,$$

where $c = c(p, q)$ is some constant depending only on p and q. Hence, if $K \in \mathcal{S}_{p,q}$ then $(\lambda_n(K)) \in l_{p,q}$. In case K belongs even to $\overset{\circ}{\mathcal{S}}_{p,q}$ ($0 < p < \infty$), we have $|\lambda_n(K)| = o(n^{-1/p})$.

Thus, the problem of finding estimates for the eigenvalues and s-numbers has been reduced to establishing criteria for an operator to belong to a Triebel class.

2.1.3. The kernels of the integral operators studied here can be conveniently handled from the view point of the theory of abstract functions.

Let (Y, τ) be a measure space with σ-finite measure τ and let E be a Banach space with norm $\|\cdot\|$. We denote by $[L_p(Y, \tau), E]$ ($1 \leqq p \leqq \infty$) the space of all measurable functions $f: Y \to E$ such that $\|f(\cdot)\| \in L_p(Y, \tau)$. Analogously we define the space $[B^\alpha_{p,u}(Q), E]$ ($-\infty < \alpha < \infty, 1 \leqq p, u \leqq \infty$), where $B^\alpha_{p,u}(Q)$ is the Besov space of functions on $Q \subseteqq \mathbb{R}^m$ (sufficiently interesting special cases are $Q = \mathbb{R}^m$ or $Q = [0, 1]^m$).

Recall that for non-integral α the space $B^\alpha_{p,p}$ coincides with the Sobolev-Slobodetskiĭ space W^α_p. If α is an integer, then these two spaces are equal to each other only in the case $p = 2$, whereas $B^\alpha_{p,p} \subset W^\alpha_p$ for $p < 2$ and $B^\alpha_{p,p} \supset W^\alpha_p$ for $p > 2$. The classes $B^\alpha_{p,\infty}$ coincide with S.M. Nikol'skiĭ's spaces H^α_p. Notice that $H^\alpha_p \supset W^\alpha_p$.

2.2. Integral Operators on $L_2(X, \mu)$

2.2.1. M.Sh. Birman and M.Z. Solomyak [1977] consider integral operators $K: L_2(X, \varrho) \to L_2(Y, \tau)$ of the form

$$(Kg)(y) := \int_X K(x, y)g(x)d\varrho(x), \tag{2.1}$$

where (X, ϱ) and (Y, τ) are separable measure spaces. Let $H_1 = L_2(X, \varrho)$ and $H_2 = L_2(Y, \tau)$ denote the corresponding Hilbert spaces of square-integrable functions. It is usually assumed that X and Y are subset of the Euclidean $\mathbb{R}^n$. If the kernel of the operator is sufficiently smooth, then ϱ and τ may be arbitrary finite (or at least locally finite) Borel measures. In the case where the kernel is not smooth enough, it is assumed that one of the measures, ϱ say, is absolutely continuous and that its density $d\varrho/dx$ belongs to an appropriate class $L_p(X, \varrho)$ or to an integral Lorentz space $L_{p,r}(X, \varrho)$. The classes $\mathscr{S}_{q,s}$ whose member the operator K is are described in terms of the properties of the kernel and the measures.

It turns out that for operators which are not Hilbert-Schmidt estimates for their s-numbers are not connected with the smoothness of the kernel but depend only on its integrability properties. Sharp "individual" estimates can be given for kernels in the Nikol'skiĭ classes H_p^α. Precise conditions for an operator K to belong to $\mathscr{S}_p$ may be stated in terms of the Besov classes. In particular, the classes $B_{2,1}^\alpha$ are of great relevancy in connection with nuclearity criteria. Here are two results from the survey by Birman and Solomyak [1977] pertaining to this subject.

Theorem. *Let $X = [0, 1]^m$, let ϱ be a finite Borel measure on X, and suppose K belongs to $[L_2(Y, \tau), B_{p,q}^\alpha(X)]$, where $2 \leq p < \infty, 2 \leq q \leq \infty$, and $p\alpha > m$. Then $K \in \mathscr{S}_{\delta,q}$ for $\delta := (\alpha/m + 1/2)^{-1}$.*

Corollary. *Under the hypothesis of the preceding theorem with $q = \infty$, we have $s_n(K) = O(n^{-1/\delta})$. If $K \in [L_2(Y, \tau), H_p^\alpha(X)]$, then $K \in \mathscr{S}_{\delta,\infty}^\circ$, i.e., $s_n(K) = o(n^{-1/\delta})$.*

We remark that in the case $p = q = 2$ the conditions on the kernel in the above theorem can be restated in terms of the space W_2^α. In this case the result is sharp.

Analogous theorems hold in the case $p\alpha < m$. Notice that the limiting case $p\alpha = m$ was studied as well.

To prove the results cited above, the authors apply a method by Weyl, which is based on approximating the given kernel by degenerate ones. The construction of the approximating kernels relies on a special technique, the approximation by piecewise polynomial functions. In this way one obtains individual estimates for the s-numbers of kernels from the classes W_p^α. The systematical use of interpolation techniques then leads to an improvement of the estimates and yields extensions to more general classes of kernels and measures.

Moreover, the above results can be generalized to certain classes of integration domains different from the unit cube of $\mathbb{R}^m$ and also to smooth compact m-dimensional manifolds or manifolds with boundary. Furthermore, certain cases of kernels with "mixed" smoothness were studied, too (see 2.3).

2.2.2. A great deal of effort has been expended in finding asymptotic formulas for the distribution function of the s-numbers,

$$n(s, K) := \sum_{k \in K_s} 1, \qquad K_s := \{k : s_k(K) > s\}, \qquad s > 0.$$

We are now going to describe one result on this subject (see Birman and Solomyak [1977]). Let X and Y be measurable subsets of $\mathbb{R}^m$ and suppose the measures ϱ and τ are both absolutely continuous with respect to Lebesgue measure. Let $|a(x)|^2 = d\varrho/dx$, $|b(y)|^2 = d\tau/dy$, and consider the operator $K_{ab}: L_2(X) \to L_2(Y)$ defined by

$$(K_{ab}g)(y) := \int_X b(y)K(x, y)a(x)g(x)\, dx. \tag{2.2}$$

Note that the operators (2.1) and (2.2) have the same s-numbers.

Denote by $\mathcal{O}_m^{k,\beta}$ $(k > -m, \beta \geq 0)$ the class of all functions which are positively homogeneous of the order k and belong to $C^\beta(\mathbb{S}^{m-1})$, where $\mathbb{S}^{m-1}$ is the unit sphere of $\mathbb{R}^m$; $C^0(\mathbb{S}^{m-1})$ is defined as $L_\infty(\mathbb{S}^{m-1})$. Let $F \in \mathcal{O}_m^{k,\beta}$ or let F be of the form $F(z) = P(z)\log f(z)$, where $f \in \mathcal{O}_m^{1,\beta}$, $f(z) > 0$ for $|z| = 1$, and P is a homogeneous polynomial of the degree $k \geq 0$. Denote by $\tilde{F}$ the Fourier transform of F, understood in the sense of Riesz,

$$\tilde{F}(\xi) := \lim_{r \to \infty} \int_{|x| < r} (1 - |x|^2 r^{-2})^j F(x) e^{-ix\xi}\, dx.$$

If j is sufficiently large (the infimum of all candidates for j depends of course on k), then this limit exists for all $\xi \neq 0$ and represents a continuous function positively homogeneous of order $-(m + k)$. Finally put $\gamma(F) := (2\pi)^{-m} \operatorname{mes}\{\xi : |\tilde{F}(\xi)| > 1\}$.

Then if $K(x, y) = \varphi(x, y)F(x - y)$ with some function φ which is continuous on the diagonal $x = y$, the asymptotic behavior of $n(s, K)$ for $K = K_{ab}$ can be described by the formula

$$\lim_{s \to 0} s^\delta n(s, K) = \gamma(F) \int_{X \cap Y} |a(x)\varphi(x, x)b(x)|^\delta\, dx,$$

where $\delta^{-1} = 1 + km^{-1}$. The general case, in which φ is not continuous on the diagonal, can be tackled by appropriately regularizing the integral.

2.3. Integral Operators on Banach Spaces

2.3.1. In the late seventies, the basic results on compact operators acting between Hilbert spaces (see 7.3 of Chap. 1 and 2.1 of the present chapter) were generalized to operators on Banach spaces. Let E and F be any Banach spaces. Given an operator $S \in \mathcal{L}(E, F)$, we define its *approximation numbers* $\alpha_n(S) := \inf\{\|S - T\| : T \in \mathcal{F}_{n-1}\}$ and denote by $\mathcal{A}_{p,q}$ the class of all $S \in \mathcal{L}(E, F)$ for which $(\alpha_n(S))$ lies in the Lorentz space $l_{p,q}$ $(0 < p, q < \infty)$. The classes $\mathcal{A}_{p,q}$ are a "good" generalization of the classes $\mathcal{S}_{p,q}$, because for operators $S \in \mathcal{A}_{p,q}$ the theorem that the eigenvalue sequence $(\lambda_j(S))$ belongs to $l_{p,q}$ remains in force (H. König, 1977). Moreover, for the s-numbers defined *in the sense of Weyl*,

$$\varkappa_n(S) := \sup\{\alpha_n(SX) : X \in \mathcal{G}(H, E), \|X\| \leq 1\},$$

the classical Weyl inequality can be generalized (Pietsch, 1979). Here H ranges over all Hilbert spaces, and $\mathcal{G}(H, E)$ denotes the class of all operators $A \in \mathcal{L}(H, E)$ such that there are finite-rank operators K_n with $\lim_n \|A - K_n\| = 0$. A survey of these topics, built upon the theory of operator ideals, was given in

Pietsch [1980] (also see Pietsch [1978, 1987]). In recent time these results have found important applications to the problem of estimating the eigenvalues of integral operators.

2.3.2. Integral operators of the form

$$(Kg)(t) := \int_0^1 k(t, s)g(s)\, ds$$

whose complex-valued kernel k is defined on the square $[0, 1]^2$ and is subject to the "mixed" smoothness condition $k \in [B_{p,u}^\alpha, B_{q,v}^\beta]$ were thoroughly studied by Pietsch [1980–83].

Note that, by definition, $k \in [B_{p,u}^\alpha, B_{q,v}^\beta]$ if the abstract function f_k given by $f_k \colon t \to k(t, s)$ (the values of which are functions of s) belongs to $[B_{p,u}^\alpha, B_{q,v}^\beta]$ in the sense of 2.1.3. Via

$$S_k \colon g \to h, \qquad h(t) = \int_0^1 k(t, s)g(s)\, ds$$

an operator $S_k \in \mathscr{L}(B_{q',v'}^{-\beta}, B_{p,u}^\alpha)$ is defined, where $B_{q',v'}^{-\beta} = (B_{q,v}^\beta)'$.

Now suppose that $-\infty < \alpha, \beta < \infty$, $1 \leq p, q \leq \infty$, $1 \leq u, v \leq \infty$,

$$\alpha + \beta > (1/p + 1/q - 1)_+, \tag{2.3}$$

where $(x)_+ := x$ for $x \geqslant 0$ and $(x)_+ := 0$ for $x < 0$. In this case there exists a canonical embedding I of $B_{p,u}^\alpha$ into $B_{q',v'}^{-\beta}$.

A Banach space M will be called *admissible* if $B_{p,u}^\alpha \subseteq M \subseteq B_{q',v'}^{-\beta}$. If M is admissible, we let $I_0 \in \mathscr{L}(B_{p,u}^\alpha, M)$ and $I_1 \in \mathscr{L}(M, B_{q',v'}^{-\beta})$ refer to the natural embeddings. Note that $I = I_1 I_0$. Then put $S_k^M := I_0 S_k I_1$, which is an operator in $\mathscr{L}(M)$. Since the operators S_k^M and $S_k^0 := I_1 I_0 S_k$ are related to each other, the nonzero eigenvalues of S_k^M do actually not depend on M. Therefore it makes sense to speak of the eigenvalue sequence $(\lambda_n(k))$ of the kernel k without mentioning the space M. Notice that the space $L_2(0, 1)$ is admissible for $\alpha + 1/2 > 1/p$ and $\beta + 1/2 > 1/q$.

2.3.3. The following basic result was established by Pietsch [1980–83].

Theorem. *Suppose* (2.3) *holds and put* $q^+ := \max(2, q')$. *If* $K \in [B_{p,u}^\alpha, B_{q,v}^\beta]$ *and* $1/r := \alpha + \beta + 1/q^+$, *then* $(\lambda_n(k)) \in l_{r,u}$.

Pietsch's proof is founded upon his theories of s-numbers in the Weyl sense and of eigenvalues of operators on Banach spaces, on the one hand, and on the method of approximating functions from Besov spaces by splines, on the other.

We remark that the above theorem can be generalized to the case where the interval $[0, 1]$ is replaced by the cube $[0, 1]^m$: all conclusions remain valid if the $\alpha + \beta$ occuring in condition (2.3) and in the definition of $1/r$ is replaced by $(\alpha + \beta)/m$ (see Pietsch [1980–83]). In the second of those papers by Pietsch, analogous results are proved for the case that the integration domain is $\mathbb{R}$. In this situation some additional weight conditions must be imposed on the kernel in order to guarantee that $k(t, s)$ decreases sufficiently rapidly as $|t|, |s| \to \infty$ (for more about this subject see the recent book Pietsch [1987]).

2.3.4. S. Heinrich and T. Kühn (1985) studied integral operators of the form (2.1) with Hölder continuous kernels. Given a compact metric space X, let $C^\alpha(X)$ denote the space of all complex-valued functions on X satisfying a Hölder condition with the exponent α and put

$$C^{\alpha,0}(X, X) := \{k \in C(X \times X): k(\cdot, y) \in C^\alpha(X) \; \forall y \in X$$

$$\text{and } \sup_{y \in X} \|k(\cdot, y)\|_{C^\alpha(X)} < \infty\}.$$

We let $\varepsilon_n(X)$ $(n = 2, 3, \ldots)$ denote the *entropy numbers* of X, that is,

$$\varepsilon_n(X) = \inf\{\varepsilon > 0: X \text{ has an } \varepsilon\text{-net of cardinality less than } n\}.$$

Theorem (S. Heinrich and T. Kühn, 1985). *Let X be a compact metric space and let $0 < \alpha \leq 1$ and $0 < \gamma < \infty$. Then the following are equivalent:*
(i) $\lambda_n(K_\mu) = O(n^{-1/2 - \alpha\gamma})$ *for all kernels $k \in C^{\alpha,0}(X, X)$ and all finite Radon measures μ;*
(ii) $\varepsilon_n(X) = O(n^{-\gamma})$.
In particular, if $X = [0, 1]^m$ and $k \in C^{\alpha,0}(X, X)$, then $\lambda_n(K_\mu) = O(n^{-1/2 - \alpha/m})$ (also see Theorem 2.2.1).

To prove this theorem, Heinrich and Kühn determined the asymptotic behavior of the entropy numbers and of the approximation numbers of the embedding operator $C^\alpha(X) \to C(X)$, thus also solving a problem posed by A. Pietsch.

If, in addition to the hypothesis of the theorem, the kernel is positive-definite, i.e. if k is Hermitian and

$$\sum_{i,j=1}^{l} \xi_i \bar{\xi}_j k(t_i, t_j) \geq 0 \qquad \forall l \in \mathbb{N}, \forall t_i \in X, \forall \xi_i \in \mathbb{C},$$

then even $\lambda_n(K_\mu) = O(n^{-1-\alpha\gamma})$ (J.B. Reade (1983) and T. Kühn (1987)). In case X is a compact m-dimensional C^∞-manifold, μ is a finite Borel measure on X (with supp $\mu = X$), and $k \in C^{\alpha,0}(X, X)$ $(0 < \alpha < 1)$ is positive-definite, then the (sharp) estimate $\lambda_n(K_\mu) = O(n^{-1-\alpha/m})$ holds (T. Kühn (1987)).

2.3.5. The asymptotic behavior of the eigenvalues of convolution operators over finite intervals, i.e. operators of the form

$$(Kg)(t) := \int_{-T}^{T} k(t - s)g(s)\, ds, \qquad -T < t < T,$$

has been the subject of profound investigations by many people, including M. Kac (1955–1956), U. Grenander and G. Szegö (1961), H. Widom (1961–1964), D. Slepian and H. Pollack (1961), M. Rosenblatt (1963), W.H.I. Fuchs (1964), and H.I. London (1965). These authors primarily considered the case where the Fourier transform of the kernel,

$$\mathscr{K}(\lambda) := \int_{-\infty}^{\infty} e^{i\lambda x} k(x)\, dx \qquad (\lambda \in \mathbb{R}),$$

is non-negative (in which case K is a positive operator). A significant result in this direction, due to H. Widom (1963), can be stated as follows.

Theorem. *Let $\mathscr{K}$ be subject to the following conditions:*
(i) *$\mathscr{K}(\lambda)$ is non-negative, even, and monotonically decreasing for $\lambda > 0$;*
(ii) *$\mathscr{K}(\lambda) \sim \mathscr{K}(\mu)$ as $\lambda, \mu \to \infty$ and $\lambda \sim \mu$;*
(iii) *$\mathscr{K}(\lambda) = o(\mathscr{K}(\mu))$ for $\lambda = o(\mu)$.*
Then the asymptotic behavior of the eigenvalues of the operator K, thought of as acting on $L_p(-T, T)\,(1 \leq p \leq \infty)$, is given by the formula

$$\lambda_n \sim \mathscr{K}(\pi n/2T).$$

A detailed survey of much of what is known about this topic can be found in Chap. VIII, § 11, of the book by Zabreĭko et al. [1968]. The monograph Widom [1985] is a modern account of both the classical results and the recent developments concerning asymptotic expansions for the spectrum of pseudodifferential operators (see Sec. 9 of Chap. 4) on compact domains of $\mathbb{R}^n$ as well as their relations to the theory of Toeplitz operators (see Sec. 5 of Chap. 3).

2.4. Hankel Operators
A prominent class of integral operators for which estimates of the s-numbers and $\mathscr{S}_p$ criteria are known is the class of *Hankel operators*. These are intimately related to *Toeplitz operators* (see Chapter 3), and both classes of operators emerge from plenty of problems of analysis or probability theory, among others.

Let H_2 and H_∞ be the familiar Hardy spaces on the unit circle, put $H_2^- := L_2 \ominus H_2$, and denote by P and P^- the orthogonal projections of L_2 onto its subspaces H_2 and H_2^-, respectively. For a function φ in L_∞, the Hankel operator H_φ and the Toeplitz operator T_φ are defined by

$$H_\varphi f = P^- \varphi f, \qquad T_\varphi f = P \varphi f \qquad (f \in H_2).$$

Note that clearly $H_\varphi \in \mathscr{L}(H_2, H_2^-)$ and $T_\varphi \in \mathscr{L}(H_2, H_2)$.

Theorem (V.M. Adamyan, D.Z. Arov, M.G. Kreĭn, 1971). *Let $\mathscr{R}_n\,(n \geq 0)$ refer to the set of all rational functions such that their poles are located in the interior of the unit disk and the sum of the multiplicities of these poles does not exceed n. Then for every $\varphi \in L^\infty$,*

$$s_{n+1}(H_\varphi) = \inf_{H_\psi \in \mathscr{F}_{n-1}} \|H_\varphi - H_\psi\| = \mathrm{dist}_{L_\infty}(\varphi, \mathscr{R}_n + H_\infty).$$

Theorem (P. Hartman, 1958). *Let $\varphi \in L_\infty$. Then $H_\varphi \in \mathscr{S}_\infty$ if and only if φ is in $C + H_\infty$.*

Note that Hartman's theorem, if given an independent proof, implies Sarason's famous theorem that the sum $C + H_\infty$ is closed in L_∞ (also see Sec. 5 of Chap. 3).

Theorem (V.V. Peller, 1980). *Let* $\varphi \in L_\infty$ *and suppose* $1 \leqq p < \infty$. *Then* $H_\varphi \in \mathscr{S}_p$ *if and only if* $\varphi \in B_{p,p}^{1/p} + H_\infty$.

§3. Fredholm Integral Equations

In this section we apply the general results on compact operators recorded above to Fredholm integral equations of the first, second, and third kinds. In this more concrete setting some of the above results can be made considerably precise.

3.1. Hilbert-Schmidt Integral Operators. We exploit the results of Sec. 7 of Chap. 1 for studying integral operators of the form

$$(Ku)(x) = \int_X k(x, y)u(y)dy, \tag{3.1}$$

where X is a compact subset of $\mathbb{R}^m$ with positive Lebesgue measure and k belongs to $L_2(X \times X)$, i.e.

$$A^2 := \int_X \int_X |k(x, y)|^2 dx\, dy < \infty. \tag{3.2}$$

Then $K \in \mathscr{K}(L_2(X))$ and $\|K\| \leq A$ (see 1.2.2).

The material of this and the next sections can be found in the following books: Mikhlin [1959], Tricomi [1957], Jörgens [1970], Fenyö and Stolle [1982–84].

3.1.1. We first consider *Hermitian kernels*, that is, kernels satisfying $k(x, y) = \overline{k(y, x)}$. The series (7.2) of Chap. 1 now assumes the form

$$(Ku)(x) = \sum_j \lambda_j e_j(x) \int_X \overline{e_j(y)}\, u(y)dy. \tag{3.3}$$

This series is called the *Hilbert-Schmidt series* of K; it converges in the norm of $\mathscr{L}(L_2(X))$.

Taking into account that the functions $\{e_j(x)e_j(y)\}$ form an orthonormal system in $L_2(X \times X)$, we obtain from (3.3) that the kernel $k \in L_2(X \times X)$ has the following Fourier series with respect to that system:

$$k(x, y) = \sum_j \lambda_j e_j(x)\overline{e_j(y)}. \tag{3.4}$$

The series (3.4) is referred to as the *bilinear series* of the kernel k. The bilinear series of the nth iterated kernel is of the form

$$k_n(x, y) = \sum_j \lambda_j^n e_j(x)\overline{e_j(y)}. \tag{3.5}$$

Notice that (3.5) in particular implies that (λ_j^n) and (e_j) are the sequences of the eigenvalues and eigenfunctions of the operator K^n. The same formula (3.5) also gives the equality

$$\sum_j \lambda_j^{2n} = \int_X \int_X |k_n(x, y)|^2 dx\, dy \qquad (n = 1, 2, \ldots).$$ (3.6)

In general, the series (3.3), (3.4), (3.5) converge in the mean. If, in addition to (3.2), the inequality

$$\int_X |k(x, y)|^2 dy \leqq B^2 < \infty \qquad \forall x \in X$$ (3.7)

is satisfied, then the series (3.3) converges uniformly and absolutely on X, and we also have $\sum_j \lambda_j^2 |e_j(x)|^2 \leqq B^2$ for $x \in X$; an analogous remark applies to the Neumann series (for $|\lambda| > B$). Furthermore, if (3.2) and (3.7) are in force, then the series (3.5) converges uniformly in both x and y for $n > 2$, and it converges uniformly in one of the variables for almost each fixed value of the other variable in case $n = 2$. On condition that the eigenvalues are all positive and the kernel k is continuous, the bilinear series (3.4) converges uniformly on $X \times X$ (*Mercer's theorem*, established by J. Mercer in 1909). In particular, if n is an even integer and k is continuous, then the series (3.5) is uniformly convergent on $X \times X$.

3.1.2. In case k is a non-Hermitian kernel satisfying (3.2), we deduce from (7.4) and (7.6) of Chap. 1 that

$$k(x, y) = \sum_j s_j f_j(x)\overline{e_j(y)}, \qquad \sum_j \lambda_j^2 \leqq A^2.$$ (3.8)

The series in (3.8) may be regarded as the bilinear series of the non-Hermitian kernel k; it converges in the $L_2(X \times X)$ norm. The inequality in (3.8) is usually called the *Schur inequality*.

3.2. Equations of the Second Kind. Let K be a nonzero compact operator on a separable Hilbert space H and let $\lambda \in \mathbb{C}$. We consider the equation

$$Ku - \lambda u = f,$$ (3.9)

which is called an *equation of the second kind* in case $\lambda \neq 0$ and an *equation of the first kind* if $\lambda = 0$. Throughout this section we assume that $\lambda \neq 0$.

3.2.1. Equation (3.9) (for $\lambda \neq 0$) can be solved by reducing it to a system of linear algebraic equations. This approach, proposed and originally applied to Fredholm integral equations by E. Schmidt in 1908, consists in the following.

Fix any positive number R, put $\mu = 1/\lambda$ for $|\lambda| > 1/R$, and decompose the operator K into a sum $K = C + D$, where C is a finite rank operator and $\|D\| < 1/R$ (see 7.1.2 of Chap. 1). On letting the operator $(I - \mu D)^{-1}$ act on both sides of (3.9), we obtain the equation

$$u - \mu[I + \mu D(\mu)]Cu = -\mu[I + \mu D(\mu)]f =: g,$$ (3.10)

which is equivalent to equation (3.9); here $D(\mu)$ is the Fredholm resolvent of D

(see Sec. 4 of Chap. 1). The operator $T := [I + \mu D(\mu)]C$ has finite rank, and therefore it can be written in the form $Tu = \sum_{k=1}^{n}(u, b_k)a_k$, where (a_k) and (b_k) are orthonormal systems in H. Note that $a_k = a_k(\mu)$ depends on μ. Scalar multiplication of (3.10) by b_j gives the system of algebraic equations

$$c_j - \mu \sum_{k=1}^{n} \alpha_{jk}c_k = g_j \qquad (j = 1, \ldots, n), \tag{3.11}$$

in which $\alpha_{jk} = (a_k, b_j)$, $g_j = (g, b_j)$, $c_j = (u, b_j)$.

It is readily seen that the determinant of the system (3.11) is holomorphic in the disk $|\mu| < R$. This determinant takes on the value 1 at 0, hence it does not vanish identically, and thus it has at most countably many zeros in the disk $|\mu| < R$. These zeros are the characteristic values of the operator T; all remaining μ's are regular values of T.

Once a solution of the system (3.11) has been found, we have a solution of equation (3.10) in the form

$$u = g + \mu \sum_{k=1}^{n} c_k a_k.$$

Note that Schmidt's approach is not only a method of solving equation (3.9) but also produces all Fredholm theorems for equation (3.9) (including the theorem on the structure of the kernel; see 5.2 of Chap. 1).

If $|\mu| < 1/|\lambda_1|$, where λ_1 is the first eigenvalue of the operator K (note that $|\mu| < 1/|\lambda_1|$ if only $|\mu| < 1/\|K\|$), then the above arguments apply with $C = 0$ and thus the unique solution of (3.9) is given by

$$u = -\mu[I + \mu K(\mu)]f = -\mu \sum_{n=0}^{\infty} \mu^n K^n f.$$

The latter equality tells us that in the case at hand the *method of successive approximation* is applicable to equation (3.1) (for every right-hand side f). In particular, if K is the integral operator (3.1), then

$$u(x) = -\mu f(x) - \mu^2 \int_X \Gamma(x, t; \mu) f(t)\, dt,$$

where $\Gamma(x, t; \mu) := \sum_{n=1}^{\infty} \lambda^{n-1} k_n(x, t)$ is the so-called *resolvent kernel* of k. Note that if (3.9) is a Volterra integral equation (recall 1.1.3), then $\lambda_1 = 0$ and, consequently, μ may be any complex number.

3.2.2. Now suppose that K is both compact and normal. Taking into consideration Hilbert's spectral theorem (7.2.4), we may write down a formula for the solutions of equation (3.9) provided the eigenvalues λ_j and the corresponding eigenvectors e_j are known:

1) If $\lambda \neq \lambda_j$ for all j (equivalently: if $\lambda \in \varrho(K)$), then the resolvent $R_\lambda(K)$ of K has the form

$$R_\lambda(K)f = -\frac{1}{\lambda}f - \frac{1}{\lambda}\sum_j \frac{\lambda_j(f, e_j)}{\lambda - \lambda_j}e_j, \tag{3.12}$$

and thus (3.12) represents the unique solution of equation (3.9).

2) If $\lambda = \lambda_p = \cdots = \lambda_q$ is an eigenvalue (of algebraic multiplicity $q - p + 1$), then for equation (3.9) to have a solution it is necessary and sufficient that

$$(f, e_k) = 0 \qquad \text{for all } k = p, \ldots, q; \tag{3.13}$$

if (3.13) is satisfied, the general solution of equation (3.9) is given by

$$u = -\frac{1}{\lambda} f - \frac{1}{\lambda} \sum_{j \neq p, \ldots, q} \frac{\lambda_j (f, e_j)}{\lambda - \lambda_j} e_j + \sum_{k=p}^{q} c_k e_k, \tag{3.14}$$

where $c_p, \ldots, c_q$ are arbitrary constants.

To see this, notice first that (3.9) implies that

$$(Ku, e_j) - \lambda (u, e_j) = (\lambda_j - \lambda) (u, e_j) = (f, e_j). \tag{3.15}$$

Hence, in the first case, (3.12) is an immediate consequence of the equalities (7.2) of Chap. 1, (3.9), and (3.15). In the second case, (3.13) is clearly a necessary condition for the solvability of (3.9) (again by (3.15)); on the other hand, if (3.13) holds, then (3.14) can be shown to represent a particular and thus the general solution of equation (3.9).

3.3. Equations of the First Kind.

The theory of equations of the first kind,

$$Ku = f, \tag{3.16}$$

is more complicated than its second kind counterpart. The main reason for this circumstance is that infinite-dimensional compact operators are not normally solvable and so the solvability conditions for equation (3.16) are of a nature that is completely different from that of equation (3.9). In particular, the inverse of the operator K, if it exists, is unbounded and therefore equation (3.16) is an ill-posed problem (in the sense of J. Hadamard).

3.3.1. First assume that K is a nonzero compact normal operator on a Hilbert space H whose eigenvalues λ_j and corresponding eigenvectors e_j are available. In that case the following corollary of Hilbert's spectral theorem may be useful.

Theorem. *The equation (3.16) has a solution $u \in H$ if and only if*
a) $(f, w) = 0$ for all $w \in H$ such that $K^* w = 0$;
b) $\sum_j \lambda_j^{-2} |(f, e_j)|^2 < \infty$.

In case both a) and b) are fulfilled, all solutions of equation (3.16) are of the form

$$u = v + \sum_j \lambda_j^{-1} (f, e_j) e_j,$$

v being an arbitrary solution of the homogeneous equation $Kv = 0$.

3.3.2. We now consider equation (3.16) without requiring that K be normal, but we retain the assumption that K be a nonzero compact operator on H. Theorem 7.3.1 of Chap. 1 gives the following.

Theorem. *For equation (3.16) to be solvable it is necessary and sufficient that*
a) $(f, w) = 0$ *for all solutions* $w \in H$ *of the equation* $K^*w = 0$,
b) $\sum_j s_j^{-2} |(f, f_j)|^2 < \infty$.

On condition that a) *and* b) *are satisfied, the general solution of equation* (3.16) *can be written in the form*

$$u = v + \sum_j s_j^{-1} (f, f_j) e_j,$$

where v *is an arbitrary solution of the homogeneous equation* $Kv = 0$.

3.3.3. One of the most important methods of studying integral equations of the first kind (and of the third kind as well, see 3.3.4) is the so-called *method of unbounded regularization* (see Sec. 3.8 of Chap. 1).

We shall exemplify this method by the Volterra equation of the first kind with a smooth kernel,

$$\int_a^x k(x, y) u(y)\, dy = f(x). \tag{3.17}$$

The equation is considered in $C[a, b]$. We assume that the derivatives $\partial k/\partial x$, $\partial^2 k/\partial x^2, \ldots, \partial^{n+1} k/\partial x^{n+1}$ do exist, that

$$\frac{\partial^j k}{\partial x^j}(x, x) \equiv 0 \quad (j = 0, \ldots, n - 1), \qquad \frac{1}{r(x)} := \frac{\partial^n k}{\partial x^n}(x, x) \neq 0,$$

and that $1/r$ is continuous on $[a, b]$. It is readily seen that (3.17) cannot have a solution unless

$$f \in C_0^{n+1}[a, b] := \{f \in C^{n+1}[a, b] : f(a) = \ldots = f^{(n)}(a) = 0\}.$$

So assume that $f \in C_0^{n+1}[a, b]$. Then an (unbounded) equivalent left regularizer of K is given by the operator

$$(Bf)(x) = r(x) f^{(n+1)}(x)$$

whose domain is $C_0^{n+1}[a, b]$, and the corresponding equivalent left regularization of (3.17) is the following equation of the second kind:

$$u(x) + \int_a^x r(x) \frac{\partial^{n+1}}{\partial x^{n+1}} k(x, y) u(y)\, dy = r(x) f^{(n+1)}(x).$$

The latter equation has a unique solution, which can be represented by a Neumann series.

Now suppose that $1/r(x) := k(x, x) \neq 0$ for $a \leq x \leq b$ and that k is differentiable with respect to x and y. In that case the operator

$$(Cf)(x) = [r(x) f(x)]',$$

defined on $C_0^1[a, b]$, is an equivalent right regularizer, and if $f \in C_0^1[a, b]$, then (3.17) is equivalent to the equation

$$v(x) - \int_a^x r(y)k_y'(x, y)v(y)\, dy = f(x);\qquad\qquad(3.18)$$

the solutions of (3.17) and (3.18) are connected by the equality $u(x) = [r(x)v(x)]'$.

In an analogous fashion one may construct equivalent regularizers of the operator (3.1) if only its kernel is subject to some additional conditions. Various methods of solving integral equations of the first kind are discussed in detail in Imanaliev [1981], Fenyö and Stolle [1982–84], Schmeidler [1950].

3.3.4. *Integral equations of the third kind* may also be handled by the method of regularization. Such an equation is of the form

$$a(x)u(x) + \int_X k(x, y)u(y)f(y) = f(x),\qquad\qquad(3.19)$$

where a is a function which vanishes on some subset of X. For simplicity, we assume that $X = [0, 1]$, $a(x) = (x - \alpha)^m$ $(0 \le \alpha \le 1, m \in \{1, 2, \dots\})$, and that $k \in C([0, 1] \times [0, 1])$ possesses continuous derivatives $\partial k/\partial x, \dots, \partial^m k/\partial x^m$ in a neighborhood of the point $x = \alpha$. Furthermore, we assume that f belongs to the set $C(\alpha, m)$ of all functions $f \in C[0, 1]$ representable in the form

$$f(x) = \sum_{j=0}^{m-1} c_j(x - \alpha)^j + g(x)(x - \alpha)^m$$

with a function $g \in C[0, 1]$ and certain complex numbers $c_0, \dots, c_{m-1}$. The numbers $f^{\{j\}}(\alpha) := j! c_j$, which are uniquely determined by f, are referred to as the *Taylor derivatives* of f at α. Under the above assumptions, we may take the operator

$$(Bf)(x) := g(x) = \left[f(x) - \sum_{j=0}^{m-1} \frac{f^{\{j\}}(\alpha)}{j!}(x - \alpha)^j \right](x - \alpha)^{-m},$$

defined on $C(\alpha, m)$, as a left regularizer of equation (3.19). Any solution $u \in C[0, 1]$ of (3.19) is a solution of the equation of the second kind

$$u(x) + \int_0^1 k_1(x, y)u(y)\, dy = g(x)\qquad\qquad(3.20)$$

whose kernel is

$$k_1(x, y) = \left[k(x, y) - \sum_{j=0}^{m-1} \frac{k_x^{(j)}(\alpha, y)}{j!}(x - \alpha)^j \right](x - \alpha)^{-m}.$$

We emphasize that, in general, the equations (3.19) and (3.20) are not equivalent to each other.

A detailed treatment of the method of unbounded regularization and its application to several classes of integral equations can be found in Prössdorf [1974].

3.3.5. The theory of Fredholm and Volterra type integral equations of the first kind was developed to an advanced stage in the work of A.N. Tikhonov, M.M.

Lavrentyev, V.K. Ivanov, M.I. Imanaliev, their collaborators and the students of theirs. A thorough account of these matters is in Imanaliev [1981] and Imanaliev et al. [1982]. We here confine ourselves to citing a few results closely related to the method of regularization.

On condition that k and f are continuous and possess sufficiently many derivates, one can show that equation (3.17) has a generalized solution of the form

$$u(x) = v(x) + \sum_{j=0}^{m} a_j \delta^{(j)}(x - a) \tag{3.21}$$

and that equation (3.1) (with $X = [a, b]$) has a generalized solution of the form

$$u(x) = w(x) + \sum_{j=0}^{m} [b_j \delta^{(j)}(x - a) + c_j \delta^{(j)}(x - b)], \tag{3.22}$$

where $v, w \in C[a, b]$, δ refers to the Dirac δ-function, and a_j, b_j, c_j are certain constants. That (3.21) and (3.22) represent generalized solutions of (3.17) and (3.1), respectively, means that the following equalities hold:

$$\int_a^x k(x, y)v(y)\, dy + \sum_{j=0}^{m} (-1)^j a_j k_y^{(j)}(x, a) = f(x),$$

$$\int_a^b k(x, y)w(y)\, dy + \sum_{j=0}^{m} (-1)^j [b_j k_y^{(j)}(x, a) + c_j k_y^{(j)}(x, b)] = f(x).$$

In addition to the equations (3.17) and (3.1), one has also studied so-called *singularly perturbed equations*, which have the form

$$c(x, \varepsilon)u(x) + \int_a^x k(x, y)u(y)\, dy = f_\varepsilon(x),$$

$$c(x, \varepsilon)u(x) + \int_a^b k(x, y)u(y)\, dy = f_\varepsilon(x),$$

where ε is a small positive parameter, while $c(x, \varepsilon)$ and $f_\varepsilon(x)$ are given functions such that $c(x, \varepsilon) \to 0$ and $f_\varepsilon(x) \to f(x)$ $(x \in [a, b])$, in some sense, as $\varepsilon \to 0$. One has found conditions ensuring that the perturbed equations have continuous solutions in the ordinary sense which converge on $[a, b]$ to some generalized solution of the form (3.21) or (3.22) as ε goes to zero.

Chapter 3
One-Dimensional Singular Equations

The basic problems of the theory of singular integral operators concern boundedness, invertibility, Noether criteria and index computation. The solution of these problems involves the description of the spectrum of the operators under consideration. Wiener-Hopf factorization provides a method for the explicit inversion of the simplest singular operator. The algebraic approach embeds the problem of studying a single singular operator with coefficients from an appropriate subclass of L_∞ into the problem of directly investigating the whole algebra generated by all the operators with coefficients in that subclass.

The first section is devoted to weighted norm inequalities for the one-dimensional singular operator. The second and third sections concentrate on singular equations with piecewise continuous coefficients and Banach algebras generated by singular operators. The problem of factorizing matrix functions is considered in Section 4. In the Sections 5 and 6 we study Toeplitz operators and integral equations of convolution type generated by various subclasses of L_∞. Section 7 focusses on equations with degenerate symbols. The concluding Section 8 contains a concise survey of several further results and applications.

§ 1. Boundedness of the Singular Integral Operator

1.1. The Hilbert Transform. Weighted Norm Estimates

1.1.1. The *Hilbert transform* is the integral

$$(Hf)(x) = \frac{1}{\pi i} \int_{-\infty}^{\infty} \frac{f(y)}{y - x} \, dy, \tag{1.1}$$

which does exist in the Cauchy principal-value sense if, for example, f is in $C_0^\infty(\mathbb{R})$ (the infinitely differentiable functions with compact support). That H induces a bounded operator on $L_p(\mathbb{R})$ $(1 < p < \infty)$ is a classical result, which was established by N.N. Lusin in 1915 for $p = 2$ and by M. Riesz in 1927 for arbitrary p. Note that H is not bounded on $L_1(\mathbb{R})$ or $L_\infty(\mathbb{R})$. Since $H^2 = I$, the operator H fails to be compact on $L_p(R)$ $(1 < p < \infty)$.

1.1.2. In 1973, R.A. Hunt, B. Muckenhoupt, and R.L. Wheeden gave a complete description of those weights ϱ on $\mathbb{R}$, i.e. of those non-negative, measurable, and locally integrable functions on $\mathbb{R}$ for which H is in $\mathscr{L}(L_p(\mathbb{R}, \varrho))$ $(1 < p < \infty)$. This clearly amounts to describing all weights ϱ such that

$$\int_{\mathbb{R}} |(Hf)(x)|^p \varrho(x) \, dx \leq C \int_{\mathbb{R}} |f(x)|^p \varrho(x) \, dx \qquad \forall f \in C_0^\infty(\mathbb{R}). \tag{1.2}$$

Here and in what follows C denotes any positive constant which depends only on the explicitly indicated parameters.

Theorem. *For (1.2) to hold it is necessary and sufficient that*

$$\sup_{Q} (1/|Q|)\|\varrho^{1/p}\|_{L_p(Q)}\|\varrho^{-1/p}\|_{L_{p'}(Q)} < \infty, \qquad (A_p)$$

the supremum over all bounded intervals $Q \subset \mathbb{R}$; here $|Q|$ stands for the length of Q, and $p' = p/(p-1)$.

The necessity of the condition (A_p) results immediately from inserting the test function $f = \varrho^{-1/(p-1)}\chi_Q$ into (1.2); the proof of the sufficiency part is much more difficult.

The condition (A_p) first appeared in a 1972 paper of Muckenhoupt, who showed that it is equivalent to the boundedness of the Hardy-Littlewood maximal operator on $L_p(\mathbb{R}, \varrho)$ (see also 1.2 of Chap. 4).

For power weights of the form

$$\varrho(x) = (1 + |x|)^\beta \prod_{k=1}^m |x - x_k|^{\beta_k}, \qquad (1.3)$$

where $x_1, \ldots, x_m$ are pairwise distinct points on $\mathbb{R}$ and $\beta, \beta_1, \ldots, \beta_m$ are real numbers, condition (A_p) is easily seen to be equivalent to the conditions

$$-1 < \beta_k < p - 1 \quad (k = 1, \ldots, m), \qquad -1 < \beta + \sum_{k=1}^m \beta_k < p - 1. \quad (1.4)$$

For such weights the boundedness of H on $L_p(\mathbb{R}, \varrho)$ was first proved by G.H. Hardy and J.E. Littlewood in 1936.

Sufficient conditions for so-called two weight norm inequalities for the operator H to hold were obtained by Muckenhoupt and Wheeden (1976), M. Cotlar and C. Sadosky (1975–1982), and E.T. Sawyer (1981); for a detailed discussion of this subject see the survey Dyn'kin, Osilenker [1983] (and also Sec. 2 of Chap. 4).

1.2. The Cauchy Singular Integral

1.2.1. Let Γ be a rectifiable plane Jordan curve and let $f \in L_1(\Gamma)$. The *Cauchy singular integral* is the integral

$$(S_\Gamma f)(t) = \frac{1}{\pi i} \int_\Gamma \frac{f(\tau)}{\tau - t}\, d\tau := \lim_{\varepsilon \to 0} \int_{\Gamma_\varepsilon} \frac{f(\tau)}{\tau - t}\, d\tau, \qquad t \in \Gamma, \qquad (1.5)$$

where $\Gamma_\varepsilon := \{\tau \in \Gamma : |\tau - t| > \varepsilon\}$.

It is well known that the boundedness problem for the operator (1.5) on $L_p(\Gamma)$ (integration with respect to arc measure) can be reduced to the following question: For what curves Γ is S_Γ bounded on $L_2(\Gamma)$? It is not difficult to see that the boundedness of S_Γ on $L_2(\Gamma)$ necessitates Γ being a *Carleson curve*, i.e. a curve such that, for each disk $B(z, r)$ which radius r and center $z \in \mathbb{C}$, the length of the arc $\Gamma \cap B(z, r)$ does not exceed $C \cdot r$.

Until recent time all known sufficient conditions for the boundedness of S_Γ had been far away from being necessary ones. For instance, one had shown that S_Γ belongs to $\mathscr{L}(L_p(\Gamma, \varrho))$ if Γ is a Lyapunov or Radon curve and ϱ is a power weight on Γ of the form

$$\varrho(t) = \prod_{k=1}^{m} |t - t_k|^{\beta_k}, \quad -1 < \beta_k < p - 1 \quad (k = 1, \ldots, m) \qquad (1.6)$$

(see Danilyuk [1975], Khvedelidze [1956, 1975], Mikhlin, Prössdorf [1980]). Remember that a curve Γ is said to be a *Lyapunov* (resp. *Radon*) *curve* if it has a parametric representation

$$x(s) = x(0) + \int_0^s \cos \vartheta(\sigma) d\sigma, \qquad y(s) = y(0) + \int_0^s \sin \vartheta(\sigma) d\sigma$$

with respect to arc length s $(0 \le s \le \gamma)$ such that the angle $\vartheta(s)$, which is only determined up to a multiple of 2π at a given point s, can be chosen so as to satisfy a Hölder condition (resp. to be of bounded variation) throughout $[0, \gamma]$.

Only very recently the boundedness problem for S_Γ was wrapped up.

In this connection, first of all mention must be made of the following remarkable result of A.P. Calderón (1977), concerning Lipschitz curves with sufficiently small Lipschitz constant. To state his result, let Γ be a curve given by $z(x) = x + i\varphi(x)$ $(x \in \mathbb{R})$ with some real-valued function φ such that $|\varphi(x) - \varphi(x')| \le N|x - x'|$ all $x, x' \in \mathbb{R}$.

Calderón's Theorem. *For $\varepsilon > 0$, define the operator*

$$(A_{\varphi,\varepsilon} f)(x) := \int_{|x-y|>\varepsilon} \frac{f(y)}{x - y + i(\varphi(x) - \varphi(y))} dy, \qquad x \in \mathbb{R}.$$

There exists a constant $M > 0$ such that whenever $N < M$ the following statements are true:

a) *The operator $A_\varphi f := \lim_{\varepsilon \to 0} A_{\varphi,\varepsilon} f$ is bounded on $L_p(\mathbb{R})$ for all $p \in (1, \infty)$;*

b) *if $f \in L_p(\mathbb{R})$, $1 \le p < \infty$, then the limit $(A_\varphi f)(x) = \lim_{\varepsilon \to 0} (A_{\varphi,\varepsilon} f)(x)$ exists for almost all $x \in \mathbb{R}$;*

c) *A_φ is an operator of the weak type $(1, 1)$, i.e., for all $f \in L_1(\mathbb{R})$ and all $\lambda > 0$,*

$$m\{x \in \mathbb{R} : |(A_\varphi f)(x)| > \lambda\} \le \frac{C}{\lambda} \int_{\mathbb{R}} |f(x)| dx,$$

where m designates Lebesgue measure.

Calderón's restriction to sufficiently small Lipschitz constants N was subsequently removed by R.R. Coifman, A. McIntosh, Y. Meyer (1982), G. David (1982, 1984) (also see Coifman, Meyer, Stein [1983]), and T. Murai (1983). An essential issue of the proof of the Calderón theorem is L_2-estimates for so-called *Calderón commutators*:

$$T(h, \varphi) f(x) := \lim_{\varepsilon \to 0} \int_{|x-y|>\varepsilon} h\left[\frac{\varphi(x) - \varphi(y)}{x - y}\right] \frac{f(y)}{x - y} dy.$$

Such commutators (with $h(x) = x^n$) are arising when the operator $A_{\lambda\varphi}$ is expanded into a power series in λ, but they are also of importance in the theory of pseudodifferential operators. The following result is due to Coifman, David, and Meyer (1981, 1983).

Theorem (on the commutator). *If φ has a bounded derivative, $\varphi' \in L_\infty(\mathbb{R})$, and if $h \in C^\infty(\mathbb{R})$, then $T(h, \varphi) \in \mathcal{L}(L_p(\mathbb{R}))$ for all $p \in (1, \infty)$.*

1.2.2. Resorting to Calderón's theorem, David (1982) found a simple proof of the fact that S_Γ is bounded on $L_2(\Gamma)$ for every Carleson curve Γ. This result combined with standard techniques yields the following criterion.

Theorem. *Let Γ be any Carleson curve. Then $S_\Gamma \in \mathcal{L}(L_p(\Gamma, \varrho))$ $(1 < p < \infty)$ if and only if ϱ and Γ satisfy the condition*

$$\sup_{z,r} \frac{1}{r} \|\varrho^{1/p}\|_{L_p(\Gamma(z,r))} \|\varrho^{-1/p}\|_{L_{p'}(\Gamma(z,r))} < \infty, \qquad (A_p^\Gamma)$$

the supremum taken over all $z \in \Gamma$ and $r > 0$, and $\Gamma(z, r)$ standing for $\Gamma \cap B(z, r)$.

Note that if (A_p^Γ) holds and Γ is a closed curve, then $S_\Gamma^2 = I$.

For more about the results of Subsections 1.2.1 and 1.2.2 see the surveys Dyn'kin, Osilenker [1983], Journé [1983] and Murai [1988].

1.2.3. Let Γ be a contour (=curve) consisting of a finite number of non-intersecting Lyapunov curves and let Γ_0 denote the complex unit circle. Then

$$\|S_{\Gamma_0}\|_p = \|H\|_p = \||S_\Gamma\||_p = \begin{cases} \cot \dfrac{\pi}{2p} & \text{for } 2 \leqq p < \infty, \\[2mm] \tan \dfrac{\pi}{2p} & \text{for } 1 < p \leqq 2, \end{cases}$$

where $\|S_{\Gamma_0}\|_p$ is the norm of S_{Γ_0} on $L_p(\Gamma_0)$, $\|H\|_p$ the norm of H on $L_p(\mathbb{R})$, and $\||S_\Gamma\||_p$ the essential norm of S_Γ on $L_p(\Gamma)$ (recall 2.6.5). Moreover, if $\Gamma \neq \Gamma_0$, then $\|S_\Gamma\|_2 > \|S_{\Gamma_0}\|_2 = 1$. Hence, unlike the essential norm $\||S_\Gamma\||_p$, the norm $\|S_\Gamma\|_p$ depends on the shape of the contour Γ essentially. Results of this type were established by M. Cotlar (1955), I. Gohberg, N.Ya. Krupnik (1968), and S.K. Pichorides (1972); they were generalized to spaces $L_p(\Gamma, \varrho)$ with power weight by I.E. Verbitskiĭ and N.Ya. Krupnik (1980) (also see Gohberg, Krupnik [1973] and Krupnik [1984], Böttcher, Krupnik, Silbermann [1988]).

1.2.4. If $f \in C^\alpha(\Gamma)$ $(0 < \alpha < 1)$, i.e. if $|f(t) - f(\tau)| \leqq C|t - \tau|^\alpha$ for all $t, \tau \in \Gamma$, then the definition of the singular integral S_Γ can be directly applied to see that $(S_\Gamma f)(t)$ exists at each interior point t of the curve Γ and that

$$(S_\Gamma f)(t) = \frac{1}{\pi i} \int_\Gamma \frac{f(\tau) - f(t)}{\tau - t} d\tau + f(t) (S_\Gamma 1)(t)$$

in this case. Note that in the formula obtained the first integral to the right is absolutely covergent. Furthermore, we have $S_\Gamma 1 = 1$ if Γ is a closed and positively oriented curve, whereas $(S_\Gamma 1)(t) = 1 + (1/\pi i)[\log(b - t) - \log(a - t)]$ if Γ is a simple non-closed curve with endpoints a and b; in the latter case $\log(z - t)$ denotes any fixed branch of the function $z \to \log(z - t)$ which is continuous in the complex plane cut along any line joining t $(t \neq a, t \neq b)$ and infinity and lying on the right of Γ.

In particular, if Γ is a closed Lyapunov curve, then $S_\Gamma \in \mathscr{L}(C^\alpha(\Gamma))$ for $0 < \alpha < 1$ (this is the *J. Plemelj–I.I. Privalov theorem*). The latter result was extended to weighted Hölder spaces by R.V. Duduchava (1970).

1.2.5. An operator which is closely related to the Cauchy singular integral is the so-called *Hilbert singular integral* (also referred to as the *conjugation operator*) defined as

$$(\tilde{H}g)(s) := \frac{1}{2\pi} \int_{-\pi}^{\pi} \cot\frac{\sigma - s}{2} g(\sigma) d\sigma, \qquad -\pi \leq s \leq \pi, \tag{1.7}$$

the integral again understood in the Cauchy principal-value sense. This integral plays a prominent role in the convergence theory of trigonometric series and also in the theory of general orthogonal series.

If $g \in C^\alpha[-\pi, \pi]$ $(0 < \alpha \leq 1)$ is 2π-periodic, then the integral (1.7) does exist for each s and can be represented as an absolutely convergent integral:

$$(\tilde{H}g)(s) = \frac{1}{2\pi} \int_{-\pi}^{\pi} \cot\frac{\sigma - s}{2} [g(\sigma) - g(s)] d\sigma.$$

Let Γ_0 be the complex unit circle and let $t = e^{is}$, $\tau = e^{i\sigma}$ $(-\pi \leq s, \sigma \leq \pi)$ be points on Γ_0. Given a function f on Γ_0, put $\tilde{f}(s) := f(e^{is})$. We then have the following equality relating S_{Γ_0} to $\tilde{H}$:

$$(S_{\Gamma_0} f)(t) = \frac{1}{i} (\tilde{H}\tilde{f})(s) + \frac{1}{2\pi} \int_{-\pi}^{\pi} \tilde{f}(\sigma) d\sigma.$$

As an immediate consequence of this we obtain that $\tilde{H}$ is bounded on the spaces $C^\alpha(\Gamma_0)$ $(0 < \alpha < 1)$ and $L_p(\Gamma_0, \varrho)$ $(1 < p < \infty)$, in the latter case provided that ϱ and Γ_0 satisfy the condition $(A_p^{\Gamma_0})$.

1.2.6. Finally notice that the intimate connection between the Cauchy singular integral and the theory of analytic functions also comes to light in the following well-known theorem by I.I.Privalov (1950) on the boundary values of Cauchy type integrals.

Privalov's Theorem. *Let $f \in L_1(\Gamma)$. Then for the Cauchy integral*

$$F(z) = \frac{1}{2\pi i} \int_\Gamma \frac{f(\tau)}{\tau - z} d\tau \qquad (z \notin \Gamma) \tag{1.8}$$

to possess finite nontangential boundary values $F^+(t)$ (resp. $F^-(t)$) from the left (resp. right) a.e. on Γ it is necessary and sufficient that (1.5) exist and be finite a.e. on Γ. In that case the Yu.V. Sokhotskiĭ–J. Plemelj formulas

$$F^\pm(t) = \pm \tfrac{1}{2} f(t) + \tfrac{1}{2} (S_\Gamma f)(t)$$

hold a.e. on Γ.

It is worth emphasizing that this theorem is of fundamental import in many applications of one-dimensional singular integrals.

More about the topics touched upon in 1.2.3–1.2.6 can be found in Gohberg and Krupnik [1973], Khvedelidze [1975], Mikhlin and Prössdorf [1980], Krupnik [1984].

§2. Singular Integral Equations

We now consider equations of the form

$$(Au)(t) := c(t)u(t) + d(t)(S_\Gamma u)(t) + (Tu)(t) = f(t) \tag{2.1}$$

in the spaces $L_p(\Gamma, \varrho)$ $(1 < p < \infty)$. Throughout, we assume Γ and ϱ are subject to the condition (A_p^Γ) (see 1.2.2), c and d are given functions belonging to $L_\infty(\Gamma)$ (they are referred to as the *coefficients* of the equation (2.1)), T is a given compact operator on $L_p(\Gamma, \varrho)$, and f is a given function in $L_p(\Gamma, \varrho)$.

The operator $A \in \mathscr{L}(L_p(\Gamma, \varrho))$ defined by the left-hand side of (2.1) is called a *general* (or *complete*) *singular integral operator*. In case $T = 0$, the operator A is said to be *simplest* (or *characteristic*). Frequently it is more convenient to rewrite the simplest singular operator $A_0 = cI + dS_\Gamma$ in the form

$$A_0 = aP_\Gamma + bQ_\Gamma; \qquad P_\Gamma = \tfrac{1}{2}(I + S_\Gamma), \qquad Q_\Gamma = I - P_\Gamma,$$

where, of course, $a = c + d$ and $b = c - d$.

At the present time the theory of singular integral equations with piecewise continuous coefficients has been worked out almost completely, and this theory will be our concern in the Sections 2 and 3. Other classes of coefficients will be dealt with in the subsequent sections.

2.1. The Case of Continuous Coefficients. We now assume that Γ is a closed curve and that c, d belong to $C(\Gamma)$. So $S_\Gamma^2 = I$, P_Γ and Q_Γ are complementary projections, and, as already mentioned, none of the operators S_Γ, P_Γ, Q_Γ is compact.

2.1.1. A crucial role in the theory of singular equations is played by the fact that the commutator K_c given by $K_c u := cS_\Gamma u - S_\Gamma cu$ is compact. To see that K_c is compact note first that $\|K_c - K_{c_n}\| \to 0$ as $n \to \infty$ whenever the functions $c_n \in C^\alpha(\Gamma)$ $(0 < \alpha < 1)$ converge uniformly to c on Γ, and then take into account that $K_{c_n} \in \mathscr{K}(L^p(\Gamma, \varrho))$.

The undoubtedly most fundamental concept in the theory of singular equations is the concept of the symbol, first introduced by S.G. Mikhlin [1948] for the case considered here. The *symbol* of the general singular operator A is the function defined by

$$A(t, \vartheta) = c(t) + \vartheta d(t) \qquad (\vartheta \in \{-1, 1\}, t \in \Gamma). \tag{2.2}$$

This function may also be identified with the pair of functions $\{a, b\}$. Since $aP_\Gamma + bQ_\Gamma$ is never compact unless both a and b vanish identically, the decomposition of a general singular operator into a simplest operator and a compact operator is unique. This implies that the definition (2.2) is correct. Using the compactness of the commutators K_c and K_d it is easy to verify that the symbol of the sum and the product of two operators A_1 and A_2 (with continuous coefficients) is equal to the sum and the product of their symbols; we also remark that the commutar $[A_1, A_2]$ is compact. In particular, if the symbol of A does not *degenerate* (in that case A is usually said to be of *normal type* or to be *elliptic*), i.e. if $A(t, \vartheta) \neq 0$ for all $t \in \Gamma$ and $\vartheta \in \{-1, 1\}$, or, what is the same, if

$$a(t)b(t) = c^2(t) - d^2(t) \neq 0 \qquad \forall t \in \Gamma, \tag{2.3}$$

then the singular operator $B := a^{-1}P_\Gamma + b^{-1}Q_\Gamma$ is a regularizer of A and, consequently, A is a Noether operator. An important addition to the facts listed above is the following theorem, which was established in the work of I. Gohberg, S.G. Mikhlin, I.B. Simonenko, and B.V. Khvedelidze (see, e.g., Mikhlin and Prössdorf [1980]).

Theorem. *For the operator $A_0 = aP_\Gamma + bQ_\Gamma$ to be semi-Noetherian on the space $L_p(\Gamma, \varrho)$ it is necessary and sufficient that (2.3) be satisfied. If this condition is fulfilled, then A_0 is invertible, only left-sided invertible or only right-sided invertible in dependence on whether the integer $\varkappa = \mathrm{ind}\, a/b$ is zero, positive or negative; moreover,*

$$\dim \ker A_0 = \max(-\varkappa, 0), \qquad \dim \ker A_0^* = \max(\varkappa, 0). \tag{2.4}$$

Here $\mathrm{ind}\, a/b$ *denotes the winding number of the positively oriented curve $V(a/b)$* $:= \{a(t)/b(t) : t \in \Gamma\}$ *about the origin; more precisely* $\mathrm{ind}\, a/b := \dfrac{1}{2\pi i} \displaystyle\int_V z^{-1}\, dz.$

Notice that (2.4) combined with 3.5(iii) of Chap. 1 gives the following *index formula for the general singular operator*:

$$\mathrm{Ind}\, A = \mathrm{Ind}\, A_0 = -\varkappa.$$

This is the celebrated formula found by F. Noether in 1921.

All results stated above remain valid for the space $C^\alpha(\Gamma)$ $(0 < \alpha < 1)$ provided one requires that a and b themselves belong to that space, that T be compact on that space, and that Γ be a closed Lyapunov curve (see Muskhelishvili [1968]).

2.1.2. Let $\tilde{C}(\Gamma)$ denote the Banach algebra of all functions of the form (2.2) with $c, d \in C(\Gamma)$, the norm being given by

$$\|A(t, \vartheta)\| = \max_{t \in \Gamma,\, \vartheta = \pm 1} |c(t) + \vartheta d(t)|.$$

We have $\|A(t, \vartheta)\| \leq \|A\|$, where $\|A\|$ refers to the norm of A as element of $\mathscr{L}(L_p(\Gamma, \varrho))$. Indeed, if there would exist t_0, ϑ_0 such that $|A(t_0, \vartheta_0)| > \|A\|$, then $B = I - (1/A(t_0, \vartheta_0))A$ would be invertible, contradicting the fact that the symbol of B vanishes at (t_0, ϑ_0).

The inequality we have just proved along with the results of 2.1.1 yields the following theorem, which is due to I. Gohberg (1952, 1964).

Theorem. *The collection of all singular operators of the form (2.1) with continuous coefficients constitutes a closed subalgebra $\mathfrak{A}$ of the Banach algebra $\mathscr{L}(L_p(\Gamma, \varrho))$. The quotient algebra $\tilde{\mathfrak{A}} = \mathfrak{A}/\mathscr{K}(L_p(\Gamma, \varrho))$ is isomorphic to the algebra $\tilde{C}(\Gamma)$.*

The isomorphism of $\tilde{\mathfrak{A}}$ into $\tilde{C}(\Gamma)$ sends the coset $A + \mathscr{H}(L_p(\Gamma, \varrho))$ to the symbol of A; it is nothing else than the Gelfand transform of $\tilde{\mathfrak{A}}$ (see 1.4 and 1.6 of Chap. 1). In the case $p = 2$ and $\varrho = 1$ it is an isometrical *-isomorphism.

2.2. The Case of Piecewise Continuous Coefficients. Now assume that Γ is an oriented curve consisting of a finite number of nonintersecting closed Lyapunov curves. We let $PC(\Gamma)$ denote the closed algebra of all piecewise continuous functions on Γ. Our subject here is the investigation of equation (2.1) with coefficients $c, d \in PC(\Gamma)$ in the space $L_p(\Gamma, \varrho)$ $(1 < p < \infty)$ with a power weight of the form (1.6). Note that the results of 2.1 break down in the situation considered here; the reason is that the commutator K_c is no longer compact if $c \in PC(\Gamma)$ is discontinuous.

Before stating generalizations of the results of 2.1 to the case of piecewise continuous coefficients we must introduce some notation. Put $\delta(t) = 2\pi/p$ for $t \in \Gamma \setminus \{t_1, \ldots, t_m\}$, $\delta(t_k) = 2\pi(1 + \beta_k)/p$, $\gamma(t) = \pi - \delta(t)$, and

$$f(t, \mu) = \begin{cases} \sin(\gamma\mu)\sin^{-1}\gamma\exp(i\gamma(\mu - 1)) & \text{if } \delta(t) \neq \pi \\ \mu & \text{if } \delta(t) = \pi. \end{cases}$$

With each function $a \in PC(\Gamma)$ we associate the function $a_{p\varrho}: \Gamma \times [0, 1] \to \mathbb{C}$ defined by

$$a_{p\varrho}(t, \mu) := a(t + 0)f(t, \mu) + a(t - 0)(1 - f(t, \mu))$$

where $t \in \Gamma$ and $\mu \in [0, 1]$. The range $V_{p\varrho}(a)$ of the function $a_{p\varrho}$ is a naturally oriented curve, which is obtained from the (essential) range of the function a by filling in certain circular arcs or straight line segments joining $a(t - 0)$ to $a(t + 0)$ for each discontinuity. Note that the shape of the arc linking up $a(t - 0)$ and $a(t + 0)$ (line segments are nothing but degenerate arcs) depends on the value of p and the behavior of the weight ϱ at (or in a neighborhood of) the point t.

Now Theorem 2.1.1 can be converted into a theorem on singular equations with piecewise continuous coefficients: one has only to replace condition (2.3) by the condition

$$b(t \pm 0) \neq 0, \quad (a/b)_{p\varrho}(t, \mu) \neq 0 \quad \forall(t, \mu) \in \Gamma \times [0, 1] \tag{2.5}$$

and to define $\varkappa$ as $\varkappa = \text{ind}_{p\varrho}(a/b) := \text{ind } V_{p\varrho}(a/b)$.

In this full generality the preceding result was first proved by I. Gohberg and N.Ya. Krupnik (1969). For several special cases it had been previously established by B.V. Khvedelidze (1956), E. Shamir (1960), H. Widom (1960), and A. Devinatz (1964). Singular integral operators on $L_p(\Gamma, \varrho)$ of the more general form

$$(Au)(t) = a(t)u(t) + \int_\Gamma b(t, \tau)(\tau - t)^{-1}u(\tau)\, d\tau$$

with $a, b \in PC$ were studied by A.P. Soldatov (1978) and N.Ya. Krupnik (1984).

It turns out that it is impossible to introduce a scalar-valued symbol on algebras generated by singular integral operators with piecewise continuous coefficients (see Krupnik [1984]). However, such algebras may be equipped with a matrix-valued symbol (see 3.4).

2.3. The Case of Non-Closed Curves. We now consider the case where Γ is an oriented contour comprising a finite number of nonintersecting closed and non-closed Lyapunov curves. This case can be reduced to the one treated in 2.2 by means of the following argument. We construct any closed contour $\tilde{\Gamma}$ containing Γ as a part and continue the coefficients a, $b \in PC(\Gamma)$ of the operator $A := aP_\Gamma + bQ_\Gamma$ to functions $\tilde{a}$, $\tilde{b} \in PC(\tilde{\Gamma})$ by defining

$$\tilde{a}(t) = \begin{cases} a(t) & \text{for } t \in \Gamma, \\ 1 & \text{for } t \in \tilde{\Gamma} \backslash \Gamma, \end{cases} \qquad \tilde{b}(t) = \begin{cases} b(t) & \text{for } t \in \Gamma \\ 1 & \text{for } t \in \tilde{\Gamma} \backslash \Gamma. \end{cases} \tag{2.6}$$

One can show that the operator $A \in \mathscr{L}(L_p(\Gamma, \varrho))$ is normally solvable (is a $\Phi_\pm$-operator, is at least one-sided invertible) if and only if the singular operator $\tilde{A} := \tilde{a}P_{\tilde{\Gamma}} + \tilde{b}Q_{\tilde{\Gamma}} \in \mathscr{L}(L_p(\tilde{\Gamma}, \tilde{\varrho}))$ has the corresponding property; here $\tilde{\varrho}$ is the weight which coincides with ϱ on Γ and equals 1 on $\tilde{\Gamma} \backslash \Gamma$. Moreover,

$$\ker A = \ker \tilde{A}, \qquad \operatorname{coker} A = \operatorname{coker} \tilde{A}.$$

That this is so can be easily seen as follows. The space $L_p(\tilde{\Gamma}, \tilde{\varrho})$ clearly decomposes into the direct sum $L_p(\Gamma, \varrho) \dotplus L_p(\tilde{\Gamma} \backslash \Gamma)$. Letting $\tilde{P}$ denote the projection of $L_p(\tilde{\Gamma}, \tilde{\varrho})$ onto $L_p(\Gamma, \varrho)$ along $L_p(\tilde{\Gamma} \backslash \Gamma)$, we have

$$A = \tilde{A}|L_p(\Gamma, \varrho), \qquad \tilde{A} = (I + \tilde{P}\tilde{A}\tilde{Q})(A\tilde{P} + \tilde{Q}), \tag{2.7}$$

where $\tilde{Q} = I - \tilde{P}$, and the operator $I + \tilde{P}\tilde{A}\tilde{Q}$ is invertible, its inverse being $I - \tilde{P}\tilde{A}\tilde{Q}$. From (2.7) we infer in particular that if $\tilde{B}$ is a left (right) inverse or regularizer of $\tilde{A}$, then $B := \tilde{P}\tilde{B}|L_p(\Gamma, \varrho)$ is a left (right) inverse or regularizer of A, respectively.

We finally remark that all results stated hitherto for the singular operator $aP_\Gamma + bQ_\Gamma$ can be literally maintained for operators of the form $P_\Gamma aI + Q_\Gamma bI$. Effective methods for solving singular integral equations with piecewise Hölder-continuous coefficients are discussed in detail in the books Gakhov [1977] and Muskhelishvili [1968] (also see Sec. 4).

2.4. The Riemann-Hilbert Boundary Value Problem. Let Γ be a closed contour. By the Sokhotskiĭ-Plemelj formulas, equation (2.1), with $T = 0$, can be rewritten in the form

$$a(t)\Phi^+(t) = b(t)\Phi^-(t) + f(t) \qquad (t \in \Gamma), \tag{2.8}$$

where $\Phi^\pm \in L_p(\Gamma, \varrho)$ are the boundary values of the Cauchy integral (1.8) with f replaced by u. Hence the simplest singular integral equation (2.1) is equivalent to the following classical problem of function theory (which is nowadays known under a babel of names, e.g. Riemann-Hilbert, Riemann alone, solely Hilbert, or is even called the "problem of conjugation"):

Find a function Φ which is representable by the Cauchy integral (1.8) with a density $u \in L_p(\Gamma, \varrho)$ such that its boundary values $\Phi^{\pm}$ satisfy (2.8) almost everywhere on Γ.

The solution of this problem (in a more general setting) is deferred to Section 4 (see also the survey Khvedelidze [1975]).

§3. Singular Operators with Matrix Coefficients and Banach Algebras Generated by Them

Throughout this section we shall assume that Γ is an oriented contour which is composed by a finite number of nonintersecting closed or non-closed Lyapunov curves and that ϱ is a weight of the form (1.6)

3.1. We first introduce some notation.

Given a linear space E, we let E^n (n being a natural number) refer to the linear space of all column-vectors of length n with components from E, and we let $E^{n \times n}$ denote the linear space of all square n by n matrices whose entries are from E.

Now suppose E is a Banach (Hilbert) space. Then E^n can be converted into a Banach (Hilbert) space by defining, for example, the norm (scalar product) of vectors in E^n as the sum of the norms (scalar products) of the components. The norm of a matrix $A = (a_{jk})_1^n \in E^{n \times n}$ may be defined by $\|A\| = n \max_{j,k} \|a_{jk}\|$. If E is a Banach algebra, then $E^{n \times n}$ endowed with this norm is also a Banach algebra.

It can be verified without difficulty that $\mathscr{L}(E^n) = \mathscr{L}(E)^{n \times n}$ and $\mathscr{K}(E^n) = \mathscr{K}(E)^{n \times n}$; we shall usually think of an operator $A \in \mathscr{L}(E^n)$ as a matrix $A = (a_{jk})_1^n$ with $a_{jk} \in \mathscr{L}(E)$.

The following theorems reveal that, under some additional conditions, the determinant $\det A \in \mathscr{L}(E)$ answers for the Noether properties of an operator $A \in \mathscr{L}(E^n)$. Note that if the entries of A commute pairwise modulo compact operators, then the arrangement of the factors in the terms of the sum defining the determinant may be neglected, since all possible results differ only by a compact operator.

Theorem 1. *Let $A \in \mathscr{L}(E^n)$ and suppose the entries of A commute pairwise up to compact operators. Then:*

1. $A \in \Phi_{\pm}(E^n) \Leftrightarrow \det A \in \Phi_{\pm}(E)$.
2. *A has a left (right) regularizer if and only if $\det A \in \mathscr{L}(E)$ owns this property.*

Theorem 2. *Suppose $\mathfrak{A}$ is a subalgebra of $\mathscr{L}(E)$ possessing the following properties: $\mathscr{K}(E) \subseteq \mathfrak{A}$; $[A, B] \in \mathscr{K}(E)$ for all $A, B \in \mathfrak{A}$; the set $\Phi(E) \cap \mathfrak{A}$ is dense in $\mathfrak{A}$. Then if $A \in \mathfrak{A}^{n \times n} \cap \Phi(E^n)$, we have $\mathrm{Ind}\, A = \mathrm{Ind}\, \det A$.*

Multidimensional singular integral operators on $L_p(\mathbb{R}^m)$ with continuous symbols exemplify that Theorem 2 is no longer true if $\Phi(E) \cap \mathfrak{A}$ is not required to be dense in $\mathfrak{A}$ (see Chap. 4). The Φ-version of Theorem 1 is due to N.Ya. Krupnik (1965); in the form stated here the theorem was established by U. Köhler and B.

Silbermann (1973). Theorem 2 is Silbermann's (1973). The next theorem was first proved by A.S. Markus and I.A. Feldman (1977).

Theorem 3. *Let H be a Hilbert space. If A belongs to $\Phi(H^n)$ and the entries of A commute pairwise modulo nuclear operators, then* Ind A = Ind det A.

3.2. We now consider singular integral operators $A = aP + bQ$ with coefficients $a, b \in [L_\infty(\Gamma)]^{n \times n}$ on the spaces $L_p^n(\Gamma, \varrho) := [L_p(\Gamma, \varrho)]^n$; here $P = (P_\Gamma \delta_{jk})_1^n$ and $Q = I - P$.

By virtue of the Riesz representation theorem, every continuous linear functional on $L_p^n(\Gamma, \varrho)$ may be written in the form

$$\langle u, v \rangle = \int_\Gamma u(t)v(t)\, dt; \qquad u \in L_p^n(\Gamma, \varrho), \qquad v \in L_{p'}^n(\Gamma, \varrho^{1-p'}). \tag{3.1}$$

In other words, (3.1) generates a dual system (recall 2.5.1 of Chap. 1). It is not difficult to check that $\langle Au, v \rangle = \langle u, A^T v \rangle$, where $A^T := Pb^T + Qa^T$ and a^T designates the matrix transposed to a. Thus, A^T is the transposed operator of A with respect to the dual system (3.1).

If, in particular, the coefficients a and b are continuous or piecewise continuous and if A is Noetherian on $L_p^n(\Gamma, \varrho)$, then Ind $A = -$ Ind A^T. In this case the necessary and sufficient condition for the solvability of the equation $Au = f$ quoted in 3.7 of Chap. 1 takes the form

$$\langle f, v_k \rangle = 0 \qquad \text{for} \qquad k = 1, 2, \dots, \alpha(A^T),$$

where $v_k \in L_{p'}^n(\Gamma, \varrho^{1-p'})$ are the linearly independent solutions of the equation $A^T v = 0$.

Theorem (I.B. Simonenko, 1964). *Let $a, b \in [L_\infty(\Gamma)]^{n \times n}$. If $aP + bQ$ (or $Pa + Qb$) is a Φ_+-or Φ_--operator, then* ess inf$|\det a(t)| > 0$ *and* ess inf$|\det b(t)| > 0$. *On condition that $n = 1$, the operator $aP + bQ$ (or $Pa + Qb$) is invertible from the left (resp. right) if it is a Φ_+-(resp. Φ_--) operator.*

3.3. Now assume that Γ consists only of closed curves and that $a, b \in [C(\Gamma)]^{n \times n}$. In that case it is natural to call the matrix function defined by (2.2), with $a = c + d$ and $b = c - d$, the *symbol* of the complete singular operator $A = aP + bQ + T$, $T \in \mathcal{K}(L_p^n(\Gamma, \varrho))$.

Theorem. *For the operator $A = aP + bQ + T$ to be a Φ_+- or Φ_--operator on $L_p^n(\Gamma, \varrho)$ it is necessary and sufficient that*

$$\det a(t) \ne 0, \qquad \det b(t) \ne 0 \quad \forall t \in \Gamma. \tag{3.2}$$

If (3.2) is in force, then the operator $B = a^{-1}P + b^{-1}Q$ is a regularizer of A and we have

$$\text{Ind } A = \text{ind}[\det b/\det a]. \tag{3.3}$$

Notice that there is also a matrix analogue of the theorem stated in 2.1.2.

For operators acting on $[C^\alpha(\Gamma)]^{n \times n}$ ($0 < \alpha < 1$) the index formula (3.3) was first given by N.I. Muskhelishvili and N.P. Vekua (1943). The theorem as it is cited here grew out of the work of G. Giraud (1939), S.G. Mikhlin (1948), I. Gohberg (1952, 1964), B.V. Khvedelidze (1956), G.F. Mandzhavidze (1958), and I.B. Simonenko (1961).

3.4. We now proceed to the case that the contour is composed by both closed and non-closed Lyapunov curves. We let $\mathscr{K}$ denote the ideal of compact operators on $L_p^n(\Gamma, \varrho)$ and $\mathfrak{A}$ refer to the smallest closed subalgebra of $\mathscr{L}(L_p^n(\Gamma, \varrho))$ containing all singular operators $aP + bQ$ with coefficients $a, b \in [PC(\Gamma)]^{n \times n}$. One can show that $\mathscr{K}$ is a subset of $\mathfrak{A}$, and it turns out that the algebra $\mathfrak{A}/\mathscr{K}$ is isomorphic to some algebra of $2n$ by $2n$ matrix functions. As will be shown below, this isomorphism depends significantly on p and the weight ϱ.

In order to state precise results, we need some more notation. Suppose Γ contains exactly l non-closed curves (l may be zero) and denote by τ_j and τ_{j+l}, respectively, the initial and final points of the jth curve ($j = 1, \dots, l$). We so obtain a set of points $\Delta := \{\tau_1, \dots, \tau_l, \tau_{l+1}, \dots, \tau_{2l}\}$. Given a weight ϱ of the form (1.6), we define the function $f(t, \mu)$ as in 2.2, and we put $h(t, \mu) = \sqrt{f(t, \mu)(1 - f(t, \mu))}$.

The *symbol* of an operator $A = aP + bQ$ with a, b in $[PC(\Gamma)]^{n \times n}$ considered as acting on $L_p^n(\Gamma, \varrho)$ is the matrix function $\mathscr{A}$ defined on $\Gamma \times [0, 1]$ as follows: for $k = 1, \dots, l$ set

$$\mathscr{A}(\tau_k, \mu) = \begin{pmatrix} f(\tau_k, \mu)a(\tau_k + 0) + (1 - f(\tau_k, \mu))b(\tau_k + 0) & 0 \\ 0 & b(\tau_k + 0) \end{pmatrix};$$

for $k = l + 1, \dots, 2l$ put

$$\mathscr{A}(\tau_k, \mu) = \begin{pmatrix} (1 - f(\tau_k, \mu))a(\tau_k - 0) + f(\tau_k, \mu)b(\tau_k - 0) & 0 \\ 0 & b(\tau_k - 0) \end{pmatrix};$$

and for $t \in \Gamma \backslash \Delta$ let $\mathscr{A}(t, \mu)$ equal

$$\begin{pmatrix} f(t, \mu)a(t + 0) + (1 - f(t, \mu))a(t - 0) & h(t, \mu)(b(t + 0) - b(t - 0)) \\ h(t, \mu)(a(t + 0) - a(t - 0)) & (1 - f(t, \mu))b(t + 0) + f(t, \mu)b(t - 0) \end{pmatrix}.$$

Notice that p and ϱ enter into the symbol via f and h.

Before dealing with the algebra $\mathfrak{A}$, let us consider the algebra $\mathfrak{A}_0$ of all operators of the form

$$A = \sum_{j=1}^{k} A_{j1} A_{j2} \dots A_{jr}, \tag{3.4}$$

where $A_{jl} = a_{jl}P + b_{jl}Q$ and $a_{jl}, b_{jl} \in [PC(\Gamma)]^{n \times n}$. The *symbol* of the operator $A \in \mathfrak{A}_0$ given by (3.4) is defined as the matrix function $\mathscr{A} = \sum_{j=1}^{k} \mathscr{A}_{j1} \mathscr{A}_{j2} \dots \mathscr{A}_{jr}$, where $\mathscr{A}_{jl}$ refers to the symbol of A_{jl} in the above sense. For the entries α_{uv} of the symbol $\mathscr{A}(t, \mu) = (\alpha_{uv}(t, \mu))_{u,v=1}^{2n}$ the following (basic!) estimates can be proved (compare with 2.1.2):

$$\max_{t \in \Gamma, 0 \leq \mu \leq 1} |\alpha_{uv}(t, \mu)| \leq C_{p\varrho} |||A|||, \tag{3.5}$$

where $|||A||| = \inf\{\|A + T\| : T \in \mathscr{K}\}$. The inequality (3.5) implies in particular that the symbol of an operator $A \in \mathfrak{A}_0$ is independent of its representation in the form (3.4). Moreover, the symbol of the sum or the product of two operators in $\mathfrak{A}_0$ equals the sum or the product of the symbols of the operators.

The algebra $\mathfrak{A}$ is the closure of $\mathfrak{A}_0$ in $\mathscr{L}(L_p^n(\Gamma, \varrho))$. Hence, given $A \in \mathfrak{A}$ there exists a sequence (A_v) of operators in $\mathfrak{A}_0$ such that $A_v \to A$ in the norm, and by virtue of (3.5) the entries of the symbols of A_v converge uniformly on $\Gamma \times [0, 1]$ to the entries of some $2n$ by $2n$ matrix function on $\Gamma \times [0, 1]$. The latter matrix function is called the *symbol* of the operator $A \in \mathfrak{A}$; it is easily seen that it is independent of the choice of the approximating sequence (A_v). The algebra $\mathfrak{A}$ can be shown to contain $\mathscr{K}$. Moreover, one can prove that the correspondence between operators in $\mathfrak{A}$ and their symbols induces an isomorphism of the quotient algebra $\mathfrak{A}/\mathscr{K}$ onto the algebra $\mathscr{S}$ comprised by the symbols of all operators in $\mathfrak{A}$. In the case $p = 2$, the algebra $\mathfrak{A}/\mathscr{K}$ is a (non-commutative) C^*-algebra and the symbol mapping is in fact an isometrical *-isomorphism.

The next theorem shows how a Noether criterion for operators in $\mathfrak{A}$ can be stated in terms of their symbol. Its formulation still requires one more definition.

Write the symbol of $A \in \mathfrak{A}_0$ as $\mathscr{A}(t, \mu) = (c_{uv}(t, \mu))_{u, v=1}^n$ with n by n matrices $c_{uv}(t, \mu)$. It can be proved that if $\det \mathscr{A}(t, \mu) \neq 0$ for all $(t, \mu) \in \Gamma \times [0, 1]$, then $\det(c_{22}(t, 1)c_{22}(t, 0))$ does not vanish on Γ and the function

$$\varDelta_A(t, \mu) := \det \mathscr{A}(t, \mu)/\det(c_{22}(t, 1)c_{22}(t, 0))$$

traces out a continuous, closed, and naturally oriented curve in the complex plane. The winding number of that curve about the origin is denoted by ind $\varDelta_A$. More generally, let now $A \in \mathfrak{A}$, $\det \mathscr{A}(t, \mu) \neq 0$ for all $(t, \mu) \in \Gamma \times [0, 1]$, and $A_v \to A$ ($A_v \in \mathfrak{A}_0$). Then the functions $\varDelta_{A_v}$ converge uniformly to some function $\varDelta_A$, and ind $\varDelta_{A_v}$ takes on a constant value for all sufficiently large v. This value will be denoted by ind $\varDelta_A$.

Theorem. *For an operator $A \in \mathfrak{A}$ to be a Φ_+- or Φ_--operator on $L_p^n(\Gamma, \varrho)$ it is necessary and sufficient that its symbol do not degenerate:*

$$\det \mathscr{A}(t, \mu) \neq 0 \qquad \forall(t, \mu) \in \Gamma \times [0, 1]. \tag{3.6}$$

If (3.6) is satisfied, then A is a Noether operator on $L_p^n(\Gamma, \varrho)$ and its index is given by Ind $A = -$ ind $\varDelta_A$.

Here are three easy but important consequences of this theorem. If Γ is a closed contour and $a, b \in PC(\Gamma)$, then

1. $[aI, S_\Gamma] = \frac{1}{2}[aI, P_\Gamma] \in \mathscr{K}(L_p(\Gamma, \varrho)) \Leftrightarrow a \in C(\Gamma)$;
2. $P_\Gamma a P_\Gamma b P_\Gamma - P_\Gamma ab P_\Gamma \in \mathscr{K}(L_p(\Gamma, \varrho)) \Leftrightarrow a$ and b have no common points of discontinuity.
3. $P_\Gamma a P_\Gamma b P_\Gamma - P_\Gamma b P_\Gamma a P_\Gamma \in \mathscr{K}(L_p(\Gamma, \varrho))$.

The above constructions and results as well as their generalizations to the case of piecewise Lyapunov contours with a finite number of self-intersection points are due to I. Gohberg and N.Ya. Krupnik (1971). They were extended to the case of piecewise Lyapunov curves possessing a finite number of corner points in the work of R.V. Duduchava (1976), N.Ya. Krupnik and V.I. Nyaga (1975), and M.

Costabel (1980) (see also Elschner [1985b], where systematic use is made of the apparatus of Mellin transform). Similar results for the spaces C^{α} were obtained by R.V. Duduchava (1970–1973). We finally wish to point out that N.Ya. Krupnik [1984] succeeded in describing all the Banach algebras of operators that have a scalar or matrix symbol; he also found an ingenious matrix analogue of the Gelfand homomorphism. In connection with the results of this section we also refer to the recent books Simonenko, Chin' Ngok Min' [1986], Prössdorf, Silbermann [1990, Chap. 5] and to the papers Roch, Silbermann [1988, 1990].

§4. Factorization of Matrix Functions and the Solution of Singular Integral Equations

The problem of explicitly solving singular integral equations and systems of such equations is intimately related to the so-called problem of Wiener-Hopf factorization of (matrix) functions given on a contour.

Throughout this section we assume that Γ is an oriented closed Lyapunov contour bounding a connected region $G_+ \subset \mathbb{C}$. Without loss of generality we shall always suppose that $0 \in G_+$.

4.1. Let $\mathfrak{B} = \mathfrak{B}(\Gamma)$ be any Banach algebra consisting of complex-valued bounded measurable functions on Γ and containing the functions $\varphi_1(t) = t$ and $\varphi_2(t) = t^{-1}$. We denote by P_Γ and $Q_\Gamma = I - P_\Gamma$ the projections on $L_p(\Gamma)$ ($1 < p < \infty$) introduced in Section 2.

The algebra $\mathfrak{B}$ is said to be *decomposing* if the subspaces $P_\Gamma\mathfrak{B}$ and $Q_\Gamma\mathfrak{B}$ ($\subset L_p(\Gamma)$) are subalgebras of $\mathfrak{B}$.

Given a decomposing algebra $\mathfrak{B}$, we put $\mathfrak{B}_+ = P_\Gamma\mathfrak{B}$, $\mathring{\mathfrak{B}}_- = Q_\Gamma\mathfrak{B}$, and $\mathfrak{B}_- = \mathfrak{B}_0 + \mathring{\mathfrak{B}}_-$, where $\mathfrak{B}_0$ is the one-dimensional subspace of constant functions. We so have $\mathfrak{B} = \mathfrak{B}_+ \dotplus \mathring{\mathfrak{B}}_-$, and it is not difficult to check that P_Γ and Q_Γ are in $\mathscr{L}(\mathfrak{B})$.

A matrix function $a \in G(\mathfrak{B}^{n \times n})$ is said to admit a (right) *factorization in* $\mathfrak{B}$ if it can be represented in the form

$$a = a_- d a_+, \quad d(t) = \operatorname{diag}(t^{\varkappa_1}, \ldots, t^{\varkappa_n}), \tag{4.1}$$

where $a_\pm$ belong to $G(\mathfrak{B}_\pm^{n \times n})$ and $\varkappa_1 \geq \cdots \geq \varkappa_n$ are integers. The integers $\varkappa_j$ ($j = 1, \ldots, n$) can be shown to be uniquely determined by the matrix function a; they are referred to as the *partial* (or *factorization*) *indices*. If every matrix function $a \in G(\mathfrak{B}^{n \times n})$ admits a factorization in $\mathfrak{B}$, then $\mathfrak{B}$ is called an algebra with the *factorization property*.

The following algebras are classical examples of algebras with the factorization property:
1) $C^{\alpha}(\Gamma)$, $0 < \alpha < 1$ (N.I. Muskhelishvili and N.P. Vekua, 1943); 2) W (I. Gohberg and M.G. Krein, 1958); 3) decomposing R-algebras, i.e. decomposing algebras of continuous functions containing the set of all rational functions with poles off Γ as a dense subset (I. Gohberg, 1964). More general decomposing algebras with the factorization property were studied by M.S.

Budyanu and I. Gohberg (1968). A survey of the classical factorization theory is
given in Clancey, Gohberg [1981] and in Litvinchuk, Spitkovskiĭ [1987] (also
see Mikhlin, Prössdorf [1980]).

4.2. Factorizing matrix functions is an effective tool for constructing a generalized inverse of singular operators and thus for solving singular integral
equations (see 3.2 of Chap. 1).

Let $\mathfrak{B} = \mathfrak{B}(\Gamma)$ be an algebra with the factorization property. We consider the
singular integral operator $A = aP + bQ \in \mathscr{L}(L_p^n(\Gamma))$ $(1 < p < \infty)$ whose coefficients a, b are in $G(\mathfrak{B}^{n \times n})$. For the sake of convenience, put $c = b^{-1}a$.

Assume we are given a factorization $c = c_- d c_+$, $d(t) = (t^{\varkappa_j}\delta_{jk})_1^n$. Then the
operator

$$A^{(-1)} = (c_+^{-1}P + c_- Q)(d^{-1}P + Q)c_-^{-1}b^{-1}$$

is a generalized inverse of A. Moreover, we have

$$\dim \ker A = -\sum_{\varkappa_j < 0} \varkappa_j, \qquad \dim \operatorname{coker} A = \sum_{\varkappa_j > 0} \varkappa_j.$$

The equation

$$Au = f \tag{4.2}$$

is solvable in $L_p^n(\Gamma)$ if and only if the vector $g = c_-^{-1}b^{-1}f =: (g_1, g_2, \ldots, g_n)^T$
satisfies the conditions

$$\int_\Gamma t^k g_j(t)\, dt = 0 \qquad (k = 0, 1, \ldots, \varkappa_j - 1) \tag{4.3}$$

for all $j \in \{1, \ldots, n\}$ such that $\varkappa_j > 0$. If the conditions (4.3) are fulfilled, then
$u_0 = A^{(-1)}f$ is a particular solution of equation (4.2).

The general solution of the homogeneous equation $Au = 0$ is given by

$$u = (c_+^{-1} - c_-)\, dq, \qquad q = (q_{-\varkappa_1 - 1}, \ldots, q_{-\varkappa_n - 1})^T,$$

where $q_{-\varkappa_j - 1}$ denotes an arbitrary polynomial whose degree does not exceed
$-\varkappa_j - 1$ $(q_{-\varkappa_j - 1}(t) \equiv 0$ in case $\varkappa_j \geq 0)$.

Analogous results hold for the operator $A = Pa + Qb$ (in that case put $c = ab^{-1}$). Moreover, all the above results remain in force for the spaces $L_p^n(\Gamma, \varrho)$.

The method of solving singular integral equations by means of a factorization
of the coefficients (or, what is the same, by solving the corresponding problem
of conjugation) was proposed by T. Carleman in 1922, who also gave birth to
the idea of constructing the factorization explicitly with the help of Cauchy
type integrals (for $n = 1$). Carleman's method was subsequently widely utilized
and generalized in the work of F.D. Gakhov, N.I. Muskhelishvili, I.N. Vekua,
N.P. Vekua, D.A. Kveselava, B.V. Khvedelidze, I.B. Simonenko, and others
(see Vekua [1970], Gakhov [1977], Muskhelishvili [1968], Khvedelidze [1956,
1975]).

4.3. A characterization of all algebras with the factorization property was
given by G. Heinig and B. Silbermann in 1983 (also see Böttcher, Silbermann
[1983]).

Let $\mathfrak{B} = \mathfrak{B}(\Gamma)$ be a decomposing algebra. With every matrix function $a \in \mathfrak{B}^{n \times n}$ we associate two operators $T_+(a) \in \mathscr{L}(\mathfrak{B}^n_+)$ and $T_-(a) \in \mathscr{L}(\mathring{\mathfrak{B}}^{1 \times n}_-)$ defined by

$$T_+(a)f_+ = P_\Gamma a f_+ \quad (f_+ \in \mathfrak{B}^n_+), \qquad T_-(a)g_- = Q_\Gamma g_- a \quad (g_- \in \mathring{\mathfrak{B}}^{1 \times n}_-).$$

Theorem. *A decomposing algebra $\mathfrak{B}$ possesses the factorization property if and only if for every $a \in G(\mathfrak{B}^{n \times n})$ the following two conditions are satisfied:*
a) $T_+(a)$ *and* $T_-(a)$ *are Noether operators;*
b) Ind $T_+(a) = -$ Ind $T_-(a)$.

Besides the algebras 1) by 3) instanced in 4.1, there is an ample list of interesting algebras satisfying the conditions of this theorem, including the following algebras of bounded functions on the unit circle Γ_0: 4) $C^\alpha + H_\infty$ $(0 < \alpha < 1)$, where $C^\alpha = C^\alpha(\Gamma_0)$ and $H_\infty = L_\infty(\Gamma) \cap P_{\Gamma_0} L_\infty(\Gamma_0)$ (i.e. H_∞ is the familiar Hardy space); 5) $B^1_{1,1} + H_\infty$, where $B^1_{1,1}$ is the Besov space.

4.4. The algebra $C(\Gamma)$ is not decomposing, since the operator S_Γ is not bounded on $L_\infty(\Gamma)$. Nevertheless, this algebra owns the following so-called generalized factorization property (G.F. Mandzhavidze and B.V. Khvedelidze (1958), I.B. Simonenko (1961)). Put $L^+_p(\Gamma) = P_\Gamma L_p(\Gamma)$ and $L^-_p(\Gamma) = Q_\Gamma L_p(\Gamma)$.

Theorem. *Every matrix function $a \in G(C(\Gamma)^{n \times n})$ admits a factorization (4.1) such that, for all $p \in (1, \infty)$,*
1) $a^{\pm 1}_+ \in [L^+_p(\Gamma)]^{n \times n}$, $a^{\pm 1}_- \in [L^-_p(\Gamma)]^{n \times n}$,
2) $a_- P a^{-1}_- \in \mathscr{L}(L^n_p(\Gamma))$.

An important consequence of this theorem is that all results of 4.2 remain valid under the sole assumption that $a, b \in G(C(\Gamma)^{n \times n})$.

4.5. A *generalized factorization* of a matrix function $a \in G(L_\infty(\Gamma)^{n \times n})$ in the space $L^n_p(\Gamma, \varrho)$ is a factorization (4.1) in which $a_- \in [L^-_p(\Gamma, \varrho)]^{n \times n}$, $a_+ \in [L^+_{p'}(\Gamma, \varrho^{1-p'})]^{n \times n}$, $a^{-1}_+ \in [L^+_p(\Gamma, \varrho)]^{n \times n}$, $a^{-1}_- \in [L^-_{p'}(\Gamma, \varrho^{1-p'})]^{n \times n}$, and $a_- P a^{-1}_- \in \mathscr{L}(L^n_p(\Gamma, \varrho))$, where $p' = p/(p-1)$.

Theorem. *A matrix function $a \in G[L_\infty(\Gamma)^{n \times n}]$ admits a generalized factorization in $L^n_p(\Gamma, \varrho)$ if and only if $aP + Q$ is a Noether operator on $L^n_p(\Gamma, \varrho)$.*

The concept of generalized factorization permits carrying over all results of 4.2 to operators on $L^n_p(\Gamma, \varrho)$ with coefficients $a, b \in G[L_\infty(\Gamma)^{n \times n}]$. This was done for the first time by I.B. Simonenko (1964, 1968).

The simplest example of a function on Γ_0 which has no factorization in the sense of 4.4 is the function $a(t) = t^{1/2} (\in G(L_\infty(\Gamma)))$. Its generalized factorization in the space $L_p(\Gamma_0)$ $(1 < p < 2)$ is given by $a(t) = (t-1)^{1/2}(1 - 1/t)^{-1/2}$.

Sufficiently large classes of functions $a \in L_\infty(\Gamma)$ generating a Noether operator $aP_\Gamma + Q_\Gamma$ on $L_p(\Gamma, \varrho)$ were studied by I.B. Simonenko (1964, 1968), F.D. Frolov (1970), N.Ya. Krupnik and V.I. Nyaga (1974) (for more about this see also Krupnik [1984] and the following section).

§5. Sufficient Noether Conditions and Index Computation for Singular Integral Equations with Bounded Measurable Coefficients. Toeplitz Operators

The topics referred to in the heading are subject of active research in our days, and we here have the opportunity to present some recent success along these lines. For simplicity, we content ourselves to the case of the unit circle $\Gamma = \Gamma_0$ and the space $L_2 = L_2(\Gamma_0)$.

Because of Theorem 3.2, it is enough to consider only operators of the form $T_a := aP + Q$, where $a \in G(L_\infty^{n \times n})$. Moreover, given $A = aP + bQ$, where b is also in $G(L_\infty^{n \times n})$, we may put $c = b^{-1}a$ and then have

$$aP + bQ = b(cP + Q) = b(PcP + Q)(I + QcP),$$

the operator $I + QcP$ being invertible (its inverse is $I - QcP$). Hence the operator A is (left, right, or generalized) invertible or is a $\Phi_\pm$-operator if and only if the operator $T(c) := PcP|H_2$ given on the Hardy space $H_2 = \text{im } P$ possesses the corresponding property. Also notice that

$$\dim \text{der } A = \dim \ker T(c), \qquad \dim \text{coker } A = \dim \text{coker } T(c).$$

Similar remarks apply to the operator $Pa + Qb$ as well.

The operator $T(c)$ is referred to as the *Toeplitz operator* generated by the function c. By what has been said above, $T(c)$ may be identified with the singular integral operator T_c.

5.1. A matrix function $a \in L_\infty^{n \times n}$ is said to be *locally sectorial* if for each point $\tau \in \Gamma_0$ there exist an open neighborhood $U_\tau \subset \Gamma_0$ and functions $b_\tau, c_\tau \in G(C^{n \times n})$, $g_\tau \in L_\infty^{n \times n}$ such that $a(t) = b_\tau(t)g_\tau(t)c_\tau(t)$ for $t \in U_\tau$ and $\text{Re } g_\tau(t) \geq \delta > 0$ for almost all $t \in \Gamma_0$.

In 1968, I.B. Simonenko showed that every locally sectorial matrix function generates a Noetherian Toeplitz operator.

5.2. We let $\mathbb{H}f$ denote the *harmonic extension* of a function $f \in L_\infty$. This is a function defined in the open unit disk, and it is given at $z = re^{i\varphi}$ $(0 \leq r < 1, 0 \leq \varphi \leq 2\pi)$ by

$$(\mathbb{H}f)(re^{i\varphi}) := \frac{1}{2\pi} \int_0^{2\pi} \frac{(1 - r^2)f(e^{i\psi})}{1 - 2r \cos(\varphi - \psi) + r^2} d\psi.$$

Put $\mathbb{H}_r f(t) := \mathbb{H}f(rt)$ for $t \in \Gamma_0$, and in case $a = (a_{jk}) \in L_\infty^{n \times n}$ let $\mathbb{H}_r a := (\mathbb{H}_r a_{jk})_1^n$.

Theorem. *If $a \in L_\infty^{n \times n}$ is locally sectorial or if a belongs to $G(C + H_\infty)^{n \times n}$, then $T_a \in \Phi(L_2^n)$ and*

$$\text{Ind } T_a = -\lim_{r \to 1} \text{ind det } \mathbb{H}_r a. \tag{5.1}$$

Under the hypothesis that $a \in G(C + H_\infty)^{n \times n}$, this theorem was established by R.G. Douglas [1973], who also proved the following reverse of the theorem:

If $a \in C + H_\infty$ and $T_a \in \Phi(L_2)$, then necessarily $a \in G(C + H_\infty)$. The algebra $C + H_\infty$ is frequently called the *Douglas algebra* (it is the smallest algebra among all closed algebras $\mathfrak{A}$ such that $H_\infty \subset \mathfrak{A} \subsetneqq L_\infty$, which are also termed Douglas algebras). Douglas [1972] was also the first to realize the important role played by the harmonic extension in the study of Toeplitz operators generated by $C + H_\infty$ functions. For locally sectorial matrix functions, the index formula (5.1) was only very recently proved by B. Silbermann (1985). His proof also works in the case where $a \in G(C + H_\infty)^{n \times n}$. A very simple new proof of the above theorem was recently also found by A. Böttcher (1987).

5.3. A natural generalization of the algebra PC is the class V of all functions on Γ_0 which have at most two essential limit values at every point of Γ_0.

If $a \in V^{n \times n}$, then $T_a \in \Phi(L_2^n)$ if and only if a is locally sectorial (K.F. Clancey, 1974). So in this case formula (5.1) applies (N.S. Faour, 1977).

The very complicated problem of studying Toeplitz operators generated by functions of $V^{n \times n}$ on the spaces $L^p (1 < p < \infty)$ was considered by I.M. Spitkovskiĭ (1983), who established Noether criteria and provided a method for computing the index.

5.4. Let PQC denote the smallest C^*-subalgebra of L_∞ containing PC and $QC := (C + H_\infty) \cap \overline{(C + H_\infty)}$ (here the bar refers to complex conjugation). The functions in PQC and QC are called the *piecewise quasicontinuous* and *quasicontinuous functions*, respectively. The theory of Toeplitz operators induced by PQC functions was developed by D.E. Sarason (1977); he also studied the C^*-algebra generated by all Toeplitz operators T_a with $a \in PQC$. In this theory, the harmonic extension again proved to play a prominent role. Recently, in 1985, Sarason's results were generalized by Silbermann to the case $n > 1$. A central result of this theory reads as follows.

Theorem. *Let* $A = \sum_j \prod_k T_{a_{jk}}$ *with* $a_{jk} \in (PQC)^{n \times n}$. *Then the following are equivalent:*

1. *A is a Noether operator on L_2^n;*
2. *there exist numbers $\delta, \varepsilon > 0$ such that*

$$|\det \sum_j \prod_k \mathbb{H}_r a_{jk}| \geq \varepsilon \qquad for \; 1 - \delta < r < 1.$$

If either of these conditions is satisfied, then

$$\text{Ind } A = - \lim_{r \to 1} \text{ind} \left(\det \sum_j \prod_k \mathbb{H}_r a_{jk} \right).$$

5.5. The great effort that has been expended in studying Toeplitz operators with more general coefficients has revealed that their spectral theory is the more difficult, the more the coefficient is allowed to "oscillate". These difficulties have been overcome in a series of important special cases, notably in the case of almost periodic coefficients.

Let AP be the classical algebra of *almost periodic functions*, i.e. the closure in $L_\infty(\mathbb{R})$ of the linear span of all functions of the form $e^{i\delta x}$ ($\delta \in \mathbb{R}$). Given $\zeta \in \Gamma_0$, we put $AP_\zeta := \{f \circ \omega : f \in AP\}$, where $\omega(z) = -i(z + \zeta)(z - \zeta)^{-1}$ is the standard mapping of Γ_0 onto $\mathbb{R}$ sending ζ to the point at infinity. If $f \in AP$ and $\inf_\mathbb{R}|f| > 0$, then, by a well-known theorem of Bohr, the limit

$$\omega_f := \lim_{l \to \infty} \frac{1}{2l}(\arg f(l) - \arg f(-l))$$

exists and is finite. For functions $\varphi = f \circ \omega \in AP_\zeta$, the number ω_f serves as a substitute for the "winding number" of φ about the point ζ.

Theorem. *Let $\varphi \in AP_\zeta$, $\zeta \in \Gamma_0$, $\varphi = f \circ \omega$ ($f \in AP$).*

(a) *The following are equivalent:* 1. T_φ *is invertible;* 2. T_φ *is Noetherian;* 3. $\inf|\varphi| > 0$ *and* $\omega_f = 0$.

(b) *The operator* T_φ *is one-sided invertible if and only if* $\inf|\varphi| > 0$; *if this situation prevails, then* $\dim \ker T_\varphi = \infty$ *if* $\omega_f < 0$, *whereas* $\dim \ker T_\varphi^* = \infty$ *in case* $\omega_f > 0$.

This theorem was established by I. Gohberg, I.A. Feldman (1968) and L.A. Coburn, R.G. Douglas (1969). It was generalized to a variety of other settings by I. Gohberg and A.A. Sementsul (1970) (see also Gohberg, Krupnik [1973] and Gohberg, Feldman [1971]), Sarason (1977), A.I. Saginashvili (1979), M.B. Abrahamse (1979), S.C. Power (1980), V.B. Dybin (1976, 1985), and S.M. Grudskiĭ (1980, 1985).

A series of papers concerns several other, sometimes rather exotic classes of coefficients (e.g. M. Lee and D. Sarason (1971), R.G. Douglas (1978), A. Böttcher (1986)).

5.6. We wish to conclude this section by stating the following simple but important formula relating Toeplitz and Hankel operators (recall 2.4 of Chapter 2):

$$T_{\varphi\psi} - T_\varphi T_\psi = H_{\bar\varphi}^* H_\psi \qquad (\varphi, \psi \in L_\infty). \tag{5.2}$$

Formula (5.2), which in this form first appeared in papers by D. Sarason (1973) and H. Widom (1976), is of great relevancy in the study of algebras generated by Toeplitz operators. On the one hand, it allows us to "estimate" *semicommutators* $[T_\varphi, T_\psi] := T_\varphi T_\psi - T_{\varphi\psi}$ and thus also commutators (which are the difference of two semicommutators) of Toeplitz operators in the language of Hankel operators. On the other hand, it involves much information on the symbols of operators belonging to Toeplitz algebras. For example, formula (5.2) combined with Hartman's theorem (2.4 of Chap. 1) yields the following results:

1. *If at least one of the functions $\bar\varphi$ and ψ belongs to $C + H_\infty$, then $[T_\varphi, T_\psi) \in \mathscr{S}_\infty$, and if both φ and ψ are in $C + H_\infty$, then $[T_\varphi, T_\psi] \in \mathscr{S}_\infty$.*

2. *If at each point of Γ_0 at least one of the functions φ and ψ is continuous, then $[T_\varphi, T_\psi) \in \mathscr{S}_\infty$.*

Let us finally cite a remarkable compactness criterion for semicommutators of Toeplitz operators, whose sufficiency part was first proved by S. Axler, S.-Y.A.

Chang, and D. Sarason (1978) and the necessity portion of which was settled by A.L. Vol'berg (1982). By $\mathcal{H}_\infty(a)$ $(a \in L_\infty)$ is meant the smallest closed subalgebra of L_∞ containing H_∞ and a.

Theorem. $[T_\varphi, T_\psi] \in \mathcal{S}_\infty \Leftrightarrow \mathcal{H}_\infty(\bar{\varphi}) \cap \mathcal{H}_\infty(\psi) \subseteqq C + H_\infty$.

For more about the matter of this subsection see Nikol'skiĭ [1985] (and also Böttcher, Silbermann [1983, 1989] and Heinig, Rost [1984]).

§6. Wiener-Hopf Integral Equations

These are equations on the half-line $\mathbb{R}_+ = (0, \infty)$ given by

$$(W_k\varphi)(x) := c\varphi(x) + \int_0^\infty k(x - t)\varphi(t)\, dt = f(x), \qquad x \in \mathbb{R}_+. \qquad (6.1)$$

N. Wiener and E. Hopf (1931) were the first to study equations of this form. Note that such equations emerge from various problems of mechanics, physics, and mathematics. There exist intimate connections between Wiener-Hopf and Toeplitz operators, which cause much likeness between the properties of equation (6.1) (considered, for instance, in the space $L_p(\mathbb{R}_+)$ under the condition that $c \neq 0$ and $k \in L_1(\mathbb{R})$) and their counterparts for singular integral equations.

6.1. In all situations in which it is possible to make use of the Fourier transform, $\mathcal{F}$, the operator W_k can be converted into a Toeplitz operator T_a. To do an example, consider the operator W_k on $L_2(\mathbb{R}_+)$. Let $\hat{k} \in L_\infty(\mathbb{R})$ and put $k := \mathcal{F}\hat{k}$ (understood in the distribution sense), define $\omega : \Gamma_0 \to \mathbb{R}$ by $\omega(z) = i(1 + z)(1 - z)^{-1}$, and put $a := c + \hat{k} \circ \omega$ and $Uf = \pi^{-1/2}(x + i)^{-1}f \circ \omega^{-1}$ (U is a unitary mapping of H_2 onto the Hardy space of the upper half-plane). Then $W_k = (\mathcal{F}U)T_a(U^{-1}\mathcal{F}^{-1})$, a relation which was first recognized by Devinatz in 1967.

The latter formula is in particular true if $k \in L_1(\mathbb{R})$ and $\hat{k}(x) := \int_{-\infty}^\infty e^{ixt}k(t)\, dt$. If this case happens, the operators W_k and T_a are connected by sharing the same matrix representation with respect to the bases constituted by the functions $\{z^n\}_{n \geq 0}$ and the Laguerre functions, respectively (see Gohberg, Feldman [1971]). The matrix representing both operators is the *Toeplitz matrix* $(a_{j-k})_{j,k=0}^\infty$ composed by the Fourier coefficients a_i $(i = 0, \pm 1, \ldots)$ of the function a.

6.2. Another way of relating the operators $W_k \in \mathcal{L}(L_p(\mathbb{R}))$ $(k \in L_1)$ and $T_a = (a_{j-k}) \in \mathcal{L}(l_p^+)$ $(a \in W)$, $1 \leq p \leq \infty$, has its origin in the observation that both operators can be interpreted as functions of certain one-sided invertible operators of convolution type. In the fundamental paper Krein [1958], M.G. Krein took up this idea and developed a unique theory for both the operators W_k and T_a on a scale of Banach spaces, including those of functions and sequences summable in the pth power.

Theorem (M.G. Krein). *Let $k \in L_1(\mathbb{R})$. Then for the operator W_k defined by (6.1) to be a Φ_+-or Φ_--operator on $L_p(\mathbb{R}_+)$ $(1 \leq p \leq \infty)$ it is necessary and sufficient that*

$$A(x) := c + \int_{-\infty}^{\infty} e^{ixt} k(t)\, dt \neq 0, \qquad -\infty \leqq x \leqq \infty. \tag{6.2}$$

If (6.2) is fulfilled, then W_k is invertible, only left-sided invertible or only right-sided invertible in dependence on whether the integer $x = \text{ind } A$ is zero, positive or negative. Furthermore, one has

$$\dim \ker W_k = \max(-x, 0), \qquad \dim \text{coker } W_k = \max(x, 0).$$

Theorems 2.1 and 3.2 have analogues for the so-called *paired operators* $A_1 P + A_2 Q$ and $PA_1 + QA_2$ on $L_p(\mathbb{R})$ ($1 \leqq p \leqq \infty$) and also for systems of such operators; here A_j ($j = 1, 2$) is given by

$$(A_j\varphi) = c_j\varphi(x) + \int_{-\infty}^{\infty} k_j(x - t)\varphi(t)\, dt, \qquad x \in \mathbb{R},$$

where $k_j \in L_1$, P is the operator of multiplication by the function $(1 + \text{sgn } x)/2$, and $Q = I - P$. Moreover, the results of 4.1 and 4.2 apply to the decomposing algebra $W(\mathbb{R})$ (see 1.1 of Chapter 1). All these facts stem from I. Gohberg and M.G. Krein (1958) (see Gohberg, Feldman [1971]).

6.3. The Wiener-Hopf analogue of equations with Toeplitz operators as in 5.5 is the so-called *Wiener-Hopf integro-difference equation*

$$(PAP\varphi)(x) = f(x), \qquad x \in \mathbb{R}_+,$$

where

$$(A\varphi)(x) = \sum_{j=-\infty}^{\infty} a_j\varphi(x - \delta_j) + \int_{-\infty}^{\infty} k(x - t)\varphi(t)\, dt, \qquad x \in \mathbb{R}, \tag{6.3}$$

$$\sum_{-\infty}^{\infty} |a_j| < \infty, \quad \delta_j \in \mathbb{R} \quad (j = 0, \pm 1, \ldots), \quad k \in L_1(\mathbb{R}).$$

We remark that there is a close connection between these equations and Wiener-Hopf integral equations of the first kind ($c = 0$); see 7.3.

The function

$$\mathscr{A} = a + \hat{k}, \qquad a(x) = \sum_{-\infty}^{\infty} a_j e^{i\delta_j x} \qquad (x \in \mathbb{R}) \tag{6.4}$$

is referred to as the *symbol* of the operators A and $A_+ := PAP$, belonging to $\mathscr{L}(L_p(\mathbb{R}))$ and $\mathscr{L}(L_p(\mathbb{R}_+))$ ($1 \leqq p \leqq \infty$), respectively. The functions (6.4) form a Banach algebra, $\mathfrak{G}$, under the norm $\|\mathscr{A}\| := \sum_{-\infty}^{\infty} |a_j| + \|k\|_{L_1}$. One has $\|A_+\|_{\mathscr{L}(L_p)} \leqq \|\mathscr{A}\|$, and the mapping which assigns the symbol $\mathscr{A}$ to each operator A of the form (6.3) is an isomorphism of the commutative algebra of all operators (6.3) onto $\mathfrak{G}$. The correspondence between the symbols (6.4) and the operators A_+ is one-to-one and linear, but fails to be multiplicative (note that the set of all operators A_+ is not an algebra).

It is a simple matter to verify that $\inf_{\mathbb{R}}|\mathscr{A}| \leqq \inf_{\mathbb{R}}|a|$, from which we infer that for each function $\mathscr{A} = a + \hat{k} \in \mathfrak{G}$ such that $\inf|\mathscr{A}| > 0$ the numbers ω_a (recall 5.5) and $n_{\mathscr{A}} := \text{ind}(1 + a^{-1}\hat{k})$ are well-defined.

Theorem. *The operator* $A_+ = PAP$ *is a* Φ_+*- or* Φ_-*-operator on* $L_p(\mathbb{R}_+)$ $(1 \leqq p \leqq \infty)$ *if and only if* $\inf|\mathscr{A}| > 0$. *In that case* A_+ *is only left-invertible if* $\omega_a > 0$ *and only right-invertible if* $\omega_a < 0$. *If the case* $\omega_a = 0$ *prevails, the operator* A_+ *is invertible if* $n_{\mathscr{A}} = 0$, *only left-invertible if* $n_{\mathscr{A}} > 0$ *and only right-invertible if* $n_{\mathscr{A}} < 0$. *Furthermore,*

$$\dim \ker A_+ = \begin{cases} \infty & \text{if } \omega_a < 0, \\ -n_{\mathscr{A}} & \text{if } \omega_a = 0, n_{\mathscr{A}} \leqq 0, \end{cases}$$

$$\dim \operatorname{coker} A_+ = \begin{cases} \infty & \text{if } \omega_a > 0, \\ n_{\mathscr{A}} & \text{if } \omega_a = 0, n_{\mathscr{A}} \geqq 0. \end{cases}$$

A similar theorem also holds for paired operators. All these results are due to I. Gohberg and I.A. Feldman [1971].

6.4. In a cycle of papers by R.V. Duduchava (1973–1982) (see especially the book Duduchava [1979]), Wiener-Hopf integral equations with discontinuous "presymbols" were studied. These equations include singular integral equations on the real line and paired convolution integral operators as well.

Given $a \in L_\infty(\mathbb{R})$, define $W_a^0 \varphi = \mathscr{F} a \mathscr{F}^{-1} \varphi$ for $\varphi \in L_p(\mathbb{R}) \cap L_2(\mathbb{R})$. On condition that a is piecewise constant (with only a finite number of jumps), W_a^0 can be extended by continuity to an operator on all of $L_p(\mathbb{R})$ $(1 < p < \infty)$. Let $\Pi C_p(\mathbb{R})$ denote the closure of the algebra of all piecewise constant functions with respect to the norm $\|a\|_p^0 := \|W_a^0\|_p$. One can show that $\Pi C_p(\mathbb{R})$ contains both all the functions of bounded variation on $\mathbb{R}$ and the algebra $W(\mathbb{R})$. Note that if, for example,

$$a(x) = c_0 + \hat{k}_0(x) + c_1 \operatorname{sgn}(x + \delta); \qquad k_0 \in L_1(\mathbb{R}), \delta \in \mathbb{R},$$

then W_a^0 acts by the rule

$$(W_a^0 \varphi)(x) = c_0 \varphi(x) + \int_{-\infty}^\infty k(x - t)\varphi(t)\, dt,$$

where $k(x) := k_0(x) + c_1 e^{i\delta t}/(\pi i t)$. In particular, $-W_{\operatorname{sgn}}^0 = H$ is just the Hilbert transform.

Now put $W_a := PW_a^0|L_p(\mathbb{R}_+)$. In case $a \in \Pi C_p(\mathbb{R})$, we associate a function a_p with the operator W_a defined for $x \in \tilde{\mathbb{R}} := \mathbb{R} \cup \{\infty\}$ and $\xi \in \mathbb{R}$ as follows:

$$a_p(x, \xi) := \frac{a(x + 0)}{2}\left[1 + \coth \pi\left(\frac{i}{p} + \xi\right)\right] + \frac{a(x - 0)}{2}\left[1 - \coth \pi\left(\frac{i}{p} + \xi\right)\right],$$

with the appointment that $a(\infty \pm 0) := a(\pm \infty)$. If $\inf|a_p(x, \xi)| > 0$ $(x \in \tilde{\mathbb{R}}, \xi \in \mathbb{R})$, we denote by $\operatorname{ind} a_p$ the increment of the function $(2\pi)^{-1} \arg a_p(x, \xi)$ as x runs through $\tilde{\mathbb{R}}$ and, at the points where a is discontinuous (possibly including ∞), as ξ ranges over $\mathbb{R}$.

Theorem. *Let* $a \in \Pi C_p(\mathbb{R})$ $(1 < p < \infty)$. *Then for* W_a *to be normally solvable on* $L_p(\mathbb{R}_+)$ *it is necessary and sufficient that* $\inf|a_p(x, \xi)| > 0$ $(x \in \tilde{\mathbb{R}}, \xi \in \mathbb{R})$. *In this case the operator* W_a *is invertible, only left-invertible or only right-invertible in* *dependence on whether the integer* ind a_p *is zero, positive or negative; the index of* W_a *is given by* Ind $W_a = -$ ind a_p.

Duduchava also studied Banach algebras generated by operators of the form

$$\sum_{j=1}^{n} a_j W_{b_j}^0 c_j I; \quad a_j, c_j \in [PC(\mathbb{R})]^{m \times m}, \quad b_j \in [\Pi C_p(\mathbb{R})]^{m \times m}$$

on the spaces $L_p^m(\mathbb{R})$ ($1 < p < \infty$). He developed a theory of such algebras that is, in many respects, analogous to the corresponding theory for singular integral operators due to Gohberg and Krupnik (see 3.3). The approach of Duduchava makes essential use of the local principle proposed by I.B. Simonenko [1965] and modified by I. Gohberg and N.Ya. Krupnik [1973] (also Sec. 8 of Chap. 4). The above results are significant generalizations of earlier results by V.A. Fock, I.M. Rapoport, F.D. Gakhov, Yu.I. Cherskiĭ, M.G. Krein, I. Gohberg, L.S. Rakovshchik, M. Kremer, N.K. Karapetyants, S.G. Samko, and others. Recently R. Schneider (1984) extended these results to the spaces $L_p(\mathbb{R}, \varrho)$ with a power weight ϱ of the form (1.3). A part of the afore-mentioned results were generalized to integro-differential equations of the Wiener-Hopf type on the half-line by H.O. Cordes (1969), E. Gerlach (1969), J. Elschner (1979), and G. Thelen (1985).

Detailed surveys of Wiener-Hopf and convolution type integral equations as well as their applications to problems of mathematical physics are in Arabadzhyan, Engibaryan [1984], Duduchava [1979], Sakhnovich [1980], Meister, Speck [1980], Talenti [1973], Meister [1987], Böttcher, Silbermann [1989]. Almost each of these works is provided with an ample list of further references. Notice that Talenti [1973] develops a theory of such equations in Sobolev spaces.

§7. Equations with Degenerate Symbols

There is an extensive literature on this topic, dealing with the subject from several points of view and by means of quite different methods. An account of the work done on singular integral equations, discrete and integral Wiener-Hopf equations, and also systems of such equations with degenerate symbols until the middle of the seventies can be found in the monograph Prössdorf [1974], where a theory of these equations is developed in a unified framework using extensively functional analysis methods. The basic outcomes of the last ten years were summed up in the English edition of the book Mikhlin, Prössdorf [1980]. In this connection we wish also to turn attention to the recent book Elschner [1985a], which treats integro-differential and one-dimensional pseudodifferential equations with degenerate symbols as well as their applications to plane boundary value problems for partial differential equations.

We here limit ourselves to considering some simple examples which will illuminate the characteristic properties of equations with degenerate symbols and shed light on the methods of their investigation.

7.1. A non-elliptic singular integral operator or Wiener-Hopf operator A, thought of as acting on a Banach space E of the sort considered above (e.g. on L_p), is not semi-Noetherian and need not even be normally solvable. So the basic methods of studying such operators are aimed at their "normalization", i.e. at reducing them to normally solvable operators. One method of doing so is the method of unbounded regularization (see 3.3 of Chap. 1 and 3.3, 3.4 of Chap. 2). Another of these methods, which, however, is closely related to the method of unbounded regularization, consists in appropriately choosing Banach spaces $\bar{E}$ and $\tilde{E}$ such that $\bar{E} \subseteq E \subseteq \tilde{E}$, the embeddings being continuous, and in simultaneously extending the operator A to a normally solvable operator $\tilde{A} \in \mathscr{L}(\tilde{E}, \bar{E})$. The latter construction is especially easy performable in the following situation.

Suppose a representation of A in the form

$$A = BCD \tag{7.1}$$

is available, in which $C \in \mathscr{L}(E)$ is normally solvable and the operators B and D ($\in \mathscr{L}(E)$) satisfy the following conditions:

1. $\dim \ker B = 0$.
2. There exist an operator $D^{(-1)}$ defined on all of E and having a range $\tilde{E} := D^{(-1)}(E)$ containing E and a linear extension $\tilde{D}$ of the operator D to $\tilde{E}$ such that

$$D^{(-1)}Df = f, \qquad \tilde{D}D^{(-1)}f = f \qquad \forall f \in E.$$

The operator $D^{(-1)}$ occuring in condition 2 plays the role of a "formal" inverse of D. It is not hard to see that the requirements of condition 2 can be certainly met in case $\dim \ker D = \dim \operatorname{coker} D = 0$. On letting $\bar{E} = B(E)$ and introducing norms by $\|x\|_{\bar{E}} = \|B^{-1}x\|_E$ ($x \in \bar{E}$) and $\|y\|_{\tilde{E}} = \|\tilde{D}y\|_E$ ($y \in \tilde{E}$) we obtain two Banach spaces $\bar{E}$ and $\tilde{E}$ with the desired properties ($\tilde{A} := BC\tilde{D}$). Note that, by Banach's theorem, the norms $\|\cdot\|_E$ and $\|\cdot\|_{\bar{E}}$ on $\bar{E}$ are equivalent if and only if B is normally solvable. It is obvious that $\tilde{D}$ and B are isometrical isomorphisms of $\tilde{E}$ onto E and of E onto $\bar{E}$, respectively. Thus, the equation

$$A\varphi = f \qquad (\varphi \in \tilde{E}, f \in \bar{E}) \tag{7.2}$$

is equivalent to the normally solvable equation $C\psi = B^{-1}f$ ($\psi \in E$).

It turns out that a representation (7.1) with the properties 1 and 2 exists whenever the symbol of the operator A does not degenerate "too strongly", for instance, whenever it has at most finitely many (or even no more than countably many) zeros of finite orders. If this happens, the operators B and D figuring in (7.1) are not uniquely determined. Under certain assumptions on the symbol, one can even reach the two limiting cases of (7.1), namely, either $B = I$ (in which case a "generalized" solution of (7.2) with $f \in E$ is sought) or $D = I$ (which is tantamount to seeking a solution $\varphi \in E$ provided the right-hand side, $f \in \bar{E}$, is "sufficiently smooth"). The problem one is left with in any case is to choose the factors B, D in (7.1) in such a clever way that the analytic description of the spaces $\bar{E}$ and $\tilde{E}$ be as simple as possible.

7.2. We want now to illustrate the strategy outlined in 7.1 by example of a *Wiener-Hopf integral equation of the first kind.* Thus, we consider equation (6.1) with $c = 0$ and $k \in L_1(\mathbb{R})$. In this case the symbol defined by (6.2) vanishes at infinity. For the sake of simplicity, we assume that $x = \infty$ is the only zero of $\mathscr{A}$ and also that $\mathscr{A}$ can be written as $\mathscr{A}(x) = (x + i)^{-l}\mathscr{B}(x)$, where l is a natural number and $\mathscr{B} \in W(\mathbb{R})$ is a function of the form (6.2) such that $\mathscr{B}(x) \neq 0$ for all $x \in \mathbb{R} \cup \{\infty\}$. Fix any two nonnegative integers m and n such that $m + n = l$ and rewrite $\mathscr{A}$ in the form

$$\mathscr{A}(x) = (x - i)^{-m}\mathscr{C}(x)(x + i)^{-n}, \qquad \mathscr{C}(x) = \left(\frac{x - i}{x + i}\right)^m \mathscr{B}(x) \ (\in W(\mathbb{R})).$$

Then denote by B, C, D the Wiener-Hopf operators whose symbols are $(x - i)^{-m}$, $\mathscr{C}(x), (x + i)^{-n}$, respectively. Notice that B and D are of the first kind. Owing to the well-known theorem on the Fourier transform of convolutions, we have $A = W_k = BCD$. Clearly, $B = I$ if $m = 0$, while $D = I$ if $n = 0$.

In the case at hand the space $\tilde{L}_p(\mathbb{R}_+) =: \tilde{L}_p(\mathbb{R}_+; n) \ (1 \leq p \leq \infty)$ consists of all functions of the form

$$D^{(-1)}f = i^n(1 + d/dx)^n f(x) \qquad (f \in L_p(\mathbb{R}_+)), \tag{7.3}$$

the derivatives understood in the distribution sense.

The space $\bar{L}_p(\mathbb{R}_+) =: \bar{L}_p(\mathbb{R}_+; m)$ contains precisely those functions $f \in L_p(\mathbb{R}_+)$ whose derivatives $f^{(j)}$ are absolutely continuous on $\mathbb{R}_+$ for $j = 0, 1, \ldots, m - 1$ and belong to $L_p(\mathbb{R}_+)$ for $j = 0, 1, \ldots, m$. The norm in $\bar{L}_p(\mathbb{R}_+; m)$ is equivalent to the norm $\|f\| = \sum_{j=0}^m \|f^{(j)}\|_{L_p}$ and the inverse of B is given by $B^{-1} = i^m(-1 + d/dx)^m$.

One can show that the conclusions of Theorem 6.2 (with $\mathscr{A}$ replaced by $\mathscr{C}$) remain in force for the operator

$$W_k : \tilde{L}_p(\mathbb{R}_+; n) \to \bar{L}_p(\mathbb{R}_+; m) \qquad (1 \leq p < \infty).$$

Analogous results hold in the case where the symbol has a finite number of zeros (of not necessarily integral orders) and were also established, e.g., for paired convolution operators and singular integral operators (see Prössdorf [1974]).

7.3. There is a close connection between non-elliptic operators and operators with oscillating coefficients. To see this, consider the following simple example. Define W_k by (6.1) with $c = 0$ and

$$k(t) = h_\delta(t) + i \int_{-\infty}^0 e^s k_1(t - s)\, ds, \qquad h_\delta(t) = \begin{cases} ie^{t-\delta}, & t < \delta, \\ 0, & t > \delta, \end{cases}$$

where $\delta \in \mathbb{R}$ and $k_1 \in L_1(\mathbb{R})$. It is clear that $k \in L_1(\mathbb{R})$ and that the Fourier transform of k (i.e. the symbol of the operator W_k) equals

$$\hat{k}(x) = (x - i)^{-1}\mathscr{C}(x), \qquad \mathscr{C}(x) = e^{i\delta x} + \hat{k}_1(x) \qquad (x \in \mathbb{R}).$$

The case $\delta = 0$ was settled in 7.2. So let now $\delta \neq 0$. Then the function $\mathscr{C}$, which belongs to $\mathfrak{G}$, is of the form (6.4) with $a(x) = e^{i\delta x} \in AP$. Obviously, $\omega_a = \text{sgn } \delta$ (see 5.5).

From Theorem 6.3 we infer that if $\inf|\mathscr{C}| > 0$, then the operator $W_k : L_p(\mathbb{R}_+) \rightarrow \bar{L}_p(\mathbb{R}_+; 1)$ $(1 \leqq p \leqq \infty)$ is only left-invertible for $\delta > 0$ and only right-invertible for $\delta < 0$, in which cases, respectively, the cokernel and kernel dimensions are infinite. Thus, what we have constructed is an example of an operator A with a degenerate continuous symbol ($\hat{k} \in W(\mathbb{R})$) such that, when considered on the space $L_p(\mathbb{R}_+)$ $(1 \leqq p \leqq \infty)$, either dim ker A or dim coker A is infinite.

The recent book Dybin [1988] contains a theory of elliptic singular integral equations with infinite index, especially for coefficients with almost periodic discontinuities. Moreover, non-elliptic singular integral equations with symbols vanishing at a countable set of points are studied by the aid of the normalization method.

7.4. The theory of non-elliptic singular integral equations on closed contours attains an especially easy and round form in the countably normed space of infinitely differentiable functions $C^\infty(\Gamma)$ as well as in the space of generalized functions $C^{-\infty}(\Gamma) = [C^\infty(\Gamma)]^*$. In this setting we have the following result: For a singular integral operator with coefficients in $C^\infty(\Gamma)$ to be a Noether operator on $C^\infty(\Gamma)$ (or $C^{-\infty}(\Gamma)$) it is necessary and sufficient that its symbol have at most finitely many zeros of integral orders. If this situation prevails, the kernel and cokernel can be fully described and the index can be effectively computed (see Prössdorf [1974]).

Similar results are true for paired convolution equations in the space $L_p^\infty = \bigcap_{k=0}^\infty L_p^{(k)}$ and $L_p^{-\infty} = \bigcup_{k=0}^\infty L_p^{(-k)}$ $(1 \leqq p < \infty)$, where

$$L_p^{(k)} := \{ f : \| f \|_k = \|(x+i)^k f \|_{L_p(\mathbb{R})} < \infty \} \qquad (k \in \mathbb{Z}),$$

and also for their discrete analogues.

7.5. Recently H. Bart, I. Gohberg, and M.A. Kaashoek [1979] elaborated constructive methods for solving explicitly elliptic systems of Wiener-Hopf integral equations of the form (6.1), where c is a constant $n \times n$ matrix, the kernel k is an $n \times n$ matrix-valued function, and the symbol (6.2) is rational or analytic on the real line and at infinity. Their method is based on the "realization idea", i.e. on writing the symbol $\mathscr{A}$ in the form of a transfer function,

$$\mathscr{A}(x) = c + d(x - a)^{-1} b, \qquad x \in \mathbb{R},$$

where $a : X \rightarrow X$, $b : \mathbb{C}^n \rightarrow X$ and $d : X \rightarrow \mathbb{C}^n$ are linear operators, X is a finite dimensional vector space, and the operator a has no real eigenvalues. In the elliptic case c is invertible, the associate main operator $a^\times := a - bc^{-1}d$ has no real eigenvalues, and an explicit Wiener-Hopf factorization of $\mathscr{A}$ may be constructed in terms of the spectral projections of a and $a^\times$ corresponding to the eigenvalues in the open upper half-plane (in this connection see also Sec. 8.4).

In the thesis of L. Roozemond [1987] the realization idea was developed further and applied to the non-elliptic case. This allowed him to get necessary and sufficient conditions for the invertibility of the Wiener-Hopf operator and to obtain an explicit formula for its inverse in terms of the operators a, b, c, d for the cases where (i) det $c \neq 0$ and det $\mathscr{A}(x) = 0$ for some $x \in \mathbb{R}$, or where (ii) $c = 0$.

In both cases, the problem of solving equation (6.1) involves a splitting of the zeros of det $\mathscr{A}(x)$ on the real line or at infinity.

§8. Notes on Further Results

8.1. Singular integral equations of the type

$$c_0\psi(x) + c_1 \int_0^1 \frac{\psi(y)\,dy}{y-x} + \sum_{k=0}^{n} c_{k+2} \int_0^1 \frac{y^{k-n_k}\psi(y)}{(y+x)^{k+1}}\,dy = g(x), \qquad (8.1)$$

where $x \in [0, 1]$ and $0 \leq \operatorname{Re} n_k \leq k$, as well as some other equations of a similar form, can be reduced to Wiener-Hopf equations on the half-line of the form $W_a\varphi = f$ with $a \in \Pi C_p(\mathbb{R})$ (recall 6.4). To do this, it suffices to make the substitutions $x = e^{-t}$ and $y = e^{-\tau}$. Equation (8.1) is distinguished for the circumstance that its kernels in addition to the singularity along the diagonal $x = y$ have also so-called fixed singularities at the point $x = y = 0$. This peculiarity exerts a significant influence on the Noether properties and index formulas of these equations. We also remark that equation (8.1) emerges from manifold problems in elasticity theory and mathematical physics (see Duduchava [1979]).

By using local principles and the results quoted in 5.4, R.V. Duduchava [1979] established a complete theory of the equations (8.1) and equations similar to them. In particular, he found Noether criteria, an index formula and a formula for the solutions in the case where the equation is considered in the space $L_p([0, 1], x^\alpha(1-x)^\beta)$ (see also Elschner [1985b]).

8.2. Let Γ be a closed Lyapunov contour and suppose α is a diffeomorphism of Γ onto itself which preserves the orientation of Γ and is subject to the following conditions: 1) there exists a natural number $n \geq 2$ such that $\alpha_n(t) = t$ and $\alpha_i(t) \neq t$ $\forall t \in \Gamma$ for $i = 1, \ldots, n-1$, where $\alpha_i(t) := \alpha(\alpha_{i-1}(t))$ and $\alpha_0(t) := t$; 2) $\alpha' \in C^\lambda(\Gamma)$ for some $\lambda \in (0, 1)$. Such a mapping α is called a (*direct*) *Carleman shift*. Operators of the form $K = aP + bQ$, where P and Q are defined as in 3.2 and a and b are given by

$$a = \sum_{j=0}^{n-1} a_j W^j, \qquad b = \sum_{j=0}^{n-1} b_j W^j; \qquad (W\varphi)(t) = \varphi[\alpha(t)],$$

$a_j, b_j \in [L_\infty(\Gamma)]^{k \times k}$ ($j = 0, \ldots, n-1$), are referred to as *singular integral operators with shift*. They are usually considered on the spaces $L_p^k(\Gamma, \varrho)$ ($1 < p < \infty$). We associate a singular integral operator without shift, $C = AP + BQ$, with K in the following way. The operator C is defined on $L_p^{k \cdot n}(\Gamma, \varrho)$ and its $k \cdot n$ by $k \cdot n$ matrix coefficients are given by

$$A(t) = (a_{j-i}[\alpha_i(t)])_{i,j=0}^{n-1}, \qquad B(t) = (b_{j-i}[\alpha_i(t)])_{i,j=0}^{n-1},$$

with the convention that $a_{-j} := a_{n-j}$ and $b_{-j} := b_{n-j}$. The Noether theory of K is completely settled by the following theorem (see Mikhlin, Prössdorf [1980]).

Theorem. *Either both K and C are Noetherian or none of them is so. If they are Noether operators, then* $\operatorname{Ind} K = 1/n \operatorname{Ind} C$.

In the case of continuous coefficients this theorem goes back to G.S. Lit-vinchuk (1967); a more general result (which also applies to degenerate coefficients A and B) was established by Ch. Meyer and B. Silbermann (1977). Notice that the above theorem has an analogue in the situation that α reverses the orientation of Γ but maintains the properties 1 and 2 (such α's are called *inverse Carleman shifts*).

The matter is much more complicated for non-closed contours Γ or non-Carleman shifts. At the present moment only the theory of equations with Carleman shift has achieved an advanced stage. Our knowledge of operators with Carleman shift includes a complete description of algebras generated by operators of the form K with piecewise continuous coefficients and of the corresponding symbol algebras; Noether criteria and index formulas in terms of the symbol are also available. A sufficiently complete account of what is known on these subjects is contained in Karlovich, Kravchenko, Litvinchuk [1983, 1990] and in the monograph Litvinchuk [1977] (also see Przeworska-Rolewicz [1973]). As for non-Carleman shifts, we mention that it was proved that no matrix symbol exists (see Krupnik [1984]).

8.3. One field of present-day research is the study of singular integral equations on curves of infinite length or even on a countable union of curves. A bibliography of these subjects can be found in the survey Khvedelidze [1975], § 13, Chap. IV. Simple necessary and sufficient conditions for a curve $\Gamma \colon \mathbb{R} \to \mathbb{R}^n$ to ensure the validity of L_2-estimates for the operators

$$(Hf)(x) := \int_{-\infty}^{\infty} f(x - \Gamma(t)) \frac{dt}{t}, \qquad (\bar{H}f)(x) := \int_{-1}^{1} f(x - \Gamma(t)) \frac{dt}{t}$$

$(x \in \mathbb{R}^n, f \in C_0^{\infty}(\mathbb{R}^n))$ were given by A. Nagel, J. Vance, S. Wainger, and D. Weinberg (1983). For more about this see also a 1978 survey by E.M. Stein and S. Wainger.

8.4. Recently H. Bart, I. Gohberg, and M.A. Kaashoek (1984) proposed a new method of reducing integral equations of various classes to simpler equations, which frequently turn out to be equivalent to certain linear systems. The first step of this method, called coupling method by the authors, consists in the following. Suppose that $T\colon E_1 \to E_2$ and $S\colon E_2 \to E_2$ are Banach space operators related to each other by an equality of the form $\left(\begin{smallmatrix} T & * \\ * & * \end{smallmatrix}\right)^{-1} = \left(\begin{smallmatrix} * & * \\ * & S \end{smallmatrix}\right)$, where the asterisks refer to entries of the 2 by 2 operator matrices at hand which are known. Then one can explicitly construct the inverse or generalized inverse of T and one can describe the kernel and range of T provided the same things are available for the operator S, which is of course especially profitable if S is simpler than T. This program can be carried out by resorting to special representations of the operators which are connected with systems theory and lead to the notion of the "indicator" S of T. The authors apply this method to effecting the explicit solution of integral equations of the first and second kinds (on a segment or a line) with certain special kernels, of singular integral equations and Wiener-Hopf equations with rational matrix symbols, and of other equations. Moreover, in a cycle of papers by these authors published after 1979, the connection between the equations mentioned and linear systems theory has been thoroughly pointed out. The methods and results essentially rely upon the theory of minimal factorization of matrix and operator functions developed in the authors' monograph Bart, Gohberg, Kaashoek [1979] (also see Gohberg, Kaashoek [1986]).

8.5. In connection with Riemann boundary value problems on Riemann surfaces, singular integral equations were studied by W. Koppelman (1959, 1961), Yu.L. Rodin (1959), E.I. Zverovich (1971), G.S. Litvinchuk and E.I. Zverovich (1968), and others (see Rodin [1988], Meister [1973]).

8.6. The method of reducing systems of generalized Abel integral equations (cf. Chap. 1, Sec. 1.1.3, 2°) to systems of Cauchy singular integral equations was thoroughly analyzed by F. Penzel (1987).

Chapter 4
Multidimensional Singular Equations

The first subject dealt with in this chapter is multidimensional singular kernels inducing integral operators which exist in the principal-value sense. After this we shall state theorems on the boundedness of the multidimensional singular integral operator on the spaces $L_p(\mathbb{R}^n)$ and $L_p(\mathbb{R}^n, |x|^\alpha)$. In Section 2 we introduce the class of Calderón-Zygmund operators. These generalize the "classical" singular integral operators and satisfy weighted norm estimates with weights from the Muckenhoupt class. We also point out the connection between weighted norm estimates for the singular integral and the Hardy-Littlewood maximal function. Algebras of singular integral operators and their symbols are studied in the Sections 3 and 4. Section 5 concentrates upon singular integral operators on smooth manifolds. The questions arising in connection with the index of a matrix singular integral operator are considered in Section 6. The sections 7 and 8 are concise surveys on singular integral operators with discontinuous or degenerate symbols and multidimensional Wiener-Hopf equations. Finally, Section 9 is a brief account of pseudodifferential operators; we there content ourselves to the simplest facts and shall touch upon only those problems which are to a certain extent related to the basic theme of this chapter.

§ 1. The Multidimensional Singular Integral

1.1. It is well known (and can be checked without much difficulty) that the convolution integral

$$(Ku)(x) = \int_{\mathbb{R}^n} k(x - y)u(y)\, dy \tag{1.1}$$

exists for almost all $x \in \mathbb{R}^n$ and defines a bounded mapping of the space $L_p(\mathbb{R}^n)$ $(1 \leq p \leq \infty)$ into itself whenever $k \in L_1(\mathbb{R}^n)$ (see e.g. Stein, Weiss [1971]).

In the case $n = 1$, the integral (1.1) does also exist for the kernel $k(x) = 1/(\pi i x)$, which is not absolutely integrable, if it is interpreted in the principal-value sense (so that the positive and negative values of the kernel cancel each other out) and it generates a bounded operator on the space $L_p(\mathbb{R})$ if $1 < p < \infty$, which is called the Hilbert transform (see 1.1 of Chap. 3).

In the multidimensional realm one has several possibilities of picking out classes of kernels which, though being not absolutely integrable, nevertheless make the integral (1.1) exist in the principal-value sense. One of these classes, the corresponding theory of which is easy and rich at the same time, is composed by operators which commute not only with the shift but also with the dilation $x \to \varepsilon x$ $(\varepsilon > 0)$. The kernels of this class are of the form

$$k(x) = \frac{f(x)}{|x|^n}, \tag{1.2}$$

where f is a positively homogeneous function of degree 0, i.e. $f(\varepsilon x) = f(x)$ for all $\varepsilon > 0$. Imposing this condition on f is equivalent to requiring that f be constant along each of the rays starting at the origin. So f is completely determined by its restriction to the $(n - 1)$-dimensional unit sphere $\mathbb{S}^{n-1}$. Notice that in the case of the Hilbert transform f is nothing but the function $f(x) = (\operatorname{sgn} x)/\pi i$.

When the case prevails where f is an odd function, i.e. $f(-x) = -f(x)$, we can apply the M. Riesz theorem on the boundedness of the Hilbert transform on $L_p(\mathbb{R})$ (see 1.1 of Chap. 3) to deduce that if $f \in L_1(\mathbb{S}^{n-1})$ and $u \in L_p(\mathbb{R}^n)$ $(1 < p < \infty)$, then the integral (1.1) exists for almost all $x \in \mathbb{R}^n$ is the principal-value sense,

$$Ku = \lim_{\varepsilon \to 0} K_\varepsilon u, \qquad (K_\varepsilon u)(x) := \int_{|x-y| \geq \varepsilon > 0} k(x - y)u(y)\, dy, \tag{1.3}$$

and K is a bounded mapping of $L_p(\mathbb{R}^n)$ into itself (A.P. Calderón and A. Zygmund [1956]; see also Stein [1970] and Stein, Weiss [1971]). The function $Ku \in L_p(\mathbb{R}^n)$ given by (1.3) is called a *multidimensional singular integral* and the operator K is referred to as the *singular integral operator with the kernel k*. Adopting a term used in the first work on multidimensional singular integrals, Tricomi's paper [1926], the function f is now called the *characteristic* of the singular integral (1.3).

Another, more general class of kernels is obtained by merely demanding that $f \in L^1(\mathbb{S}^{n-1})$ satisfy the so-called "cancellation condition"

$$\int_{\mathbb{S}^{n-1}} f(x)\, d\sigma(x) = 0, \tag{1.4}$$

where $d\sigma$ is surface measure on $\mathbb{S}^{n-1}$. Notice that (1.4) is evidently satisfied by odd functions. Also note that, as is readily seen, the integral (1.3) need no longer exist in case (1.4) is not fulfilled. The investigation of singular integral operators with such more general kernel is a much more difficult task than the study of odd kernels. In particular, in order to guarantee that such operators be bounded on $L_p(\mathbb{R}^n)\,(1 < p < \infty)$ one must impose conditions upon f which are stronger than integrability on $\mathbb{S}^{n-1}$.

The following theorem is due to Calderón and Zygmund [1952, 1956]; see also Chap. 2 of Stein [1970].

Theorem. *Let k be a kernel given by (1.2). Suppose f is a positively homogeneous function of degree 0 which satisfies (1.4) and, in addition, is subject to the following integrability conditions:*

$$\int_{\mathbb{S}^{n-1}} |f(x)|\,d\sigma(x) < \infty, \qquad \int_{\mathbb{S}^{n-1}} |f(x)||\log^+|f(x)|\,d\sigma(x) < \infty. \qquad (1.5)$$

Assume $u \in L_p(\mathbb{R}^n)\,(1 < p < \infty)$. Then

a) *there exists a constant C_p (independent of u and ε) such that $\|K_\varepsilon u\|_p \leq C_p \|u\|_p$;*

b) *the limit $Ku = \lim_{\varepsilon \to 0} K_\varepsilon u$ exists in the $L_p(\mathbb{R}^n)$ sense and $\|Ku\|_p \leq C_p \|u\|_p$;*

c) *if $u \in L_2(\mathbb{R}^n)$, the Fourier transforms $\hat{u}$ and $\widehat{Ku}$ are related to each other by the formula*

$$\widehat{Ku}(x) = \mathcal{K}(x)\hat{u}(x), \qquad (1.6)$$

where $\mathcal{K}$ is the positively homogeneous function of degree 0 given for $|\xi| = 1$ by

$$\mathcal{K}(\xi) = \int_{\mathbb{S}^{n-1}} \left[\frac{\pi i}{2}\,\mathrm{sgn}(\xi, y) + \log\frac{1}{|(\xi, y)|}\right] f(y)\,d\sigma(y); \qquad (1.7)$$

here, in notation slightly deviating from Sec. 6 of Chap. 3, the Fourier transform of a function $v \in L^2(\mathbb{R}^n)$ is defined as

$$\hat{v}(x) = (Fv)(x) := \int_{\mathbb{R}^n} e^{2\pi i(x, y)} v(y)\,dy.$$

We remark that the proof of this theorem heavily relies on such things as L_2-estimates, inequalities of the weak type (1.1) (see also Sec. 2), and the Marcinkiewicz interpolation theorem.

Moreover, if the continuity modulus of the characteristic function f satisfies a Dini condition, then, invoking the maximal function (see Sec. 2) and using the above conclusions a) and b), one can prove that the pointwise limit $\lim_{\varepsilon \to 0}(K_\varepsilon u)(x)$ exists for almost all $x \in \mathbb{R}^n$ (see Chap. 2 of Stein [1970]).

1.2. By virtue of formula (1.6), we may write the singular integral operator K as follows:

$$Ku = F^{-1}\mathcal{K}Fu. \qquad (1.8)$$

The function $\mathcal{K}$ is referred to as the *symbol* of the operator K. It can be shown that $\mathcal{K}$ coincides with the Fourier transform (in the distribution sense) of the kernel k (Calderón, Zygmund [1952]; see also Mikhlin [1962] and Mikhlin, Prössdorf [1980]). Formula (1.7) provides an explicit expression of the symbol via the characteristic. We remark that it is condition (1.5) which guarantees the boundedness of the symbol.

From formula (1.8) we infer at once that for the singular operator K to be bounded on the space $L_2(\mathbb{R}^n, \varrho)$ with the weight $\varrho(x) = (1 + |x|^2)^{l/2}$ $(l \in \mathbb{R})$ it is necessary and sufficient that its symbol admits the estimate

$$\|\mathscr{K} v\|_{W_2^l(\mathbb{R}^n)} \leq C \|v\|_{W_2^l(\mathbb{R}^n)}, \qquad v \in W_2^l(\mathbb{R}^n), \tag{1.9}$$

where $C > 0$ is some constant independent of v.

1.3. So-called singular integrals with "*variable kernels*", i.e. entities of the form

$$(Ku)(x) = \lim_{\varepsilon \to 0} \int_{|x-y| \geq \varepsilon} k(x, x - y)u(y)\, dy \tag{1.10}$$

have been studied as well. The kernel $k(x, z)$ is assumed to possess the following properties:

1°. k is positively homogeneous of degree $-n$ with respect to z, that is, $k(x, \varepsilon z) = \varepsilon^{-n} k(x, z)$ for all $\varepsilon > 0$;

2°. $\int_{\mathbb{S}^{n-1}} k(x, z)d\sigma(z) = 0$ for all $x \in \mathbb{R}^n$.

The following result was established by Calderón and Zygmund [1978] only recently (compared with the time at which Theorem 1.2 was born).

Theorem. *Suppose* $\int_{\mathbb{S}^{n-1}} |k(x, z)|^q\, d\sigma(z) < \infty$ *for some* $q \in (1, \infty)$ *and for all* $x \in \mathbb{R}^n$. *Then the singular integral* (1.10) *exists almost everywhere for every function* $u \in C_0^\infty(\mathbb{R}^n)$ *and one has the estimate*

$$\|Ku\|_p \leq C_{p,q} \sup_x \left(\int_{\mathbb{S}^{n-1}} |k(x, z)|^q\, d\sigma(z) \right)^{1/q} \|u\|_p$$

if only q *satisfies the following condition* (CZ_p):

(i) $q > p' \dfrac{n-1}{n}$ *in case* $1 < p \leq 2$ $(p' := p/(p-1))$,

(ii) $\dfrac{1}{q} < \dfrac{1}{p} \dfrac{n}{n-1} + \left(1 - \dfrac{2}{p}\right)$ *in case* $2 \leq p < \infty$.

This result cannot be improved (is sharp) for $1 < p \leq 2$.

There is an analogue of formula (1.8) for the singular operator (1.10). Indeed, after letting

$$f(x, \vartheta) := |x - y|^n k(x, x - y), \qquad \vartheta = (y - x)/|y - x|,$$

the formulas (1.7) and (1.8) give

$$K = F_{\xi \to x}^{-1} \mathscr{K}(x, \xi) F_{x \to \xi}, \tag{1.11}$$

$\mathscr{K}$ being a function which is positively homogeneous of degree 0 in ξ. Notice that (1.11) results from (1.7) by replacing $f(y)$ with $f(x, y)$. The function $\mathscr{K}$ is called the *symbol* of the operator (1.10).

1.4. Formulas (1.8) and (1.11) suggest an extension of the concept of the singular integral operator: now by a singular integral operator will be meant any operator of the form (1.8) or (1.11) such that $\mathscr{K}$ is a bounded function which is measurable with respect to x, ξ and positively homogeneous of degree 0 with respect to ξ. In that case the function $\mathscr{K}$ is referred to as the *symbol* of K also.

Furthermore, one may consider operators of the type (1.8) or (1.11) without assuming that their symbol, $\mathscr{K}$, is positively homogeneous of degree 0 in ξ. In this way one is led to the concept of the *pseudodifferential operator* (see the original works Agranovich [1965], Kohn, Nirenberg [1965], Hörmander [1965], Seeley [1965], Cordes [1965] and the recent monographs Eskin [1973], Shubin [1978], Cordes [1979], Kumano-go [1981], Taylor [1981], Treves [1982], Journe [1983], Grubb [1986]; also see Sec. 9).

1.5. A multidimensional analogue of the Hardy-Littlewood theorem (see 1.1.2 of Chap. 3) for the spaces $L_p(\mathbb{R}^n, \varrho)$ with power weight ϱ was found by E.M. Stein (1957).

Theorem. *Let* $(Ku)(x) = \lim_{\varepsilon \to 0} \int_{|x-y| \geqq \varepsilon} K(x, y) u(y)\, dy$, *where* $|K(x, y)| \leqq C|x - y|^{-n}$. *If the operator K is bounded on $L_p(\mathbb{R}^n)$ $(1 < p < \infty)$, then it is also bounded on those spaces $L_p(\mathbb{R}^n, |x|^\alpha)$ for which*

$$-n < \alpha < n(p - 1). \tag{1.12}$$

The same result also holds for the spaces $L_p(\mathbb{R}^n, (1 + |x|)^\alpha)$ (and is an easy consequence of the theorem we have just stated).

1.6. In connection with Stein's theorem the following question arises: To what extent does the theorem extend to values α lying outside the interval (1.12)? A series of results in this direction were obtained independently by B.A. Plamenevskiĭ (1968) for $p = 2$ and by Yu.E. Haikin (1970, 1973) for $p \in (1, \infty)$. These two authors proved that there exists a dense subset $\mathfrak{M}_+$ (resp. $\mathfrak{M}_-$) of $L_p(\mathbb{R}^n, |x|^\alpha)$ consisting of sufficiently smooth functions with compact support such that, under some restrictions on the symbol, the singular operator defined by (1.11) on $\mathfrak{M}_+$ (resp. $\mathfrak{M}_-$) is bounded in $L_p(\mathbb{R}^n, |x|^\alpha)$ for all positive (resp. negative) α distinct from

$$n(p - 1) + pj \quad (\text{resp.} \ -n + pj), \qquad j = 0, 1, 2, \ldots.$$

Surveys of other results concerning boundedness of singular integral operators are in Mikhlin [1962], Calderón, Zygmund [1978], Mikhlin, Prössdorf [1980], Dyn'kin, Osilenker [1983]; there one can also find some open problems.

1.7. We finally remark that unlike the singular operators (1.3) and (1.10), the singularity of the kernel of the operator defined by (1.7) is not located on a diagonal but on the equator $(\xi, y) = 0$. The properties of such operators are not yet very well understood. Some results on the operator (1.7) were established by A.D. Gadzhiev (1981) when he was proving the implications (3.5) that will be stated below in 3.2.

§2. Weighted Norm Inequalities for Singular Integrals and Maximal Functions

2.1. The Maximal Function. There is an intimate connection between weighted norm estimates for the singular integral and such estimates for the *Hardy-Littlewood maximal function*. Given a locally integrable function u on $\mathbb{R}^n$, its maximal function Mu is defined by

$$(Mu)(x) = \sup_{Q \ni x} \frac{1}{|Q|} \int_Q |u| \, dy, \qquad x \in \mathbb{R}^n,$$

the supremum taken over all cubes $Q \subset \mathbb{R}^n$ containing the point x. The operator $M : f \to Mf$ is called the *Hardy-Littlewood maximal operator*. Notice that this operator is not linear. A classical theorem (Stein [1970]) says that M is a bounded operator on $L_p(\mathbb{R}^n) \, (1 < p \leq \infty)$ and that it is of the weak type (1.1) (see also 1.2.1 of Chap. 3).

The problem of finding weighted norm inequalities for the maximal function (operator) can be in full generality phrased as follows: Describe all the pairs (ϱ, ω) of weights such that either an estimate of the strong type

$$\int_{\mathbb{R}^n} [(Mu)(x)]^p \varrho(x) \, dx \leq C \int_{\mathbb{R}^n} |u(x)|^p \omega(x) \, dx \qquad (2.1)$$

or an estimate of the weak type

$$\int_{\{x : (Mu)(x) > \lambda\}} \varrho(x) \, dx \leq C\lambda^{-p} \int_{\mathbb{R}^n} |u(x)|^p \omega(x) \, dx \qquad (2.2)$$

hold; of course, here C is some constant that depends neither on $u \in L_p(\mathbb{R}^n, \omega)$ nor on $\lambda > 0$.

In the one weight case $(\varrho = \omega)$ this problem was solved by Muckenhoupt in 1972, who established the following fundamental result.

Theorem. *Let $\varrho = \omega$ be a weight on $\mathbb{R}^n$ and let $1 < p < \infty$. Then the following are equivalent:* (i) $\varrho \in (A_p)$ *(this is the condition introduced in 1.1.2 of Chap. 3, Q now ranging over all cubes of $\mathbb{R}^n$);* (ii) *the estimate (2.1) holds;* (iii) *the estimate (2.2) is valid.*

That (ii) or (iii) necessitate (A_p) follows easily by applying (2.1) or (2.2) to the test function $u = \varrho^{-1/(p-1)} \chi_Q$. To prove that (A_p) is also a sufficient condition is very difficult. A relatively simple proof was given by R.R. Coifman and C. Fefferman in 1974; a key observation in this proof is the implication $(A_p) \Rightarrow (A_{p-\varepsilon})$, $\varepsilon > 0$. In 1981, E.T. Sawyer provided a completely different proof and obtained the following result on two weight norm inequalities.

Theorem 2. *Let ϱ and ω be two weights on $\mathbb{R}^n$ and suppose $1 < p < \infty$. Then the following are equivalent:* 1. *the estimate (2.1) is true;* 2. *one has the inequality*

$$\int_Q |M(\omega^{-1/(p-1)}\chi_Q)|^p \varrho \, dx \leq C \int_Q \omega^{-1/(p-1)} \, dx < \infty,$$

where C is some constant independent of the cube Q.

2.2. Calderón-Zygmund Operators. All those singular integral operators for which weighted norm inequalities are known at the present time belong to the large class of Calderón-Zygmund operators. This class, introduced by Coifman and Meyer [1978], comprises the "classical" singular operators of the form (1.1) (which in the literature are also called *singular integrals of Mikhlin-Calderón-Zygmund type*), Cauchy singular integrals on Lipschitz curves, the double layer potential in Lipschitz surfaces, as well as pseudodifferential operators (see also Journé [1983], Dyn'kin, Osilenker [1983]).

An operator $K \in \mathscr{L}(L_2(\mathbb{R}^n))$ is called a *Calderón-Zygmund operator* if there is a function $K(x, y)$ $(x \neq y)$ such that

$$|K(x, y)| \leq C|x - y|^{-n}, \tag{2.3}$$

$$|K(x, y) - K(x', y)| \leq C|x - y|^{-n-\alpha}|x - x'|^\alpha \qquad \text{for } 2r_{xx'} < r_{xy}, \tag{2.4}$$

$$|K(x, y) - K(x, y')| \leq C|x - y|^{-n-\alpha}|y - y'|^\alpha \qquad \text{for } 2r_{yy'} < r_{xy} \tag{2.5}$$

$(r_{xy} := |x - y|)$ and the action of K at the functions $u \in C_0^\infty(\mathbb{R}^n)$ is given by

$$(Ku)(x) = \int K(x, y)u(y) \, dy, \qquad x \notin \text{supp } u.$$

Here α is some positive constant. The operator K is in general not uniquely determined by its *kernel* $K(x, y)$.

The simplest example of a Calderón-Zygmund operator is the *Riesz operators* R_j $(j = 1, \ldots, n)$,

$$(R_j u)(x) := \lim_{\varepsilon \to 0} \int_{|x-y| > \varepsilon} \frac{c_n(x_j - y_j)}{|x - y|^{n+1}} u(y) \, dy; \qquad c_n := \frac{\Gamma((n+1)/2)}{\pi^{(n+1)/2}}.$$

It is easily seen that the singular operator (1.3) with the kernel (1.2) is a Calderón-Zygmund operator if only $f \in C^\alpha(\mathbb{S}^{n-1})$ $(0 < \alpha \leq 1)$. The next theorem is an important generalization of Theorem 1.1.2 of Chapter 3. It is due to Coifman and Meyer [1978].

Theorem. *If K is a Calderón-Zygmund operator and $\varrho \in (A_p)$ $(1 < p < \infty)$, then*

$$\int_{\mathbb{R}^n} |(Ku)(x)|^p \varrho(x) \, dx \leq C \int_{\mathbb{R}^n} [(Mu)(x)]^p \varrho(x) \, dx \qquad \forall u \in C_0^\infty(\mathbb{R}^n). \tag{2.6}$$

Thus, under the hypothesis of this theorem, $K \in \mathscr{L}(L_p(\mathbb{R}^n, \varrho))$.

Moreover, if ϱ is a weight such that $(M\varrho)(x) \leq C\varrho(x)$ almost everywhere ("condition (A_1)"), then every Calderón-Zygmund operator K satisfies an inequality of the weak type (2.2) (with $p = 1$, $\omega = \varrho$, and M replaced by K).

For singular integrals (1.3) with kernels of the form (1.2), these results were set up independently by R. Coifman and C. Fefferman (1974) and M. Kaneko and S. Jano (1975). They also showed that condition (A_p) is necessary for the Riesz operators to be bounded on $L_p(\mathbb{R}^n, \varrho)$ $(1 < p < \infty)$. It is also known that the estimate (2.6) remains in general no longer true without assuming that $\varrho \in (A_p)$; more precisely, if ϱ is a power weight $|x|^\alpha$, then (2.6) implies that $\varrho \in (A_p)$ (which in the case at hand is equivalent to (1.12)). Of course, (2.6) in conjunction with Sawyer's theorem cited in 2.1 also yields two weight norm inequalities for Calderón-Zygmund operators.

We also remark that if K is the kernel of a Calderón-Zygmund operator satisfying (2.3)–(2.5), if the limit $\lim_{\varepsilon \to 0} \int_{\varepsilon < |x-y| < 1} K(x, y)\, dy$ exist almost everywhere, and if $\varrho \in (A_p)$ $(1 \leq p < \infty)$, then the limit

$$\lim_{\varepsilon \to 0} \int_{|x-y| > \varepsilon} K(x, y)u(y)\, dy$$

exists almost everywhere for every $u \in L_p(\mathbb{R}^n, \varrho)$.

Recently Calderón-Zygmund operators have become a powerful tool in the theory of wavelets (see Y. Meyer [1987]). Note that for most of the classical spaces, such as $L_p(1 < p < \infty)$, Sobolev or Besov spaces, the "wavelets" form a universal unconditional basis combining the advantages of the Haar system and of the trigonometric system.

§3. The Connection Between Symbol and Kernel

3.1. We now consider singular operators of the form

$$(Au)(x) = a(x)u(x) + \int_{\mathbb{R}^n} k(x, x - y)u(y)\, dy \tag{3.1}$$

where $a \in L_\infty(\mathbb{R}^n)$ and the kernel k is subject to the conditions of Section 1. The *symbol* of the operator A is defined by

$$\mathscr{A}(x, \xi) = a(x) + \hat{k}(x, \xi), \tag{3.2}$$

$\hat{k}(x, \xi)$ denoting the Fourier transform of the kernel $k(x, z)$ with respect to z.

We expand the characteristic function $f(x, \vartheta) = |x - y|^n k(x, x - y)$ into a Fourier series with respect to spherical functions,

$$f(x, \vartheta) = \sum_{m=1}^{\infty} \sum_{l=1}^{\varkappa_m} a_m^{(l)}(x) Y_m^{(l)}(\vartheta). \tag{3.3}$$

Here $Y_m^{(l)}$ $(l = 1, \ldots, \varkappa_m; m = 0, 1, 2, \ldots)$ is the orthonormal system in $L_2(\mathbb{S}^{n-1})$ constituted by the n-dimensional spherical functions, and $\varkappa_m$ refers to the number of linearly independent spherical functions of order m. Note that in virtue of condition 2° of 1.3 the zeroth Fourier coefficient does not appear.

If $K(x) := Y_m^{(l)}(x/|x|)\,|x|^{-n}$, then, by a well known formula (see e.g. Mikhlin, Prössdorf [1980], Sec. 2 of Chap. X),

$$\hat{K}(\xi) = \pi^{n/2} i^m \frac{\Gamma(m/2)}{\Gamma((n+m)/2)} Y_m^{(l)}\left(\frac{\xi}{|\xi|}\right).$$

This together with the definition (3.2) produces the following expansion of the symbol of the operator (3.1) into a series with respect to spherical functions:

$$\mathscr{A}(x,\xi) = \sum_{m=0}^{\infty} \sum_{l=1}^{\varkappa_m} \gamma_m a_m^{(l)}(x) Y_m^{(l)}\left(\frac{\xi}{|\xi|}\right), \tag{3.4}$$

where $a_0^{(1)} = a$, $\gamma_0 = 1$, $\gamma_m = \pi^{n/2} i^m \Gamma(m/2)/\Gamma((n+m)/2)$. Formula (3.4) was established by Mikhlin (1936) for $n = 2$ and generalized by Giraud (1936) to arbitrary n.

3.2. The formulas (1.7) and (3.3), (3.4) enable us to pass to and from between the characteristic and the symbol. In particular, the connection between differentiability properties of the symbol and the characteristic can be studied by resorting to these formulas. To state a result pertaining to this subject, we need one more definition.

Denote by $H_p^l(\mathbb{S}^{n-1})$ $(l \geq 0, 1 < p < \infty)$ the completion of the space $C_0^\infty(\mathbb{R}^n)$ with respect to the norm $\|u\|_{p,l} := \|(I + \delta)^{l/2} u\|_{L_p(\mathbb{S}^{n-1})}$, where δ is the Beltrami operator on the sphere (i.e. the spherical part of the Laplace operator). In the case $p = 2$ we obtain just the Sobolev-Slobodetskiĭ spaces (see e.g. Maz'ya [1985]). Furthermore, we shall say that a function f given on $\mathbb{R}^n \times \mathbb{S}^{n-1}$ belongs uniformly (on $\mathbb{R}^n$) to the space $H_p^l(\mathbb{S}^{n-1})$ and shall write this as $f \hat{\in} H_p^l(\mathbb{S}^{n-1})$ if $f(x,\cdot) \in H_p^l(\mathbb{S}^{n-1})$ for all $x \in \mathbb{R}^n$ and $\|f(x,\cdot)\|_{p,l} \leq C$ with some constant C that does not depend on x.

The following theorem is due to M.S. Agranovich (1965) and N.M. Mikhailova-Gubenko (1966) and makes more precise an earlier result of S.G. Mikhlin [1962].

Theorem. *Let f and $\mathscr{A}$, respectively, be the characteristic and the symbol of the singular integral operator (3.1). Then*

$$f \hat{\in} H_2^l(\mathbb{S}^{n-1}) \Leftrightarrow \mathscr{A} \hat{\in} H_2^{l+n/2}(\mathbb{S}^{n-1}).$$

In the plane case $(n = 2)$ the theorem remains in force for all $p \in (1, \infty)$; this follows from formula (1.7) and M. Riesz' theorem on the boundedness of the Hilbert operator on $L_p(0, 2\pi)$. Recently A.D. Gadzhiev (1981, 1982) showed that the theorem breaks down in the case $n \geq 3, p \neq 2$ but that then the implications

$$\mathscr{A} \hat{\in} H_p^{n/2+l}(\mathbb{S}^{n-1}) \Rightarrow f \hat{\in} L_p(\mathbb{S}^{n-1}) \Rightarrow \mathscr{A} \hat{\in} H_p^{n/2-l}(\mathbb{S}^{n-1}) \tag{3.5}$$

are valid $(l := (n-2)|1/p - 1/2|)$. He also showed that (3.5) cannot be improved. Notice that, of course, there is an analogue of the chain of implications (3.5) starting with the uniform membership of the characteristic function to $H_p^k(\mathbb{S}^{n-1})$.

3.3. Theorem 1.3 and the implications (3.5) involve the following boundedness criterion for the singular integral operator in terms of the smoothness of its symbol.

Theorem. Let $a \in L_\infty(\mathbb{R}^n)$, $1 < p < \infty$, suppose $\mathscr{A} \hat{\in} H_q^{n/2+l}(\mathbb{S}^{n-1})$ with $l = (n-2)|1/q - 1/2|$, and assume q satisfies the condition (CZ_p) of Theorem 1.3. Then the singular operator (3.1) is bounded on $L_p(\mathbb{R}^n)$ and one has

$$\|A\| \leq C[\|a\|_{L_\infty(\mathbb{R}^n)} + \sup_x \|\mathscr{A}(x, \cdot)\|_{q, n/2+l}]. \tag{3.6}$$

Corollary. Let $A_m^{(k)}$ denote the singular operator whose characteristic is $Y_m^{(k)}$ $(A_0^{(1)} := I)$. Then, under the hypotheses of the theorem,

$$Au = \sum_{m=0}^{\infty} \sum_{k=1}^{\kappa_m} a_m^{(k)} A_m^{(k)} u \qquad \forall u \in L_p(\mathbb{R}^n) \tag{3.7}$$

the series (3.7) converging in the norm of $L_p(\mathbb{R}^n)$.

§4. Algebras of Singular Integral Operators

4.1. It is known that $H_q^k(\mathbb{S}^{n-1})$ is a Banach algebra if and only if $qk > n - 1$. In the sequel we suppose that this condition is fulfilled. Then the collection of all functions $f \hat{\in} H_q^k(\mathbb{S}^{n-1})$ forms a Banach algebra $\mathscr{S}_p^k$ under the norm $\sup_x \|f(x, \cdot)\|_{q,k}$. We let $\overset{\circ}{\mathscr{S}}_q^k$ denote the subalgebra of $\mathscr{S}_q^k$ consisting of all those functions which are continuous on $\tilde{\mathbb{R}}^n \times \mathbb{S}^{n-1}$, where $\tilde{\mathbb{R}}^n := \mathbb{R}^n \cup \{\infty\}$. One can show that there is a one-to-one correspondence between the set of all maximal ideals of the algebra $\overset{\circ}{\mathscr{S}}_q^k$ and the set $\tilde{\mathbb{R}}^n \times \mathbb{S}^{n-1}$, viz, that each maximal ideal contains exactly those functions from $\overset{\circ}{\mathscr{S}}_q^k$ which vanish at some point $(x_0, \vartheta_0) \in \tilde{\mathbb{R}}^n \times \mathbb{S}^{n-1}$. Hence, if $f \in \overset{\circ}{\mathscr{S}}_q^k$ and $f(x, \vartheta) \neq 0$ for all $(x, \vartheta) \in \tilde{\mathbb{R}}^n \times \mathbb{S}^{n-1}$, then $f^{-1} \in \overset{\circ}{\mathscr{S}}_q^k$ (see 1.4 of Chap. 1).

4.2. Let now A and B be singular integral operators of the form (3.1) and denote by $\mathscr{A}$ and $\mathscr{B}$ their symbols. Suppose $\mathscr{A}, \mathscr{B} \in \overset{\circ}{\mathscr{S}}_q^k$, where $k = n/2 + (n-2)$ $|1/q - 1/2|$ and q satisfies condition (CZ_p) of Theorem 1.3. Denote by C the singular integral operator whose symbol is $\mathscr{A} \cdot \mathscr{B}$, and put $[A, B) := AB - C$; the latter operator is referred to as the *semi-commutator* of the operators A and B. The results of 3.3 give the following.

Theorem. If $\mathscr{A}, \mathscr{B} \in \overset{\circ}{\mathscr{S}}_q^k$ then $[A, B) \in \mathscr{K}(L_p(\mathbb{R}^n))$.

Corollary 1. If $\mathscr{A}, \mathscr{B} \in \overset{\circ}{\mathscr{S}}_q^k$ then $[A, B] \in \mathscr{K}(L_p(\mathbb{R}^n))$.

Corollary 2. If $\mathscr{A} \in \overset{\circ}{\mathscr{S}}_q^k$ and $\inf |\mathscr{A}| > 0$, then the singular operator whose symbol is $\mathscr{A}^{-1}$ is a regularizer of A and thus A is Noetherian on the space $L_p(\mathbb{R}^n)$.

4.3. Given a singular integral operator (3.1) with symbol $\mathscr{A}$ and a compact operator T on $L_p(\mathbb{R}^n)$, it is natural to call the function $\mathscr{A}$ the *symbol* of the "general" singular operator $A + T$. From the results of 4.2 we infer that the symbol of the sum and the product of two general singular operators with symbols in $\overset{\circ}{\mathscr{S}}{}_q^k$ is equal to the sum and the product of their symbols, respectively.

Theorem 1. *Let A be a general singular integral operator the symbol $\mathscr{A}$ of which belongs to $\overset{\circ}{\mathscr{S}}{}_q^k$, where $k = n/2 + (n - 2)|1/q - 1/2|$ and q satisfies the conditions (CZ_p) and $(CZ_{p'})$ $(p' = p/(p - 1))$ of Theorem 1.3. Then the adjoint operator of $A \in \mathscr{L}(L_p(\mathbb{R}^n))$ $(1 < p < \infty)$ is also a general singular operator and its symbol equals $\overline{\mathscr{A}}$, the complex conjugate of $\mathscr{A}$.*

Theorem 2. *If the conditions of Theorem 1 are fulfilled and, in addition, $\inf|\mathscr{A}| > 0$, then $\operatorname{Ind} A = 0$.*

We emphasize that Theorem 2 has no one-dimensional analogue.

The results of 4.2 and 4.3 are S.G. Mikhlin's (see also Mikhlin, Prössdorf [1980]).

4.4. The picture is completed by the following important theorem, which was established by I. Gohberg (1960) for $p = 2$ and was then generalized by R.T. Seeley (1965) and N.Ya. Krupnik (1965) to the case of arbitrary $p \in (1, \infty)$.

Theorem. *Under the hypothesis of Theorem 1 of 4.3, for the operator A to be Noetherian on $L_p(\mathbb{R}^n)$ $(1 < p < \infty)$ it is necessary that $\inf|\mathscr{A}| > 0$.*

From this theorem we may as in the one-dimensional case (see 2.1.2 of Chap. 3) deduce that

$$\max_{(x,\,\vartheta)\,\in\,\tilde{\mathbb{R}}^n\times\mathbb{S}^{n-1}} |\mathscr{A}(x, \vartheta)| \leq \|A + T\| \tag{4.1}$$

for every compact operator $T \in \mathscr{K}(L_p(\mathbb{R}^n))$. As was pointed out by Seeley (1965), in the case $p = 2$ we even have equality:

$$\max_{(x,\,\vartheta)\,\in\,\tilde{\mathbb{R}}^n\times\mathbb{S}^{n-1}} |\mathscr{A}(x, \vartheta)| = \inf_T \|A + T\|. \tag{4.2}$$

4.5. Let $\mathfrak{A}$ be the collection of all operators of the form

$$\sum_{m=0}^{j} \sum_{k=1}^{\varkappa_m} a_m^{(k)} A_m^{(k)} + T; \qquad A_0^{(1)} = I, \, T \in \mathscr{K}(L_p(\mathbb{R}^n)), \qquad j = 0, 1, \ldots$$

with $a_m^{(k)} \in C(\tilde{\mathbb{R}}^n)$. We denote by $\mathfrak{A}_p$ $(1 < p < \infty)$ the closure of $\mathfrak{A}$ in $\mathscr{L}(L_p(\mathbb{R}^n))$. For operators in $\mathfrak{A}$ estimates of the type (4.1), (4.2) are valid. Given $A \in \mathfrak{A}_p$, we may choose any sequence (A_j) of operators in $\mathfrak{A}$ converging in the norm of $\mathscr{L}(L_p(\mathbb{R}^n))$ to A; then, by (4.1), the sequence $(\mathscr{A}_j)$ of the symbols of the operators in (A_j) converges uniformly on $\tilde{\mathbb{R}}^n \times \mathbb{S}^{n-1}$ to some continuous function $\mathscr{A}$. The function $\mathscr{A}$ will be called the *symbol* of A.

If A is a singular integral operator of the form (3.1) whose symbol $\mathscr{A}$ is in $\overset{\circ}{\mathscr{S}}{}_q^k$ (with k and q as in Theorem 1 of 4.3), then $A \in \mathfrak{A}_p$. It is clear that all the above results, including (4.1) and (4.2), apply to operators in $\mathfrak{A}_p$. Moreover, the symbol mapping turns out to be an isomorphism of the quotient algebra $\mathfrak{A}_p/\mathscr{K}(L_p(\mathbb{R}^n))$

onto the algebra $C(\tilde{\mathbb{R}}^n \times \mathbb{S}^{n-1})$ (see 1.4. and 1.6 of Chap. 1). In the case $p = 2$, the algebra $\mathfrak{A}_p / \mathcal{K}(L_p(\mathbb{R}^n))$ is a C^*-algebra and the symbol is an isometrical *-isomorphism (I. Gohberg (1960), R.T. Seeley (1965)).

4.6. Analogous results are true for algebras of singular integral operators acting on the Sobolev spaces $W_p^s(\mathbb{R}^n)$ or on other spaces of generalized functions as well (N.Ya. Krupnik (1965), S.G. Mikhlin (1977); also see Mikhlin, Prössdorf [1980]). V.G. Maz'ya and T.O. Shaposhnikova [1985] found interesting applications of their theory of multipliers on spaces of differentiable functions to the theory of singular integral operators on the spaces W_p^s. They showed that the basic properties of singular operators with symbols depending smoothly on ϑ can be retained (when passing to W_p^s) by appropriately characterizing the smoothness with respect to x in terms of multiplier spaces. In a sense, the latter characterizations bear maximal generality.

§ 5. Singular Integral Operators on Manifolds

So far we have been speaking about singular integral operators on $\mathbb{R}^n$, the Euclidean space being equipped with a fixed system of coordinates. We now proceed to giving a brief account of singular integral operators on smooth manifolds.

5.1. Let Γ be an n-dimensional infinitely differentiable manifold and, for simplicity, assume Γ is compact. Further, let $\{U_j\}$ be a finite system of coordinate neighborhoods of Γ and let $\{\varphi_j\}$ be a finite partition of unity subordinate to the cover $\{U_j\}$ and constituted by infinitely smooth non-negative functions φ_j. Thus, supp $\varphi_j \subset U_j$ and $\sum_j \varphi_j \equiv 1$.

There are natural definitions of the Sobolev spaces $W_p^s(\Gamma)$ (including the spaces $L_p(\Gamma)$), of the Hölder spaces $C^\alpha(\Gamma)$ and of certain other interesting function spaces on Γ. Given a function f on Γ, we have $f = \sum \varphi_j f$, and by passing to local coordinates, we may think of $\varphi_j f$ as a function given on $\mathbb{R}^n$. This allows us to define the space $W_p^s(\Gamma)$ as the completion of the space $C^\infty(\Gamma)$ of complex-valued infinitely differentiable functions on Γ with respect to the norm

$$\|f\| = \sum \|\varphi_j f\|_{p,s},$$

where $\|\varphi_j f\|_{p,s}$ is the norm of $\varphi_j f$ as member of $W_p^s(\mathbb{R}^n)$. The norm in $C^\alpha(\Gamma)$ may be defined in an analogous fashion. Notice that different systems of coordinate neighborhoods and different partitions of unity produce equivalent norms.

5.2. The next definition was suggested by Seeley (1959).

An operator A defined on $L_p(\Gamma)$ $(1 < p < \infty)$ is said to be a *singular integral operator* on $L_p(\Gamma)$ (or simply on Γ) if the following two conditions are satisfied:

1) if φ, $\psi \in C^\infty(\Gamma)$ are any functions with disjoint support, then $\varphi A\psi$ is compact on $L_p(\Gamma)$;

2) if $\varphi, \psi \in C^\infty(\Gamma)$ are any functions supported in one and the same coordinate neighborhood U_j, with local coordinates x, then $\varphi A \psi$ can be represented in the form $\varphi A_j L \psi + T_j$, where T_j is compact on $L_p(\Gamma)$, A_j is a singular operator of the form (3.1) on $\mathbb{R}^n = \mathbb{R}^n_x$, L is the identity operator in case U_j is an inner neighborhood and L is the canonical continuation operator from $\mathbb{R}^n_+ = \{(x', x^n) \in \mathbb{R}^n : x^n \geq 0\}$ to $\mathbb{R}^n$ in case U_j is a boundary neighborhood (the latter case can of course only happen if Γ is a manifold with boundary).

This definition involves a rather convenient representation for singular integral operators A on a manifold Γ. Namely, if we let $\psi_j \in C^\infty(\Gamma)$ denote any function whose support is contained in U_j and which takes the value 1 on the support of φ_j, then

$$A = \sum \varphi_j A_j L \psi_j + T \tag{5.1}$$

with some compact operator $T \in \mathscr{K}(L_p(\Gamma))$.

The *symbol* of the singular integral operator A on Γ is a function $\mathscr{A}: T^*\Gamma \setminus \{0\} \to C$ given on the sheaf of the nonzero cotangent vectors (P, ξ_P) $(P \in \Gamma)$ of the cotangent bundle $T^*\Gamma$; for the points $P \in U_j$, the symbol $\mathscr{A}$ is, in local coordinates x, defined as $\mathscr{A}_j(x, \xi)$ (as the restriction of $\mathscr{A}_j(x, \xi)$ to $x \in \mathbb{R}^n_+$ in case U_j is a boundary neighborhood), where $\mathscr{A}_j$ is the symbol of the singular integral operator A_j appearing in (5.1). One can show that the symbol $\mathscr{A}$ defined in this way does not depend on the particular choice of the cover $\{U_j\}$ or the partition of unity $\{\varphi_j\}$ (Seeley, 1959).

The representation (5.1) can be used to carry over results on singular integral operators on $\mathbb{R}^n$ to singular integral operators on manifolds. It follows in particular that singular integral operators with sufficiently smooth symbols on a manifold Γ without boundary form a C^*-algebra. Moreover, such an operator, A, is Noetherian on $L_p(\Gamma)$ $(1 < p < \infty)$ if and only if $\inf|\mathscr{A}| > 0$; in this case Ind $A = 0$.

5.3. Now let Γ be an n-dimensional closed manifold of the class $C^{1,\alpha}$ $(0 < \alpha < 1)$. A singular integral operator on the space $C^\alpha(\Gamma)$ is an operator which is defined as in 5.2, only with $C^\infty(\Gamma)$ and $L_p(\Gamma)$ in the conditions 1 and 2 replaced by $C^\alpha(\Gamma)$. The classical Plemelj-Privalov theorem on the boundedness of the one-dimensional singular operator on $C^\alpha(\Gamma)$ (see 1.2.4 of Chap. 3) was extended by G. Giraud (1934) to the higher-dimensional case under the assumption that the characteristic be continuously differentiable. Giraud's theorem has been subsequently generalized into various directions by S.G. Mikhlin, J.J. Kohn and D.C. Spencer, T.G. Gegelia, M. Taibleson, N.M. Mikhailova-Gubenko, V.I. Shevchenko, A.A. Khvoles, and others (see Mikhlin, Prössdorf [1980], Chapters IX and XIII). We limit ourselves to formulating two very general results established by A.A. Khvoles (1974, 1978).

Theorem 1. *If* $\mathscr{A} \hat{\in} W_2^{n/2+\varepsilon}(\mathbb{S}^{n-1})$ *for any* $\varepsilon > 0$ *and*

$$\|\mathscr{A}(x + h, \cdot) - \mathscr{A}(x, \cdot)\|_{W_2^{n/2}(\mathbb{S}^{n-1})} \leq B|h|^\alpha,$$

then A *is a bounded operator on* $C^\alpha(\Gamma)$.

Theorem 2. *Suppose* $\mathscr{A} \hat{\in} W_2^l(\mathbb{S}^{n-1})$ *with* $l > n/2$ *and*

$$\|\mathscr{A}(x + h, \cdot) - \mathscr{A}(x, \cdot)\|_{W_2^l(\mathbb{S}^{n-1})} \leqq B|h|^\alpha.$$

Then if $\inf|\mathscr{A}| > 0$, *every solution* $u \in L_2(\Gamma)$ *of the equation* $Au = g$ *with* $g \in C^\alpha(\Gamma)$
belongs to $C^\alpha(\Gamma)$.

5.4. The Noether properties of matrix singular integral operators on $L_2^m(\Gamma)$,
Γ being a smooth manifold with boundary, were investigated by I.B. Simonenko
(1965) with the help of the local principle invented by him (see Simonenko [1965],
[1986]). These results were generalized by R.V. Duduchava (1981, 1983) to the
spaces $L_p^m(\Gamma)$ and $W_p^s(\Gamma)$ (the work of W. Peterhänsel (1980) is also pertinent to
these problems). M.I. Vishik and G.I. Eskin (1964–1967) as well as Agranovich
(1965) developed a theory of elliptic singular integro-differential equations, which
was subsequently extended to pseudodifferential equations in $W_2^s(\Gamma)$ for manifolds Γ with boundary, comprising the theory of multidimensional singular
integral equations in domains with smooth boundary (see Eskin [1973]). B.A.
Plamenevskiĭ and V.N. Senichkin (1981, 1985) studied algebras of singular integral and pseudodifferential operators on manifolds with conic singularities (see
also Sections 7 and 9, and Plamenevskiĭ [1986]).

§6. Systems of Singular Integral Equations. The Index Formula

Throughout this section Γ refers either to the Euclidean space $\mathbb{R}^n$ or to an
n-dimensional compact smooth manifold without boundary. We consider systems of singular integral equations of the form

$$\sum_{k=1}^m A_{jk} u_k = g_j, \qquad j = 1, \dots, m, \tag{6.1}$$

where A_{jk} ($j, k = 1, \dots, m$) are singular integral operators in the sense of Sections
1 or 5. We also assume that the symbols $\mathscr{A}_{jk}$ meet the requirements posed in the
Sections 4 and 5.

On letting $A = (A_{jk})_1^m$ and denoting by u and g the column-vectors with
components $u_1, \dots, u_m$ and $g_1, \dots, g_m$, respectively, we may rewrite the system
(6.1) as a single equation $Au = g$ (see also Sec. 3 of Chap. 3). The matrix A is
called a *matrix singular integral operator*, while the matrix $\mathscr{A} = (\mathscr{A}_{jk})_1^m$ will be
referred to as the *symbol* (or the *symbol matrix*) of the operator A.

Matrix singular operators are multiplied in accordance with the rules of usual
matrix multiplication. This multiplication is, of course, not commutative. From
the results of Sections 4 and 5 we infer that the symbol of the proudct of two
matrix singular operators equals the product of their symbols. It is clear that the
same also applies to addition. So the symbol of the singular integral operator
det A (see 3.1 of Chap. 3) is nothing but the function det $\mathscr{A}$. Thus, taking also
into account the results of Sections 4, 5 and Theorem 1 of 3.1 of Chap. 3, we
arrive at the following conclusion.

Theorem. *A matrix singular integral operator A is Noetherian on $L_p^m(\Gamma)$ if and only if*

$$\inf|\det \mathscr{A}| > 0. \qquad (6.2)$$

Now assume that (6.2) holds. Unlike the scalar case $m = 1$ (recall Theorem 2 of 4.3), the index of a matrix singular operator need not be equal to zero. This rather surprising fact was probably first noticed by A.I. Vol'pert (1960). The index computation problem for matrix singular integral operators was then studied by many people, including A.I. Vol'pert (1962), A.S. Dynin (1961, 1962), R.T. Seeley (1963, 1965), M.F. Atiyah and I.M. Singer (1963, 1968–1971), S.G. Mikhlin (1963), B.W. Bojarski (1963), M.S. Agranovich (1965), B.V. Fedosov (1970, 1974). The final solution of the problem of computing the index of the general elliptic operator was given by M.F. Atiyah and I.M. Singer (1963) for the case of manifolds without boundary and by M.F. Atiyah and R. Bott (1964) for the case of manifolds with boundary. Note that matrix singular integral operators actually constitute only a humble subclass of the class of operators considered by Atiyah, Singer and Bott. The latter authors' work laid the foundation of the subsequent enormous development, whose outcomes are sufficiently completely reflected by the books Palais [1965], Booss [1977], Rempel, Schulze [1982] (also see Mikhlin, Prössdorf [1980], Chap. XIV).

To state at least a special version of the Atiyah-Singer index formula, let Γ be an n-dimensional compact manifold without boundary which is embedded in an Euclidean space $\mathbb{R}^N$ of sufficiently large dimension N and suppose A is a matrix singular integral operator satisfying (6.2). In that case the Atiyah-Singer index formula was reduced by B.V. Fedosov (1970) to the equality

$$\text{Ind } A = \frac{(-1)^{n+1}(n-1)!}{(2\pi i)^n (2n-1)!} \int_{S^*\Gamma} \text{tr}(\mathscr{A}^{-1} d\mathscr{A})^{2n-1}, \qquad (6.3)$$

where $S^*\Gamma$ is the unit spheres bundle in the cotangent bundle $T^*\Gamma$ and $(\mathscr{A}^{-1} d\mathscr{A})^{2n-1}$ is understood in the sense of outer multiplication of differential forms.

Formula (6.3) can be further simplified in the case $\Gamma = \widetilde{\mathbb{R}}^n$. To avoid complications, assume $\mathscr{A}(x, \xi)$ equals the identity matrix outside some ball $|x| < a$. Then in (6.3) integration over $S^*\Gamma$ may be replaced by integration over the sphere $|x|^2 + |\xi|^2 = a^2$, the orientation of the sphere being the one inherited from the orientation of the $\mathbb{R}^{2n}$ defined by the ordering $x_1, \xi_1, x_2, \xi_2, \ldots, x_n, \xi_n$ of the coordinates.

§7. Notes on Further Results

7.1. C^*-algebras $\mathfrak{A}$ generated by multidimensional singular integral operators (more precisely, by pseudodifferential operators of the degree zero; see Sec. 9) on closed manifolds with "conic" singularities are studied in the work of

B.A. Plamenevskiĭ (1979, 1982) and of B.A. Plamenevskiĭ and V.N. Senichkin (1981, 1985) (also see Plamenevskiĭ [1986]). They define an operator symbol for singular operators and construct an isomorphism between the quotient algebra of $\mathfrak{A}$ by the compact operators $\mathcal{K}$ and the algebra of operator symbols. The symbols themselves are allowed to have two types of singularities: at a singularity of the first type, it is required that the values of the symbol converge to a limit as the variables approach the singularity, the limit depending possibly on the direction of approaching the singularity; in the case of a singularity of the second type, the symbol may oscillate near the singularity. The authors reveal the dependence of the spectrum of the cosets in $\mathfrak{A}/\mathcal{K}$ on the nature of the singularity of their symbol, on the singularities of the manifold, and on the choice of the underlying function space. It is shown that the symbol singularities induce, in a sense, infinite-dimensional representations of the algebra $\mathfrak{A}/\mathcal{K}$ provided dim $\Gamma \geq 2$; in the case where dim $\Gamma = 1$, singularities of the first type induce two-dimensional representations and those of the second type infinite-dimensional ones. It turns out that even in the smooth situation the algebra $\mathfrak{A}/\mathcal{K}$ may possess irreducible infinite-dimensional representations if only the singular operators are considered on spaces with weight; in the case of $L_p(\Gamma)$ as underlying space all irreducible representations are one-dimensional (see Sec. 5).

7.2. Since about 1979, N.L. Vasil'evskiĭ has published a cycle of papers devoted to the algebra $\mathfrak{R}$ generated by operators of the form $A = aI + bK_D + T$ on $L_2(\bar{D})$, where D is a bounded domain in the complex plane whose boundary is composed by a finite number of closed nonintersecting smooth curves, a and b are piecewise continuous functions defined on D, T is a compact operator, and K_D is the operator acting on $L_2(D)$ by the rule

$$(K_D u)(z) = \int_D \int K_D(z, \bar{\zeta}) u(\zeta) dD_\zeta,$$

where $K_D(z, \bar{\zeta}) = -(2/\pi)(\partial^2 G(z, \zeta)/\partial z \partial \bar{\zeta})$ is the Bergman kernel and G the Green function of D. We remark that if D is the unit disk, then $K_D(z, \bar{\zeta}) = [\pi(1 - z\bar{\zeta})^2]^{-1}$.

The singular operator K_D shares a lot of properties with the one-dimensional Cauchy singular operator (see 1.2 of Chap. 3). In particular, $K_D \in \mathcal{L}(L_2(\bar{D}))$, $K_D^2 = K_D$, and the commutator $cK_D - K_D c$ is compact whenever $c \in C(\bar{D})$. A consequence of this is that the algebra $\mathfrak{R}$ coincides with the smallest Banach algebra containing all operators of the form $aI + bS_D + T$, where $S_D = I - 2K_D$. Since obviously $S_D^2 = I$, the algebra $\mathfrak{R}$ may thus be viewed as a two-dimensional analogue of the algebra $\mathfrak{A}$ of one-dimensional singular integral operators considered in 3.4 of Chapter 3.

By invoking the local principle of Simonenko [1965], [1986] and Gohberg, Krupnik [1973], N.L. Vasil'evskiĭ succeeded in describing the C^*-algebra of all symbols which is isomorphic to the algebra $\mathfrak{R}/\mathcal{K}$. He shows that all irreducible representations of the symbol algebra have either dimension 1 or dimension 2. Noether criteria as well as index formulas are given in terms of the symbol. These results generalize earlier work of A.A. Dzhuraev (1979) pertaining to the case where $a, b \in C^1(D) \cap C^\alpha(\bar{D})$ $(0 < \alpha < 1)$ and of I.I. Komyak (1979) concerning the case of continuous coefficients a, b.

7.3. Despite the considerable progress in the theory of one-dimensional equations with degenerate symbols (see Sec. 7 of Chap. 3), the properties of multidimensional singular integral equations with degenerate symbols are not yet very well understood. Profound results are merely known for two classes of such equations: first, for equations whose symbols depend only on $\zeta \in \mathbb{S}^{n-1}$ (see 1.1.1 and 1.1.2) (V.G. Maz'ya and B.A. Plamenevskiĭ (1965), V.G. Maz'ya, B.A. Plamenevskiĭ, and Yu.E. Haikin (1977), M. Lorenz (1974–1979), K.J. Schäfer (1980)) and secondly, for equations involving a certain multidimensional analogue of the Cauchy singular integral introduced by A.V. Bitsadze in 1953 (W. Sprössig (1974)). A survey of these results is in Mikhlin, Prössdorf [1980], Chapter XVI.

§8. Multidimensional Wiener-Hopf Operators

8.1. Given an open subset Ω of $\mathbb{R}^n$, let P_Ω denote the canonical projection (restriction operator) of $L_2(\mathbb{R}^n)$ onto $L_2(\Omega)$ and denote by $L_\Omega: L_2(\Omega) \to L_2(\mathbb{R}^n)$ the operator of continuation by zero. The *multidimensional Wiener-Hopf integral operator* $W_\Omega(a)$ generated by a function $a \in L_\infty(\mathbb{R}^n)$ is the operator defined on $L_2(\Omega)$ by $W_\Omega(a) := P_\Omega \mathscr{F}^{-1} a \mathscr{F} L_\Omega$, $\mathscr{F}$ referring to the Fourier transform. If a has the form $a = c + \mathscr{F}k$ with $c \in \mathbb{C}$ and $k \in L_1(\mathbb{R}^n)$, then $W_\Omega(a)$ can be written as

$$(W_\Omega(a)\varphi)(x) = c\varphi(x) + \int_\Omega k(x - t)\varphi(t)\, dt \qquad (x \in \Omega)$$

(see Sec. 6 of Chap. 3).

8.2. For a subset G of $\mathbb{Z}^n$ (n-dimensional integer group), denote by P_G the canonical projection of $l_2(\mathbb{Z}^n)$ onto $l_2(G)$ and by $L_G: l_2(G) \to l_2(\mathbb{Z}^n)$ the operator of continuation by zero. Also let $\mathbb{T}$ refer to the complex unit circle and let V be the canonical isomorphism of $l_2(\mathbb{Z}^n)$ onto $L_2(\mathbb{T}^n)$. For $a \in L_\infty(\mathbb{T}^n)$, the operator $W_G(a) := P_G V^{-1} a V L_G$, considered on $l_2(G)$, is called the *multidimensional discrete Wiener-Hopf operator* generated by a; it may be regarded as a generalization of the one-dimensional operator on $l_2(\mathbb{Z}_+)$ induced by a Toeplitz matrix (see 6.1 of Chap. 3). If the sequence $\{a_k\}_{k \in \mathbb{Z}^n}$ of the Fourier coefficients of a belongs to $l_1(\mathbb{Z}^n)$, then $W_G(a)$ admits the representation

$$(W_G(a)\varphi)_j = \sum_{k \in G} a_{j-k}\varphi_k \qquad (j \in G).$$

The function $a \in L_\infty(\mathbb{R}^n)$ (resp. $a \in L_\infty(\mathbb{T}^n)$) is usually referred to as the *symbol* of the operator $W_\Omega(a)$ (resp. $W_G(a)$).

Under certain additional assumptions on the symbol a, the operators $W_\Omega(a)|L_2(\Omega) \cap L_p(\Omega)$ and $W_G(a)|l_2(G) \cap l_p(G)$ extend to bounded operators on the spaces $L_p(\Omega)$ and $l_p(G)$ $(1 \le p \le \infty)$.

8.3. The case where Ω or G are conic sets, i.e. sets containing with every of its point all of the ray starting at the origin and passing through that point, is of particular interest. The theory of multidimensional Wiener-Hopf operators on the half-space, $\Omega = \mathbb{R}^{n-1} \times \mathbb{R}_+$ or $G = \mathbb{Z}^{n-1} \times \mathbb{Z}_+$, was worked out by L.S. Gol'denshtein and I. Gohberg (1960, 1967); the half-space theory is in many respects akin to the theory of one-dimensional Wiener-Hopf operators (see also Sec. 4 of Chap. III of Gohberg, Feldman [1971]). "Non-classical" singular integral operators of the Calderón-Mikhlin-Zygmund type on $\mathbb{R}^{n-1} \times \mathbb{R}_+$ were studied by R.V. Duduchava and R. Schneider (1987). When dealing with more general (conic) sets Ω and G, there emerge serious difficulties, a considerable part of which has not yet been overcome at present time.

8.4. To understand the nature of multidimensional Wiener-Hopf operators, it is, in a sense, sufficient to restrict attention to the quarter-plane case: $\Omega = \mathbb{R}^2_{++} = \mathbb{R}_+ \times \mathbb{R}_+$ and $G = \mathbb{Z}^2_{++} = \mathbb{Z}_+ \times \mathbb{Z}_+$.

I.B. Simonenko (1967, 1968) and, independently, R.G. Douglas and R. Howe (1971) were the first to study the Noether properties of Wiener-Hopf operators on $L_p(\mathbb{R}^2_{++})$ and $l_p(\mathbb{Z}^2_{++})$ with continuous symbols. They established the following criterion.

Theorem. *Let* $a \in C(\mathbb{T}^2)$. *For each* $\tau \in \mathbb{T}$, *define functions* b_τ, $c_\tau \in C(\mathbb{T})$ *by* $b_\tau(t) = a(t, \tau)$ *and* $c_\tau(t) = a(\tau, t)(t \in \mathbb{T})$. *Then* $W_{\mathbb{Z}^2_{++}}(a)$ *is a Noether operator on* $l_2(\mathbb{Z}^2_{++})$ *if and only if the operators* $W_{\mathbb{Z}_+}(b_\tau)$ *and* $W_{\mathbb{Z}_+}(c_\tau)$ *are invertible for all* $\tau \in \mathbb{T}$. *In this case* Ind $W_{\mathbb{Z}^2_{++}}(a) = 0$.

This result was extended to the case of matrix symbols, to Wiener-Hopf integral operators, and to the case $p \neq 2$. Simonenko's proof was based on his local principle, by means of which he reduced the problem to the half-plane case, which had been earlier studied by Gol'denshtein and Gohberg. Douglas and Howe proved the result by resorting to C^*-algebra techniques. Subsequently several new proofs were found. In particular, G. Strang (1970) and R.V. Duduchava (1977) verified the sufficiency part of the theorem by explicitly constructing a regularizer of $W_{\mathbb{Z}^2_{++}}(a)$ (such an approach was also suggested by V.S. Pilidi (1971)). E. Meister and F.-O. Speck (1980) generalized these results to operators on three-dimensional wedge-shaped regions.

The case of piecewise continuous symbols was considered as well. By a piecewise continuous symbol $a \in L_\infty(\mathbb{R}^2)$ usually a symbol is meant which can be uniformly approximated by finite sums of the form $\sum_j b_j(x)c_j(y)$ $(x, y \in \mathbb{R})$ with $a_j, b_j \in PC(\mathbb{R})$. The Noether theory of the operators $W_{\mathbb{R}^2_{++}}(a)$ with piecewise continuous symbols on $L_p(\mathbb{R}^2_{++})$ was settled by R.V. Duduchava (1977). The corresponding problem for the discrete operators $W_{\mathbb{Z}^2_{++}}(a)$ on $l_p(\mathbb{Z}^2_{++})$ was solved by R.V. Duduchava (1977) for $p = 2$ and by A. Böttcher (1983) for general $p \in (1, \infty)$. In all these cases it was shown that the quarter-plane operator is Noetherian if and only if two appropriately chosen families of half-line operators consist only of invertible ones. For more about this subject and the "bilocalization approach" see the recent monograph Böttcher, Silbermann [1989]. Some progress in the

Noether theory of multidimensional Wiener-Hopf operators with symbols possessing discontinuities of a more general kind has recently been made by L.I. Sazonov (1985) and A. Böttcher (1987).

8.5. The question on the invertibility of quarter-plane Wiener-Hopf operators is extremely difficult and has presently been answered for only a handful of more less special classes of symbols.

It was Douglas and Howe (1971) who discovered Noetherian but non-invertible discrete Wiener-Hopf operators on the quarter-plane. To give their example, let the symbol a have the form $a(\xi, \eta) = b(\xi \eta^{-1})$ ($\xi, \eta \in \mathbb{T}$), where $b \in C(\mathbb{T})$, and denote by b_k ($k = 0, \pm 1, ...$) the Fourier coefficients of b. Then $W_{\mathbb{Z}^2_{++}}(a)$ is Noetherian (resp. invertible) on $l_2(\mathbb{Z}^2_{++})$ if and only if the matrices $B_n := (b_{j-k})^n_{j,k=0}$ are nonsingular for all sufficiently large n (resp. all $n = 0, 1, ...$) and $\lim \sup_{n \to \infty} \|B_n^{-1}\| < \infty$. Hence, if $b \in C(\mathbb{T})$ is any function such that $b(t) \neq 0$ for $t \in \mathbb{T}$, ind $b = 0, b_0 = 0$, then $W_{\mathbb{Z}^2_{++}}(a)$ is Noetherian but not invertible.

Here are some more classes of symbols for which invertibility criteria for the corresponding Wiener-Hopf operators on the quarter-plane are known: 1) the symbol a admits a factorization $a = a_{--}a_{-+}a_{++}$, the Fourier coefficient sequence of a_{++} being supported in $\mathbb{Z}_+ \times \mathbb{Z}_+$ (V.S. Rabinovich, 1967); 2) the symbol a is of the form $a(\xi, \eta) = b(\xi)c(\eta) + d(\xi)$ (S. Osher, 1970); 3) $a(\xi, \eta) = (\xi - \lambda)^{-1}(\eta - \mu)^{-1} \sum_{i,j \geq 0} a_{ij} \xi^i \eta^j$ ($|\lambda| < 1, |\mu| < 1$) (V.A. Malyshev (1971) and R. Douglas (1972)); 4) the support of the kernel $k(x)$ ($x \in \mathbb{R}^2$) resp. $\{a_k\}_{k \in \mathbb{Z}^2}$ is located in a half-plane whose boundary contains the origin (A. Böttcher and A.E. Pasenchuk, 1982); 5) the symbol $a \in L_\infty(\mathbb{T}^2)$ is locally sectorial over $C \otimes C$, i.e. each point $\xi_0 \in \mathbb{T}$ has an open neighborhood $U(\xi_0) \subseteq \mathbb{T}$ such that the set $a(U(\xi_0) \times \mathbb{T})$ can be rotated into the right open half-plane (A. Böttcher, 1987). Note that except for 4) in each of these cases Noethericity (Fredholmness) coincides with invertibility. Interesting results on the spectrum of Wiener-Hopf operators on n-dimensional regions were also obtained by E. Gerlach and N. Latz (1977).

E. Meister and F.-O. Speck (1984) succeeded in explicitly constructing the Moore-Penrose inverse for quarter-plane Wiener-Hopf operators.

For more about this topic and for a variety of applications of multidimensional Wiener-Hopf operators we refer to Malyshev [1975], Meister, Speck [1979, 1980], Böttcher, Silbermann [1989].

8.6. The theory of multidimensional Wiener-Hopf operators with degenerate symbols is much more complicated and less-developed than its one-dimensional counterpart (for which see Sec. 7 of Chap. 3). The hitherto known results in this direction mainly concern quarter-plane operators with so-called "analytic" symbols, i.e. symbols of the type a_{++} (V.B. Dybin and A.E. Pasenchuk (1978); M.B. Gorodetskiĭ (1980); A. Böttcher (1984); Yu.D. Kislitskiĭ (1985)).

8.7. We finally remark that one has also studied so-called *general* (or *abstract*) *Wiener-Hopf (Toeplitz) operators*. These are of the form $PAP|\text{Im } P$, where $A \in \mathscr{L}(X)$, X is a Hilbert (Banach or Fréchet) space, and $P \in \mathscr{L}(X)$ is a projection.

In this general setting a series of profound results were established by M. Shinbrot (1964), G.N. Chebotarev (1967), A. Devinatz, M. Shinbrot (1969), and F.-O. Speck (1985) (see Speck [1985]). Of course, the operators we have considered above fall into the class of general Wiener-Hopf operators. The various possibilities one has for specifying X (and P and A) lead to several concrete classes of operators with quite different properties; for instance, much work has been done on the case that PX is a Hardy space (of a higher-dimensional sphere, say) or that PX is a Bergman space (on a ball of $\mathbb{C}^n$, for example). In this connection we refer to papers by L.A. Coburn (1973), U. Venugopalkrishna (1972), R. Douglas (1973), A.M. Davie and J.P. Jewell (1977), McDonald and C. Sundberg (1979), L. Boutet de Monvel and V. Guillemin (1981), S. Axler (1984), H. Upmeier (1983–84), A. Dynin (1986); this list is naturally incomplete. The papers we have just cited primarily concentrate on the case that A is a multiplication operator. Connections to pseudodifferential operators are elucidated in the survey Guillemin [1984].

§9. Pseudodifferential Operators

Pseudodifferential operators (ψdo's) are natural generalizations of singular integral operators (see Sec. 1). The foundation of the theory of ψdo's was laid in 1965 by the work of Agranovich [1965], Cordes [1965], Hörmander [1965], Kohn and Nirenberg [1965], and Seeley [1965]. A detailed account of all the different aspects of the modern theory of ψdo's as well as their applications can be found in the monographs Eskin [1973], Shubin [1978], Cordes [1979], Taylor [1989], Kumano-go [1981], Journé [1983], Treves [1982], Hörmander [1985a] Grubb [1986]. We here confine our attention to some simple facts concerning ψdo's and immediately tying in with Section 1, 4, 5 of this chapter (for more details see Agranovich, Vishik [1968] and Mikhlin, Prössdorf [1980]).

9.1. Let $a(x, \xi)$ be a $\mathbb{C}$- or $\mathbb{C}^{m \times m}$-valued infinitely differentiable function on $\mathbb{R}^n \times \mathbb{R}^n \setminus \{0\}$ which is positively homogeneous of degree $\sigma \geq 0$ in ξ:

$$a(x, t\xi) = t^\sigma a(x, \xi) \qquad \forall t > 0.$$

Assume that the limit $a(\infty, \xi) = \lim_{|x| \to \infty} a(x, \xi)$ exists and that the difference $a(x, \xi) - a(\infty, \xi)$, thought of as a function of x, belongs uniformly with respect to ξ to the space $S(\mathbb{R}^n)$ of all (possibly matrix-valued) infinitely differentiable functions on $\mathbb{R}^n$ which together with all their derivatives decrease more rapidly than any (negative) power of $|x|$ as $|x| \to \infty$.

The *homogeneous pseudodifferential operator* with *symbol* $a(x, \xi)$ is the operator defined for $u \in S(\mathbb{R}^n)$ by

$$(Au)(x) = \mathscr{F}^{-1}_{\xi \to x} a(x, \xi) \mathscr{F}_{x \to \xi} u. \tag{9.1}$$

The number σ is called the *degree* of the ψdo (9.1).

There are two special cases worthwhile to be mentioned separately: 1) If $\sigma = 0$, then the ψdo (9.1) is the singular integral operator whose symbol is a (see (1.11)); 2) in case $a(x, \xi) = \sum_{|\alpha|=\sigma} a_\alpha(x)\xi^\alpha$ is a polynomial of degree σ in ξ, then A is nothing else than the differential operator given by $(Au)(x) = \sum_{|\alpha|=\sigma} a_\alpha(x)(D^\alpha u)(x)$.

9.2. In ψdo theory a prominent role is played by the Sobolev-Slobodetskiĭ spaces $H^l := W_2^l(\mathbb{R}^n)$ $(l \in \mathbb{R})$, which are defined as the completion of $S(\mathbb{R}^n)$ in the norm $\|u\|_l = (\int (1 + |\xi|^2)^l |(\mathscr{F}u)(\xi)|^2 d\xi)^{1/2}$.

A linear operator A defined on $S(\mathbb{R}^n)$ and extendible to an operator in $\mathscr{L}(H^l, H^{l-\sigma})$ for all $l \in \mathbb{R}$ and some $\sigma \in \mathbb{R}$ is said to have the *order* σ. The infimum of all numbers σ such that A has order σ is referred to as the *actual order* of A.

The so-called *boundedness theorem* says that a homogeneous ψdo of degree $\sigma \geq 0$ has the order σ.

The algebra $\mathscr{L}_{-\infty}$ of all operators of actual order $-\infty$ is of great importance for the theory of ψdo's. It is a natural substitute for the ideal of compact operators and plays the same part in ψdo theory as compact operators do in the theory of singular integral operators.

9.3. ψdo's of negative degree are defined as follows. Let a be a function as in 9.1 but suppose now that σ is negative. So a has a singularity for $\xi = 0$, as a consequence of which formula (9.1) does not make sense unless it is suitably interpreted. To overcome this inconvenience, we take any nonnegative function $\zeta \in C^\infty(\mathbb{R}^n)$ which vanishes identically in some neighborhood of the point $\xi = 0$ and is identically equal to 1 outside some other (bigger) neighborhood and put

$$(A_\zeta u)(x) = F_{\xi \to x}^{-1} a(x, \xi) \zeta(\xi) F_{x \to \xi} u. \tag{9.2}$$

The operator A_ζ defined in this way is called a *homogeneous pseudodifferential operator of negative degree σ with symbol a*.

Unlike a ψdo of nonnegative degree, defined by (9.1), a ψdo of negative degree is not uniquely determined by its symbol (notice that A_ζ depends on ζ). However, if ζ_1 and ζ_2 are any two "cut-off" functions of the above type, then $A_{\zeta_1} - A_{\zeta_1} \in L_{-\infty}$. In other words, we may think of a ψdo of negative degree as a being which is uniquely determined by its symbol up to an $\mathscr{L}_{-\infty}$ operator. To unite the definitions given in 9.1 and here, we note that the above construction makes of course also sense in the case $\sigma \geq 0$, so that a ψdo of nonnegative degree may be defined via (9.2) as well. We wish to point out that for $\sigma \geq 0$ even $A - A_\zeta \in \mathscr{L}_{-\infty}$, so that A itself is a natural representative of the "coset" of all ψdo's with symbol a.

The boundedness theorem cited in 9.2 extends to ψdo's of negative degree.

9.4. If A and B are homogeneous ψdo's with symbol $a(\xi)$ and $b(\xi)$ which do not depend on x, then the product AB is obviously again a ψdo and its symbol is $a(\xi)b(\xi)$. In the general case the following theorem applies.

Theorem. (on the multiplication of ψdo's). *Let A and B be homogeneous ψdo's of the degrees σ_a and σ_b whose symbols are a and b, respectively. Then, for every natural number ϱ, the product AB can be represented in the form*

$AB = C_0 + C_1 + \ldots + C_{\varrho-1} + T_\varrho$, where C_k $(k = 0, \ldots, \varrho - 1)$ is a homogeneous ψdo of degree $\sigma_a + \sigma_b - k$ with symbol

$$c_k(x, \xi) = \sum_{|\alpha| = k} \frac{1}{\alpha!} \partial^\alpha a(x, \xi) D^\alpha b(x, \xi) \qquad (9.3)$$

and T_ϱ is an operator which has the order $\sigma_a + \sigma_b - \varrho$.

Here and in what follows ∂^α denotes differentiation with respect to ξ, while D^α refers to differentiation with respect to x.

This theorem implies that the product AB of two homogeneous ψdo's admits an asymptotic expansion into a series of ψdo's of descending degrees $\sigma_a + \sigma_b$, $\sigma_a + \sigma_b - 1, \ldots$. In particular, the semicommutator $[A, B)$ (defined as in 4.2) is an operator of order $\sigma_a + \sigma_b - 1$.

9.5. Now let A be a homogeneous ψdo of degree σ with symbol a and let B be a (the) homogeneous ψdo with symbol a^*, where a^* is the Hermitian conjugate matrix of a. We denote by A^* the operator which is adjoint to A with respect to the scalar product of $L_2(\mathbb{R}^n)$. If the symbol $a = a(\xi)$ does not depend on x, it is easily seen that $B = A^*$. The general situation is described by the following theorem.

Theorem. *If A is a homogeneous ψdo of degree σ with symbol $a(x, \xi)$, then, for every natural number ϱ, $A^* = B_0 + B_1 + \cdots + B_{\varrho-1} + T_\varrho$, where B_k $(k = 0, \ldots, \varrho - 1)$ is a ψdo with symbol*

$$\sum_{|\alpha| = k} \frac{1}{\alpha!} D^\alpha \partial^\alpha a^*(x, \xi)$$

and T_ϱ is an operator which has the order $\sigma - \varrho$.

9.6. Theorems 9.4 and 9.5 tell us that multiplication and taking adjoints of homogeneous ψdo's is impossible within the class of homogeneous ψdo's, i.e. leads to objects which live beyond the realm of homogeneous ψdo's. On the other hand, these theorems suggest a way of constructing a C^*-algebra of ψdo's. This way consists in thinking of a ψdo as something which comprises a whole family of ψdo's of descending orders. The precise construction of such a C^*-algebra is as follows.

Let (σ_k) be a strictly decreasing finite or infinite sequence of real numbers; if the sequence is infinite, we assume that $\sigma_k \to -\infty$. Furthermore, let A_k be homogeneous ψdo's of the degrees σ_k with symbols a_k. Then any operator A defined on $S(\mathbb{R}^n)$ such that, for every natural integer N, the operator $A - \sum_{k=0}^N A_k$ has an order less than σ_N is called a *pseudodifferential operator with the asymptotic expansion*

$$A \sim \sum_0^\infty A_k. \qquad (9.4)$$

The formal series

$$a(x, \xi) \sim \sum_0^\infty a_k(x, \xi) \qquad (9.5)$$

is referred to as the *symbol* of A and $a_0(x, \xi)$ is said to be the *principal (part of the) symbol*.

We denote the class of all ψdo's with an asymptotic expansion of the form (9.4) by $\mathscr{L}$. Note that the operators $A \in \mathscr{L}$ which correspond to finite sequences (σ_k) can be written as

$$A = \sum A_k + T \text{ with } T \in \mathscr{L}_{-\infty}.$$

From the definition given above we infer that every operator $T \in \mathscr{L}_{-\infty}$ is a ψdo belonging to $\mathscr{L}$, namely, a ψdo whose symbol is zero. A moment's reflection also reveals that every operator $A \in \mathscr{L}$ is determined by its symbol up to an additive operator $T \in \mathscr{L}_{-\infty}$.

L. Hörmander (1965) has shown that for every formal series (9.5) of symbols a_k of descending degrees there exists a ψdo $A \in \mathscr{L}$ whose symbol is a.

In our new language, what the theorems 9.4 and 9.5 say is that the operation $A \rightarrow A^*$ makes $\mathscr{L}$ become an algebra with an involution; given operators A and B in $\mathscr{L}$ with symbols a and b, the symbols c of AB and d of A^* are represented by

$$c \sim \sum_\alpha \frac{1}{\alpha!} \partial^\alpha a \, D^\alpha b, \qquad d \sim \sum_\alpha \frac{1}{\alpha!} D^\alpha \partial^\alpha a^*,$$

in the sense that the formal series to the left and the right of "$\sim$" are generated by the same degree sequence (σ_k) and that the terms corresponding to the same degree are equal to each other (a^* refers to the formal series $a^* \sim \sum_0^\infty a_k^*$).

9.7. Now let Γ be an n-dimensional infinitely smooth compact manifold without boundary. In analogy to the definition of a singular integral operator on Γ (recall 5.2), an operator A acting on $C^\infty(\Gamma)$ is called a *pseudodifferential operator on Γ of degree σ* if it obeys the following two requirements:

1) If $\varphi, \psi \in C^\infty(\Gamma)$ have disjoint support, then $\varphi A \psi$ has the order $\sigma - 1$ (that is, $\varphi A \psi$ extends to an operator in $\mathscr{L}(H^l(\Gamma), H^{l-\sigma+1}(\Gamma))$ for all $l \in \mathbb{R}$).

2) Condition 2 of 5.2 is satisfied, the A_j's now being homogeneous ψdo's of the degree σ on $\mathbb{R}^n = \mathbb{R}^n_x$ and the T_j's being operators which have order $\sigma - 1$.

The *symbol* of a ψdo on Γ is defined similarly as in the case of a singular integral operator on Γ (see 5.2).

The ψdo's on Γ constitute an algebra with an involution (the involution is taking the adjoint with respect to the scalar product in $L_2(\Gamma)$). Moreover, one can show that a ψdo A on Γ of degree σ is a Noether operator between $H^l(\Gamma)$ and $H^{l-\sigma}(\Gamma)$ if and only if its symbol has no zeros (or, to include the matrix case, does not degenerate). The index of $A \in \mathscr{L}(H^l(\Gamma), H^{l-\sigma}(\Gamma))$ is independent of l.

9.8. A generalization of pseudodifferential operators is the class of so-called *Fourier integral operators*. Accounts of this theory can be found in Shubin [1978] and Treves [1982]. For applications of the theories of ψdo's and Fourier integral operators to boundary value problems for partial differential equation and to other fields we refer to Eskin [1973], Shubin [1978], Treves [1982], Grubb [1986], Hörmander [1985b].

Notes on the Bibliography

To help the reader, we here give a few notes on some of the books and surveys cited in the list of references.

The handbooks Zabreĭko et al. [1968] and Fenyö, Stolle [1982–1984] are encyclopedic in nature and provide information on all the basic branches of the theory of integral equations and their applications. The books Jörgens [1970] and Fenyö, Stolle [1982–1984] contain a detailed and modern exposition of the classical theory of linear operator equations and integral equations of Volterra and Fredholm types as well as of some approximation methods for their solution. The fundamentals of the classical theory are also in the textbooks Schmeidler [1950], Riesz, Sz.-Nagy [1952], Tricomi [1957], Mikhlin [1959].

The books Tricomi [1957], Jörgens [1970], Fenyö, Stolle [1982–1984] embark on certain types of singular equations, and the books Tricomi [1957], Zabreĭko et al. [1968] also deal with some classes of nonlinear integral equations of special import for applications. Several methods for solving Volterra and Fredholm integral equations of the first kind are discussed in Schmeidler [1950], Imanaliev [1981], Fenyö, Stolle [1982–1984].

The books Hilbert [1912] and Volterra [1913] gave an essential impetus to the development of the general theory of Volterra and Fredholm integral equations. In Hilbert's monograph [1912], the edifice of linear integral equations of Fredholm type is built up on the theory of linear and bilinear forms with infinitely many variables; the spectral theory of integral equations with Hermitian kernels is also worked out there. A detailed survey of all results on integral equations published up to 1928 is given in the excellent exposition Hellinger, Toeplitz [1928]. Furthermore, the books cited above contain discussions of the connections between the Fredholm and Hilbert-Schmidt theories on the one hand and various branches of mathematics on the other, including numerous applications to a variety of problems in physics, mechanics, and engineering sciences.

The books Riesz, Sz.-Nagy [1952], Gohberg, Krein [1965], Jörgens [1970], Kantorovich, Akilov [1982], Pietsch [1978] treat the theory of ideals of linear compact operators on abstract Hilbert and Banach spaces. The monograph Pietsch [1978] provides a complete theory of Riesz operators on Banach space, including the theory of Fredholm determinants.

The papers Birman, Solomyak [1977] and Pietsch [1980, 1983] are surveys of and the recent monograph Pietsch [1987] is dedicated to asymptotic estimates for the characteristic values and eigenvalues of integral operators. Gohberg and Krein's survey [1957] is a detailed account of the theory of unbounded Noether operators on Banach spaces; the further evolution of this subject is reflected in the books Jörgens [1970], Gohberg, Krupnik [1973], Prössdorf [1974], Mikhlin, Prössdorf [1980], Krupnik [1984].

A very systematic and clear exposition of the theory of one-dimensional singular integral equations in classes of Hölder continuous functions is given in the monograph Muskhelishvili [1968]. A major part of this book, written by one of the pioneers of that theory, is devoted to applications of singular equations to potential theory, elasticity, and other fundamental disciplines of mathematical physics. The books Vekua [1970] and Gakhov [1977] focus on boundary value problems for functions of a complex variable and on their applications to singular integral equations with kernels of Cauchy, Hilbert, power of logarithmic types in classes of Hölder continuous functions.

The books Gohberg, Feldman [1971], Gohberg, Krupnik [1973], Prössdorf [1974] are concerned with one-dimensional singular integral equations, Wiener-Hopf equations, and some of their generalizations; these equations are studied in several function spaces, and the investigation relies heavily upon modern methods of operator theory and functional analysis. Projection methods for the approximate solution of the equation listed above and considered in the books Gohberg, Feldman [1971], Prössdorf [1974], Prössdorf, Silbermann [1977], [1990], Böttcher, Silbermann [1983], [1989]. In Prössdorf [1974] and Prössdorf, Silbermann [1977] special attention is paid to non-elliptic equations, i.e. equations whose symbol has zeros.

The books Clancey, Gohberg [1981] and Litvinchuk, Spitkovskiĭ [1987] are systematic treatises of the factorization theory for both smooth and discontinuous matrix functions and its application to systems of one-dimensional singular integral equations. Various aspects of generalized factorization of matrix and operator functions as well as a series of applications to singular and Wiener-Hopf equations are illuminated in the monographs Bart, Gohberg, Kaashoek [1979], Böttcher, Silbermann [1983], Speck [1985].

The subject of the book Danilyuk [1975] is two-dimensional boundary value problems and one-dimensional singular integral equations under fairly general assumptions on the smoothness of the boundary contour and the coefficients of the equation (the class of contours admitted involves Lyapunov and Radon curves). The survey Khvedelidze [1975] is an account of the basic results (published up to the middle of the seventies) on integrals of Cauchy type and their connection with boundary value problems for analytic functions under "non-classical" hypotheses; it also dwells on applications of these results to the boundary problem of conjugation in the sense of I.I. Privalov. For the state of the art of the topic alluded to in this paragraph and for the abundance of recent results in this direction see V.G. Maz'ya's article contained in the present volume.

The book Litvinchuk [1977] concentrates on boundary value problems for analytic functions and one-dimensional singular integral equations with a shift. Several classes of integral (and other) equations with shifts are studied by means of algebraic methods in the book Przeworska-Rolewicz [1973].

The monograph Duduchava [1979] covers integral equations of convolution type with kernels whose Fourier transform is a discontinuous function, singular integral equations with fixed singularities in the kernel, and plenty of applications to problems of elasticity theory and mathematical physics. The papers Sakhnovich [1980] and Arabadzhyan, Engibaryan [1984] are surveys of special topics on integral equations of convolution type (connections with the non-linear equations emerging from the factorization problem, equations over finite intervals, applications to numerous problems of mathematics and mathematical physics).

In the book Krupnik [1984], a characterization of all the Banach algebras of operators owning a scalar- or matrix-valued symbol is given and singular integral operators with matrix-valued coefficients as well as algebras generated by such operators are studied thoroughly. The books Douglas [1972], [1973] aim at demonstrating how by a consistent recourse to Banach algebra techniques a clean and profound theory of (block) Toeplitz operators can be established; special emphasis is laid on operators with symbols in $C + H_\infty$.

The books Böttcher, Silbermann [1983], [1989] are systematic and fairly comprehensive treatments of multitudinous questions on Toeplitz operators, including the problems of invertibility, of index computation, operators on weighted l_p and H_p spaces, quarter-plane operators, projection methods, and Toeplitz determinants. Generalizations and ramifications of the problem of inverting (finite) Toeplitz and Hankel matrices are studied in the book Heinig, Rost [1984]. The English translation of the monograph Nikol'skiĭ [1980] contains an up-to-date survey of the spectral theory of Toeplitz and Hankel operators (and highlights deep relationships to the general spectral theory of the shift operator).

The book Mikhlin [1962] is the pioneering monograph on the theory of multidimensional singular integral operators. It consists primarily of results of its author, who is well-known to be one of the founders of this theory. The first part of this book reflects the state of the art of the theory of multidimensional singular integral equations in the early sixties.

The monograph Mikhlin, Prössdorf [1980] endeavours to present in a systematic manner the theory of singular integral operators, their applications, and the different methods for approximately solving singular integral equations. A significant peculiarity of this book is that both one- and multidimensional singular operators are studied from a unique point of view and that ideas and methods from functional analysis are utilized consistently.

The books Stein [1970], Stein, Weiss [1971] record the results on multidimensional singular integrals, integral transforms related to them, and Fourier multipliers obtained during the last few decades. These results form the foundation on which the theory of spaces of differentiable and harmonic functions is erected in these books.

The book Maz'ya, Shapozhnikova [1985] deals with applications of the theory of multipliers in spaces of differentiable functions developed by its authors to the theory of singular integral operators. The authors study the properties of singular operators in Sobolev spaces under fairly weak assumptions on the symbol concerning the first variable, establish two-sided estimates for the norm and essential norm of operators invariant with respect to shifts and acting on pairs of weighted L_2 spaces, and they describe the spectrum of such operators.

A large set of questions of the theory of Sobolev spaces is considered in the book Maz'ya [1985]: extensions of and new proofs to embedding theorems, necessary and sufficient conditions in terms of isoperimetric inequalities for the validity of several integral inequalities for functions possessing generalized derivatives, applications to the theory of elliptic boundary value problems.

The monograph Journé [1983] is a systematic account of the theory of Calderón-Zygmund operators. It elucidates such topics as the Hardy-Littlewood maximal operator, classes of Muckenhoupt weights, Hardy spaces and BMO, weighted norm estimates for Calderón-Zygmund operators, connections between the latter ones and pseudodifferential operators or the Littlewood-Paley theory, the Cauchy singular integral on Lipschitz curves. A similar circle of problems is also covered by the excellent treatise Garsia-Cuerva, Rubio de Francia [1985]. The article Dyn'kin, Osilenker [1983] is a detailed survey on one and two weight norm inequalities of the strong and weak types for the Hardy-Littlewood maximal function, Riesz potentials, singular integral operators and harmonic functions. This survey is written in the framework of classical function theory and is in spirit close to the book Stein [1970]. Note that Dyn'kin and Osilenker [1983] pay special attention to the works that appeared after 1980.

The articles Malyshev [1975], Meister, Speck [1979], [1980] focus on several branches of the theory of multidimensional convolution operators and their applications to probability theory and mathematical physics. A self-contained and up-to-date theory of convolution operators on the quarter-plane can be found in Böttcher, Silbermann [1989].

The book Elschner [1985a] is a systematic exposition of the theory of one-dimensional pseudo-differential equations with degenerate symbols and their applications to certain plane boundary value problems for partial differential equations. For the multitude of all the various aspects of the modern theory of ψdo's and their applications we refer to the standard books: Èskin [1973], Shubin [1978], Cordes [1979], Kumano-go [1981], Taylor [1981], Treves [1982], Journé [1983], Hörmander [1985], Grubb [1986].

The Atiyah-Singer index theory is presented in a systematic fashion in the books Palais [1965], Booss [1977], Rempel, Schulze [1982].

Several questions concerning approximation methods for integral equations are considered with different minuteness and rigour in the books Ivanov [1968], Gohberg, Feldman [1971], Prössdorf [1974], Baker [1977], Prössdorf, Silbermann [1977], [1990], Gabdulkhaev [1980], Mikhlin, Prössdorf [1980], Fenyö, Stolle [1982–1984], Böttcher, Silbermann [1983], [1989], Belotserkovskiĭ, Lifanov [1985].

The monograph Prössdorf, Silbermann [1990] contains an up-to-date convergence and error analysis of various classes of approximation methods for solving integral and operator equations, such as singular integral equations with piecewise continuous coefficients on curves with corners and on the interval, pseudodifferential equations of Mellin type, integral and discrete equations of Wiener-Hopf type etc. This analysis is based on a modern and largely unified approach using extensively functional analysis methods (e.g., Banach algebra techniques, local principles).

Bibliography*

Agranovich, M.S. [1965]: Elliptic singular integro-differential operators. Russ. Math. Surv. 20, 1–121 (Russian original: Usp. Mat. Nauk 20 (5), 3–120, 1965). Zbl.149,361

Agranovich, M.S., Vishik, M.I. [1968]: Pseudodifferential operators. Moscow State Univ.: Moscow

Arabadzhyan, L.G., Engibaryan, N.B. [1984]: Convolution equations and nonlinear functional equations. J. Sov. Math. 36, 745–791 (1987) (Russian original: Itogi Nauki Tekh., Ser. Mat. Anal. 22, 175–244, 1984). Zbl.568.45004

Baker, C.T.H. [1977]: The Numerical Treatment of Integral Equations. Clarendon Press: Oxford. Zbl.373.65060

Bart, H., Gohberg, I., Kaashoek, M.A. [1979]: Minimal Factorization of Matrix and Operator Functions. Birkhäuser Verlag: Basel, Boston, Stuttgart. Zbl.424.47001

Belotserkovskiĭ, S.M., Lifanov, I.K. [1985]: Numerical Methods for Singular Integral Equations. Nauka: Moscow. Zbl.578.65140

Birman, M.Sh., Solomyak, M.Z. [1977]: Estimates of the singular numbers of integral operators. Russ. Math. Surv. 32 (1), 15–89 (Russian original: Usp. Mat. Nauk 32 (1), 17–84, 1977). Zbl.344.47021

Booss, B. [1977]: Topologie und Analysis. Springer-Verlag: Berlin, Heidelberg, New York. Zbl.364.58017. English transl.: Springer-Verlag 1985

Böttcher, A., Krupnik, N., Silbermann, B. [1988]: A general look at local principles with special emphasis on the norm computation aspect. Integral Equations Oper. Theory 11, No. 4, 455–479. Zbl.655.46036

Böttcher, A., Silbermann, B. [1983]: Invertibility and Asymptotics of Toeplitz Matrices. Akademie-Verlag: Berlin. Zbl.578.47015

Böttcher, A., Silbermann, B. [1989]: Analysis of Toeplitz Operators. Akademie-Verlag: Berlin, Springer-Verlag: Berlin, Heidelberg, New York

Calderón, A.P. [1977]: Cauchy integrals on Lipschitz curves and related operators. Proc. Natl. Acad. Sci. USA 74: 4, 1324–1327. Zbl.373.44003

Calderón, A.P., Zygmund, A. [1952]: On the existence of certain singular integrals. Acta Math. 88 (1–2), 85–139. Zbl.47,102

Calderón, A.P., Zygmund, A. [1956]: On singular integrals. Am. J. Math. 78 (2), 289–309. Zbl.72,115

Calderón, A.P., Zygmund, A. [1978]: On singular integrals with variable kernels. Appl. Anal. 7, 221–238. Zbl.451.42012

Carleman, T. [1922]: Sur la résolution de certaines équations intégrales. Ark. Mat. 16 (26), 1–19. Jrb.48,456

Clancey, K., Gohberg, I. [1981]: Factorization of Matrix Functions and Singular Integral Operators. Birkhäuser Verlag: Basel, Boston, Stuttgart. Zbl.474.47023

Coifman, R.R., Meyer, Y. [1978]: Au dela des opérateurs pseudodifférentiels. Astérisque 57, 2–185. Zbl.483.35082

Coifman, R.R., Meyer, Y., Stein, E.M. [1983]: Un nouvel espace fonctionnel adapté à l'étude des opérateurs définis par des intégrales singulières. Harmonic Analysis, Lect. Notes Math. 992, 1–15, Springer-Verlag: Berlin, Heidelberg, New York, Tokyo. Zbl.523.42016

Cordes, H.O. [1965]: The algebra of singular integral operators in R^n. J. Math. Mech. 14 (6), 1007–1032. Zbl.137,101

Cordes, H.O. [1979]: Elliptic Pseudo-Differential Operators—An Abstract Theory. Lect. Notes Math. 756, Springer-Verlag: Berlin, Heidelberg, New York. Zbl.417.35004

Costabel, M. [1980]: An inverse for the Gohberg-Krupnik symbol map. Proc. R. Soc. Edinb., Sect. A 87, 153–165. Zbl.459.45013

---

*For the convenience of the reader, references to reviews in Zentralblatt für Mathematik (Zbl.), compiled using the MATH database, and Jahrbuch über die Fortschritte der Mathematik (Jrb.) have, as far as possible, been included in this bibliography.

Danilyuk, I.I. [1975]: Nonregular Boundary Value Problems in the Plane. Nauka: Moscow. Zbl.302.45007

Douglas, R.G. [1972]: Banach Algebra Techniques in Operator Theory. Academic Press: New York, London. Zbl.247.47001

Douglas, R.G. [1973]: Banach Algebra Techniques in the Theory of Toeplitz Operators. C.B.M.S. Reg. Conf. Ser. Math., Vol. 15, Am. Math. Soc.: Providence, R.I. Zbl.252.47025

Duduchava, R.V. [1979]: Integral Equations with Fixed Singularities. Teubner-Texte zur Mathematik, B.G. Teubner Verlagsgesellschaft: Leipzig. Zbl.439.45002

Dybin, V.B. [1988]: Correct problems for singular integral equations. Rostov Univ. Press: Rostov

Dyn'kin, E.M., Osilenker, B.P. [1983]: Weighted norm estimates of singular integrals and their applications. J. Sov. Math. *30*, 2094–2154, 1985 (Russian original: Itogi Nauki Tekh., Ser. Mat. Anal. *21*, 42–129, 1983). Zbl.568.42009

Elschner, J. [1985a]: Singular Ordinary Differential Operators and Pseudodifferential Equations. Akademie-Verlag: Berlin and Lect. Notes Math. 1128, Springer-Verlag: Berlin, Heidelberg, New York, Tokyo. Zbl.571.47038

Elschner, J. [1985b]: Asymptotics of Solutions to Pseudodifferential Equations of Mellin Type. Math. Nachr. *130*, 267–305, 1987. Zbl.663.35100

Eskin, G.I. [1973]: Boundary Value Problems for Elliptic Pseudodifferential Equations. Transl. Math. Mon., Vol. 52, Am. Math. Soc.: Providence, R.I., 1981 (Russian original: Nauka: Moscow, 1973). Zbl.262.35001, Zbl.458.35002

Fefferman, C. [1974]: Recent progress in classical Fourier analysis. Proc. Int. Congr. Math. Vancouver 1974, Vol. I, 95–118. Zbl.332.42021

Fenyö, S., Stolle, H.W. [1982–1984]: Theorie und Praxis der linearen Integralgleichungen. Bände 1–4, VEB Deutscher Verlag der Wissenschaften: Berlin. Zbl.487.45002, Zbl.504.45002, Zbl.534.45002, Zbl.534.45004.

Fichera, G. [1958]: Una introduzione a la teoria delle equazioni integrali singolari. Rend. Mat. Appl., V. Ser. *17* (1–2), 82–191

Fredholm, I. [1903]: Sur une classe d'équations fonctionelles. Acta Math. *27*, 365–390. Jrb.34,422

Gabdulkhaev, B.G. [1980]: Optimal Approximations of Solutions of Linear Problems. Kazan Univ.: Kazan

Gakhov, F.D. [1977]: Boundary Value Problems. Pergamon Press: Oxford, 1966 (Most recent Russian edition: Nauka: Moscow, 1977). Zbl.141,80, Zbl.449.30030

Garcia-Cuerva, J., Rubio de Francia, J.L. [1985]: Weighted Norm Inequalities and Related Topics. North-Holland: Amsterdam, New York, Oxford. Zbl.578.46046

Giraud, G. [1934]: Equations à intégrales principales. Ann. Sci. Ec. Norm. Supér. III. Ser. *51*, 251–372. Zbl.11,216

Giraud, G. [1936]: Sur une classe générale d'équations à intégrales principales. C.R. Acad. Sci., Paris *202* (26), 2124–2127. Zbl.14,309

Gohberg, I., Feldman, I.A. [1971]: Convolution Equations and Projection Methods for Their Solution. Transl. Math. Mon., Vol. 41, Am. Math. Soc.: Providence, R.I., 1974 (Russian original: Nauka: Moscow, 1971). Zbl.214,385

Gohberg, I., Kaashoek, M.A. (eds.) [1986]: Constructive Methods of Wiener-Hopf Factorization. Operator Theory: Advances and Applications, Vol. 21, Birkhäuser Verlag: Basel, Boston, Stuttgart. Zbl.612.47025

Gohberg, I., Krein, M.G. [1957]: The fundamentals on defect numbers, root numbers, and indices of linear operators. Am. Math. Soc., Transl. II. Ser. *13*, 185–264, 1960 (Russian original: Usp. Mat. Nauk *12* (2), 43–118, 1957). Zbl.88,321

Gohberg, I., Krein, M.G. [1965]: Introduction to the Theory of Linear Nonselfadjoint Operators in Hilbert Space. Transl. Math. Mon., Vol. 18, Am. Math. Soc.: Providence, R.I., 1969 (Russian original: Nauka: Moscow, 1965). Zbl.138,78

Gohberg, I., Krupnik, N.Ya. [1973]: Einführung in die Theorie der eindimensionalen singulären Integraloperatoren. Birkhäuser Verlag: Basel, Stuttgart, 1979 (Russian original: Shtiintsa: Kishinev, 1973). Zbl.271.47017

Grubb, G. [1986]: Functional Calculus of Pseudo-Differential Boundary Problems. Birkhäuser Verlag: Basel, Boston, Stuttgart. Zbl.622.35001

Guillemin, V. [1984]: Toeplitz operators in n dimensions. Integral Equations Oper. Theory 7 (2), 145–205. Zbl.561.47025

Heinig, G., Rost, K. [1984]: Algebraic Methods for Toeplitz-like Matrices and Operators. Akademie-Verlag: Berlin and Birkhäuser Verlag: Basel, Boston, Stuttgart. Zbl.549.15013

Hellinger, E., Toeplitz, O. [1928]: Integralgleichungen und Gleichungen mit unendlich vielen Unbekannten. In Enzyklopädie der Math. Wiss., II C 13, Verlag B.G. Teubner: Leipzig, Berlin. Jrb.53,350

Hilbert, D. [1912]: Grundzüge einer allgemeinen Theorie der linearen Integralgleichungen. Verlag B.G. Teubner: Leipzig, Berlin

Hörmander, L. [1965]: Pseudo-differential operators. Commun. Pure Appl. Math. 18 (3), 501–517. Zbl.125,334

Hörmander, L. [1985a]: The Analysis of Linear Partial Differential Operators. III. Pseudo-Differential Operators. Springer-Verlag: Berlin, Heidelberg, New York, Tokyo. Zbl.601.35001

Hörmander, L. [1985b]: The Analysis of Linear Partial Differential Operators. IV. Fourier Integral Operators. Springer-Verlag: Berlin, Heidelberg, New York, Tokyo. Zbl.612.35001

Imanaliev, M.I. [1981]: Generalized Solutions of Integral Equations of the First Kind. Ilim: Frunze

Imanaliev, M.I., Khvedelidze, B.V., Gegelia, T.G., Babaev, A.A., Botashev, A.I. [1982]: Integral equations. Differ. Equations 18, 1442–1458, 1983 (Russian original: Differ. Uravn. 18 (12), 2050–2069, 1982). Zbl.517.45001

Ivanov, V.V. [1968]: The Theory of Approximation Methods and their Applications to the Numerical Solution of Singular Integral Equations. Noordhoff Int. Publ.: Leyden, 1976 (Russian original: Naukova Dumka: Kiev, 1968). Zbl.167,161

Jörgens, K. [1970]: Lineare Integraloperatoren. Verlag B.G. Teubner: Stuttgart. Zbl.207,446. English transl.: Boston, London, Melbourne (1982)

Journé, J.-L. [1983]: Calderón-Zygmund Operators, Pseudodifferential Operators and the Cauchy Integral of Calderón. Lect. Notes Math. 994, Springer-Verlag: Berlin, Heidelberg, New York, Tokyo. Zbl.508.42021

Kantorovich, L.V., Akilov, G.P. [1982]: Functional Analysis. Pergamon Press: Oxford (Most recent Russian edition: Nauka: Moscow, 1984). Zbl.127,61, Zbl.484.46003

Karlovich, Yu.I., Kravchenko, B.G., Litvinchuk, G.S. [1983]: Noether theory of singular integral operators with a shift. Sov. Math. 27 (4), 1–34 (Russian original: Izv. Vyssh. Uchebn. Zaved., Mat. 4, 3–27, 1983). Zbl.522.45012

Karlovich, Yu.I., Kravchenko, B.G., Litvinchuk, G.S. [1990]: On Noethericity and Mikhlin's symbols of operators of singular integral operators type with shift. Z. Anal. Anwend. 9 (1), 15–32.

Khvedelidze, B.V. [1956]: Linear discontinuous boundary value problems of function theory, singular integral equations and some of their applications. Tr. Tbilis. Mat. Inst. Razmadze 23, 3–158. Zbl.83,300

Khvedelidze, B.V. [1975]: The method of Cauchy type integrals for discontinuous boundary value problems in the theory of holomorphic functions of one complex variable. J. Sov. Math. 7 (3), 309–415, 1977 (Russian original: Itogi Nauki Tekh., Ser. Sovrem. Probl. Mat. 7, 5–162, 1975). Zbl.406.30034

Khvedelidze, B.V. [1983]: Integral equations. In: Essays on the Development of Mathematics in the USSR, 444–460. Naukova Dumka: Kiev

Kohn, J.J., Nirenberg, L. [1965]: An algebra of pseudodifferential operators. Commun. Pure Appl. Math. 18 (1–2), 269–305. Zbl.171,351

Kosel, U., von Wolfersdorf, L. [1986]: Nichtlineare Integralgleichungen. In: Semin. Anal. 1985/86, S. Prössdorf, B. Silbermann (eds.), 93–128. Zbl. 603.45007

Krein, M.G. [1958]: Integral equations on a half-line with kernel depending upon the difference of the arguments. Am. Math. Soc., Transl. II. Ser. 22, 163–288, 1962 (Russian original: Usp. Mat. Nauk 13 (5), 3–120, 1958). Zbl.88,309

Krupnik, N.Ya. [1984]: Banach Algebras with a Symbol and Singular Integral Operators. Operator Theory: Advances and Applications, Vol. 26, Birkhäuser Verlag: Basel, Boston, Stuttgart, 1987 (Russian original: Shtiintsa: Kishinev, 1984). Zbl.641.47031

Kumano-go, H. [1981]: Pseudo-Differential Operators. M.I.T. Press: Cambridge. Zbl.489.35003

Larsen, R. [1971]: An Introduction to the Theory of Multipliers. Springer-Verlag: Berlin, Heidelberg, New York. Zbl.213,133

Litvinchuk, G.S. [1977]: Boundary Value Problems and Singular Integral Equations with a Shift. Nauka: Moscow. Zbl.462.30029

Litvinchuk, G.S., Spitkovskiĭ, I.M. [1987]: Factorization of Measurable Matrix Functions. Akademie-Verlag: Berlin and Birkhäuser Verlag: Basel, Boston, Stuttgart. Zbl.651.47010

Malyshev, V.A. [1975]: Wiener-Hopf equations and their applications in probability theory. J. Sov. Math. 7 (2), 129–148, 1977 (Russian original: Itogi Nauki Tekh., Ser. Teor. Veroyatn. Mat. Stat. Teor. Kibern. *13*, 5–36, 1975). Zbl.451.60047

Maz'ya, V.G. [1985]: Sobolev Spaces. Springer-Verlag: Berlin, Heidelberg, New York, Tokyo (Russian edition: Leningrad State Univ.: Leningrad, 1985)

Maz'ya, V.G., Shaposhnikova, T.O. [1985]: Theory of Multipliers in Spaces of Differentiable Functions. Pitman: Boston, London, Melbourne (Russian edition: Leningrad State Univ.: Leningrad, 1986). Zbl.645.46031

Meister, E. [1973]: Das Riemannsche Randwertproblem—Ergebnisse und Anwendungen. In: Überblicke Math. 6, 113–178. Zbl.329.30032

Meister, E. [1987]: Einige Klassen singulärer Integral—und Integro-Differentialgleichungen auf der Halbachse. Preprint Nr. 1090, Technische Hochschule Darmstadt

Meister, E., Speck, F.-O. [1979]: Some multidimensional Wiener-Hopf equations with applications. In: Trends in Applications of Pure Mathematics to Mechanics, Vol. II, Pitman: London 217–262. Zbl.415.45001

Meister, E., Speck, F.-O. [1980]: Some classes of integral and integro-differential equations of convolution type. In: Proc. Conf. Ordinary and Partial Diff. Equations, Dundee 1978, Lect. Notes Math. 827, 182–228, Springer-Verlag: Berlin, Heidelberg, New York. Zbl.444.45005

Meyer, Y. [1987]: Wavelets and Operators. Preprint, Ceremade, Université de Paris Dauphine. Published in Lond. Math. Soc. Lect. Note. Ser. 137, 256–365 (1989)

Mikhlin, S.G. [1936]: Singular integral equations with two independent variables. Mat. Sb. Nov. Ser. *1* (4), 535–551 and *1* (6), 963–964. Zbl.16,29

Mikhlin, S.G. [1948]: Singular integral equations. Usp. Mat. Nauk *3* (3), 29–112

Mikhlin, S.G. [1959]: Vorlesungen über lineare Integralgleichungen. VEB Deutscher Verlag der Wissenschaften: Berlin, 1962 (Russian original: Fizmatgiz: Moscow, 1959). Zbl.87,100

Mikhlin, S.G. [1962]: Multidimensional Singular Integrals and Integral Equations. Pergamon Press: New York, 1965 (Russian original: Fizmatgiz: Moscow, 1962). Zbl.105,303

Mikhlin, S.G., Prössdorf, S. [1980]: Singular Integral Operators. Springer-Verlag: Berlin, Heidelberg, New York, Tokyo, 1986 (German original: Akademie-Verlag: Berlin, 1980). Zbl.442.47027

Murai, T. [1988]: A real variable method for the Cauchy transform and applications to analytic capacity. Lect. Notes Math. 1307, Springer-Verlag: Berlin, Heidelberg, New York, Tokyo. Zbl.645.30016

Muskhelishvili, N.I. [1968]: Singular Integral Equations. Nordhoff: Groningen, 1953 (Last Russian edition: Nauka: Moscow, 1968). Zbl.51,332, Zbl.174,162

Naimark, M.A. [1968]: Normed Rings. Nordhoff: Groningen, 1959 (Last Russian edition: Nauka: Moscow, 1968). Zbl.89,101

Nikol'skiĭ, N.K. [1980]: Treatise on the Shift Operator. Springer-Verlag: Berlin, Heidelberg, New York, Tokyo, 1986 (Russian original: Nauka: Moscow, 1980). Zbl.508.47001

Noether, F. [1920]: Über eine Klasse singulärer Integralgleichungen. Math. Ann. *82*, 42–63. Jrb.47,369

Okikiolu, G.O. [1971]: Aspects of the Theory of Bounded Integral Operators in L^p-Spaces. Academic Press: New York. Zbl.219,440

Palais, R.S. [1965]: Seminar on the Atiyah-Singer Index Theorem. Princeton Univ. Press: Princeton, New Jersey, 1965 (Russian transl.: Mir: Moscow, 1970). Zbl.137,170

Pietsch, A. [1978]: Operator Ideals. VEB Deutscher Verlag der Wissenschaften: Berlin, 1978 (Russian transl.: Mir: Moscow, 1982). Zbl.399.47039

Pietsch, A. [1980]: Weyl numbers and eigenvalues of operators in Banach spaces. Math. Ann. *247*, 149–168. Zbl.428.47027

Pietsch, A. [1980, 1983]: Eigenvalues of integral operators. I.II. Math. Ann. *247*, 169–178, 1980 and *262*, 343–376, 1983. Zbl.428.47028 and 488.47025

Pietsch, A. [1987]: Eigenvalues and s-Numbers. Akademische Verlagsgesellschaft Geest & Portig: Leipzig. Zbl.615.47019

Plamenevskiĭ, B.A. [1986]: Algebras of Pseudodifferential Operators. Nauka: Moscow. Zbl.615.47038

Prössdorf, S. [1974]: Some Classes of Singular Equations. North-Holland: Amsterdam, New York, Oxford, 1978 (German original: Akademie-Verlag: Berlin, 1974; Russian transl.: Mir: Moscow, 1979). Zbl.302.45009

Prössdorf, S., Silbermann, B. [1977]: Projektionsverfahren und die näherungsweise Lösung singulärer Gleichungen. Teubner-Texte zur Mathematik, B.G. Teubner Verlagsgesellschaft: Leipzig. Zbl.364.65044

Prössdorf, S. Silbermann, B. [1990]: Numerical Analysis for Integral and Operator Equations. Akademie-Verlag:Berlin, and Birkhäuser-Verlag: Basel, Boston, Stuttgart

Przeworska-Rolewicz, D. [1973]: Equations with Transformed Argument—An Algebraic Approach. Elsevier Publ. Comp.: Amsterdam. Zbl.271.47008

Rempel, S., Schulze, B.-W. [1982]: Index Theory of Elliptic Boundary Problems. Akademie-Verlag: Berlin, 1982 (Russian transl.: Mir: Moscow 1986). Zbl.504.35002

Riesz, F., Sz.-Nagy, B. [1952]: Leçons d'Analyse Fonctionelle. Acad. Kiadó: Budapest, 1952 (English Transl.: Ungar: New York, 1955: Russian transl.: Mir: Moscow, 1979). Zbl.51,84

Roch, S., Silbermann, B. [1988]: Algebras generated by idempotents and the symbol calculus for singular integral operators. Integral Equations Oper. Theory *11*, No. 3, 385–419. Zbl.658.47050

Roch, S., Silbermann, B. [1990]: Algebras of convolution operators and their image in the Calkin algebra. Report R-MATH-05/90, Akad. Wiss. DDR, Karl-Weierstrass-Inst. Math.: Berlin

Rodin, Yu.L. [1988]: Generalized Analytic Functions on Riemann Surfaces. Lect. Notes Math. 1288, Springer-Verlag: Berlin, Heidelberg, New York, Tokyo. Zbl.637.30041

Roozemond, L. [1987]: Systems of Non-Normal and First Kind Wiener-Hopf Equations. Free University Press: Amsterdam

Sakhnovich, L.A. [1980]: Equations with a difference kernel on a finite interval. Russ. Math. Surv. *35* (4), 81–152 (Russian original: Usp. Mat. Nauk *35* (4), 69–129). Zbl.444.45008

Schaefer, H.H. [1971]: Topological Vector Spaces. Springer-Verlag: Berlin, Heidelberg, New York. Zbl.141,305

Schmeidler, W. [1950]: Integralgleichungen mit Anwendungen in Physik und Technik. Akademische Verlagsgesellschaft: Leipzig. Zbl.35,349

Seeley, R.T. [1965]: Integro-differential operators on vector bundles. Trans. Am. Math. Soc. *117* (5), 167–204. Zbl.133,371

Seeley, R.T. [1967]: Elliptic singular integral equations. Proc. Symp. Pure Math. *10*, 308–315. Zbl.177,386

Shubin, M.A. [1978]: Pseudodifferential Operators and Spectral Theory. Springer-Verlag: Berlin, Heidelberg, New York, Tokyo, 1987 (Russian original: Nauka, Moscow, 1978). Zbl.451.47064

Simonenko, I.B. [1965]: A new general method for investigating linear operator equations of the type of singular integral equations. I.II. Izv. Akad. Nauk SSSR, Ser. Mat. *29*, 567–586 and 757–782. Zbl.146,131

Simonenko, I.B., Chin' Ngok Min' [1986]: Local method in the theory of one-dimensional singular integral equations with piecewise continuous coefficients. Noethericity. Rostov Univ. Press: Rostov

Speck, F.-O. [1985]: General Wiener-Hopf Factorization Methods. Pitman: Boston, London, Melbourne. Zbl.588.35090

Stein, E.M. [1970]: Singular Integrals and Differentiability Properties of Functions. Princeton Univ. Press: Princeton, New Jersey (Russian transl.: Mir: Moscow, 1973). Zbl.207,135

Stein, E.M., Weiss, G. [1971]: Introduction to Fourier Analysis on Euclidean Spaces. Princeton Univ. Press: Princeton, New Jersey (Russian transl.: Mir: Moscow, 1974). Zbl.232.42007

Talenti, G. [1973]: Sulle equazioni integrali di Wiener-Hopf. Boll. Unione Mat. Ital., IV. Ser. 7, Suppl. al Fasc. 1, 18–118. Zbl.281.45007

Taylor, M.E. [1981]: Pseudodifferential Operators. Princeton Univ. Press: Princeton, New Jersey (Russian transl.: Mir: Moscow, 1985). Zbl.453.47026

Treves, F. [1980]: Introduction to Pseudodifferential and Fourier Integral Operators. Vols. 1, 2. Plenum Press: New York, London (Russian transl.: Mir: Moscow, 1984). Zbl.453.47027

Tricomi, F. [1926]: Formula d'inversione dell'ordine di due integrazioni doppio "con asterisco". Rend. Accad. Naz. Lincei, III. Ser. 6a, 9, 535–539. Jrb.52,235

Tricomi, F. [1927]: Equazioni integrale contenenti il valor principale di un integrale doppio. Math. Z. 27, 87–133. Jrb. 53,359

Tricomi, F. [1957]: Integral Equations. Interscience Publishers: New York, London (Russian transl.: Inostr. Lit.: Moscow, 1960). Zbl.78,94

Vekua, N.P. [1970]: Systems of Singular Integral Equations. Groningen (1967) (Russian original: Nauka: Moscow, 1970). Zbl.41,429

Volterra, V. [1913]: Leçons sur le Équations Intégrales et les Équations Intégro-Différentielles. Gauthier-Villars: Paris

Widom, H. [1985]: Asymptotic Expansions for Pseudodifferential Operators on Bounded domains. Lect. Notes Math., Vol. 1152, Springer-Verlag: Berlin, Heidelberg, New York, Tokyo. Zbl.582.35002

Wiener, N., Hopf, E. [1931]: Über eine Klasse singulärer Integralgleichungen. Sitzungsber. Preuss. Akad. Wiss. 30/32, 696–706. Zbl.3,307

von Wolfersdorf, L. [1974]: Integralgleichungen. In: Entwicklung Math. DDR, 453–485. Zbl.292.45001

Zabreĭko, P.P., Koshelev, A.I., Krasnosel'skiĭ, M.A., Mikhlin, S.G., Rakovshchik, L.S., Stetsenko, B.Ya. [1968]: Integral Equations. Noordhoff International Publishing, 1975 (Russian original: Nauka: Moscow, 1968). Zbl.159,410

II. Boundary Integral Equations

V.G. Maz'ya

Contents

Introduction

This article is devoted to boundary integral equations and their application to the solution of boundary and initial-boundary value problems for partial differential equations.

The essence of the boundary integral equations method for solving boundary value problems consists in the following. Its foundation lies in specific representations of the solutions of the differential equations as integrals over the boundary involving one or more unknown functions. To determine these functions one makes use of the boundary conditions. Substituting the afore-mentioned representations into the boundary operators converts the boundary conditions into integral (and occasionally also into integro-differential) equations. By solving the latter equations, one obtains the solution of the original boundary value problem in the form of an integral.

The application of integral equations to the theory of harmonic functions traces back to the pioneering work of C. Neumann, H. Poincaré, G. Robin, O. Hölder, A.M. Lyapunov, V.A. Steklov, and I. Fredholm. The further development of the method of boundary equations is tied in with the names of T. Carleman, J. Radon, G. Giraud, N.I. Muskhelishvili, V.D. Kupradze, S.G. Mikhlin, A.P. Calderón, A. Zygmund, to mention only a few of the principal figures. Detailed accounts of the history of the topic can be found in Sologub [1975], Kupradze et al. [1976], Lonseth [1977], Král [1980], Crouch, Starfield [1983], Burchuladze, Gegelia [1985]. The term "boundary integral equation" itself appeared only quite recently and has now superseded the former name "integral equation of potential theory".

We here merely dwell on the theoretical aspects of the matter. So the problem of the numerical realization of the boundary integral equations methods, which has been becoming of great importance during the last few years, is left aside.

The article consists of five chapters. In the first chapter we illuminate the physical background of the theory of harmonic potentials and present, in a systematic way, the basic properties of the integral equations arising in this theory for domains with boundaries of the class $C^{1,\alpha}$. Chapter 2 concerns similar questions for the linear elliptic (systems of) equations of static elasticity and hydromechanics. The third chapter aims at elucidating one of the two modern tendencies in the development of the boundary integral equations method, viz the extension of the method to new classes of boundary and initial-boundary value problems of mathematical physics. We do not claim completeness in this connection and so content ourselves to the oblique derivative problem for Laplace's equation, to the biharmonic equation, to the initial-boundary value problems of heat conduction and wave propagation, and to nonstationary problems of elasticity and visco-elasticity. Chapter 3 concludes by a few remarks and references to the literature on some lines of development which are intimately connected with our subject but could not be included for lack of space.

The other tendency to put forward the boundary equations method is discussed throughout the Chapters 4 and 5. It has its source in the investigations of Carleman and Radon and has now culminated in a powerful machinery for treating the classical equations on nonsmooth surfaces. Note that this direction was given the deciding impetus in very recent time. Chapter 4 concentrates upon the theory of some of the integral equations alluded to above in the spaces C and L_p on nonsmooth curves and surfaces of special classes. The same equations over piecewise smooth surfaces are studied in Chapter 5. The results of this chapter are obtained by means of a unified approach, which consists in reducing boundary integral equations to certain auxiliary boundary value problems.

I wish to express my deepest gratitude to J. Král, N.V. Grachev, G.I. Kresin, A.A. Solov'ev, N.M. Khutoryanskiĭ and T.O. Shaposhnikova, who supplied me with some material. I am especially indebted to N.G. Kuznetsov for his invaluable support at all stages of my work on this article.

Chapter 1
Theory of Harmonic Potentials

This chapter is devoted to classical results on the boundary integral equations generated by the boundary value problems for the Laplace equation. The first section contains some necessary information about simple and double layer potentials. The solvability of boundary integral equations is studied in the second section, and the spectral properties of boundary operators are the subject of the third section. In the fourth section, some more methods of reducing boundary value problems to integral equations of the first and second kinds are considered.

§ 1. Basic Concepts of the Theory of Harmonic Potentials

1.1. The Physical Background of the Method of Boundary Integral Equations. This method originates from those problems of physics and mechanics which gave birth the basic concepts of potential theory. The concept of a (harmonic) potential first appeared in connection with Newton's law of universal gravitation.

Assume we are given a particle of mass m_ξ located at a point $\xi = (\xi_1, \xi_2, \xi_3) \in \mathbb{R}^3$. This particle attracts a particle of mass m_x placed at a point $x = (x_1, x_2, x_3)$ with a force $\mathbf{F}$ whose projections onto the axes of a Cartesian system of coordinates are

$$F_i = G \frac{m_x m_\xi}{r^2} \frac{\xi_i - x_i}{r}; \qquad i = 1, 2, 3.$$

Here $r = [\sum_{i=1}^{3} (\xi_i - x_i)^2]^{1/2}$, and G is some positive constant. By choosing an appropriate system of units, we may assume that G equals 1. The components of the force of attraction are partial derivatives:

$$F_i = Gm_x \frac{\partial}{\partial x_i} \left(\frac{m_\xi}{r} \right); \qquad i = 1, 2, 3;$$

the function $U(x) = m_\xi r^{-1}$, defined on $\mathbb{R}^3 \setminus \{\xi\}$, is usually called the *potential* of the mass m_ξ concentrated at the point ξ. The potential satisfies the Laplace equation,

$$\Delta U = 0 \quad \text{in} \quad \mathbb{R}^3 \setminus \{\xi\},$$

where $\Delta U = \sum_{i=1}^{3} \partial^2 U / \partial x_i^2$.

Coulomb's law, which describes the field of electrostatic forces, can be expressed by a formula which, up to magnitude and sign of the constant G, coincides with Newton's law of gravitation. Furthermore, the quantities m_x and m_ξ, being electric charges in this context, may attain any real value (unlike the mass, which is always positive). A law of the same kind also holds in magnetostatics.

That a field (for the sake of definiteness assume we are given the gravitational field due to a point mass m_ξ) possesses a potential has an important consequence for the work done in moving a point mass m. The work done on overcoming the gravitational field when moving the particle m from a point x_1 to a point x_2 along a path γ equals

$$\int_\gamma \sum_{i=1}^{3} F_i dx_i = Gm[U(x_2) - U(x_1)],$$

and thus it does not depend on the path γ. Note that the latter property is often taken as the definition of what a potential field is.

In the case where an attracting mass is distributed over some body occupying a domain $\Omega \subset \mathbb{R}^3$ and having mass density ϱ, we infer from the superposition principle that the potential is given by

$$(U\varrho)(x) = \int_\Omega \frac{\varrho(\xi)}{r} d\xi.$$

However, in connection with the method of boundary integral equations, of much more import is the case of field sources distributed over a surface. In this case we have a so-called *simple* (or *single*) *layer potential*:

$$(V\varrho)(x) = \int_S \frac{\varrho(\xi)}{r} d_\xi S.$$

Here S is a surface in $\mathbb{R}^3$ and ϱ denotes density with respect to area measure on the surface. The property of the potential $V\varrho$ answering for its usefulness in connection with boundary equations is already revealed by some simple examples, in which the integral can be evaluated straightforwardly.

Let $S_0 = \{x \in \mathbb{R}^3 : |x| = R\}$ be a sphere centered at the origin and suppose ϱ is constant on S_0. Then

$$(V\varrho)(x) = \begin{cases} 4\pi\varrho R & \text{for } |x| \leq R, \\ 4\pi\varrho R^2/|x| & \text{for } |x| \geq R. \end{cases}$$

If $S_1 = \{x \in \mathbb{R}^3 : x_3 = 0, x_1^2 + x_2^2 \leq R^2\}$ is a disk of radius R and ϱ is constant on S_1, then, for $x^{(0)} = (0, 0, x_3)$.

$$(V\varrho)(x^{(0)}) = 2\pi\varrho[-|x_3| + (R^2 + x_3^2)^{1/2}].$$

Both examples show that when x is passing through the surface S, the simple layer potential $(V\varrho)(x)$ is continuous, whereas its derivative $(\partial V/\partial n)(x)$ in the direction of the normal n to S has a jump of size $4\pi\varrho$.

Besides the potential $V\varrho$, there is another potential of interest for the method of boundary equations. This is the so-called *double layer potential*

$$(W\chi)(x) = \int_S \chi(\xi) \frac{\partial}{\partial n_\xi} \frac{1}{r} d_\xi S.$$

The direction of the normal n is fixed by specifying which of the two sides of S is the "positive" one. In case S is a closed surface, we shall think of n as being directed outwards.

A physical object that can be described by the latter potential is the "Leiden jar", i.e. a triple layer condenser S whose middle layer is an insulator and the inner and outer layers of which are conductors. The latter two layers are assumed to have charges of equal magnitude but opposite sign (see Fig. 1). A single pair of opposite point charges lying above the same point of S is called a dipole and can be mathematically modelled as follows.

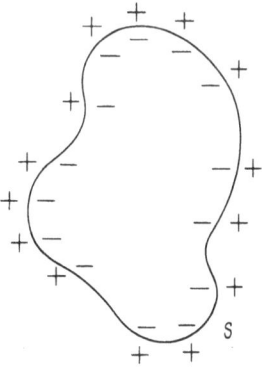

Fig. 1

Let $-\varepsilon^{-1}$ and $+\varepsilon^{-1}$ be two point charges located at the points ξ and ξ' separated by a distance ε from each other (see Fig. 2). By the superposition principle, their potential equals $\varepsilon^{-1}[(r')^{-1} - r^{-1}]$, where $r' = [\sum_{i=1}^3 (\xi_i' - x_i)^2]^{1/2}$.

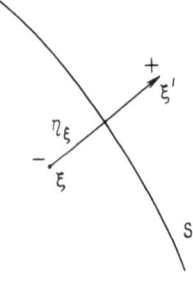

Fig. 2

As ε goes to zero, we arrive at a potential which is equal to $\dfrac{\partial}{\partial \mathbf{n}_\xi} \dfrac{1}{r}$. The so-called

dipole moment of this potential is 1; recall that the dipole moment is the product of the charge placed at the head of the vector determining the orientation of the dipole and the distance between the charges. The function χ figuring in the potential $W\chi$ characterizes the density of the distribution of the dipoles on the surface S.

Another physical interpretation of the double layer potential resorts to magnetostatics. In this context one thinks of χ as the density of the distribution of "micro-magnets" placed on the surface S in such a manner that their axes are parallel to the normal and, say, all North poles lie outside, while all South poles are located inside the surface S.

As in the case of a simple layer potential, the property of the double layer potential on which its application to boundary integral equations is based can be illustrated by some simple examples. If S_0 is the sphere introduced above and χ is constant on S_0, then

$$(W\chi)(x) = \begin{cases} 0 & \text{for } |x| > R, \\ -2\pi\chi & \text{for } |x| = R, \\ -4\pi\chi & \text{for } |x| < R. \end{cases}$$

If we are given dipoles which are distributed over the disk S_1 with constant density χ such that their direction is parallel to the x_3-axis, we have, for $x^{(0)} = (0, 0, x_3)$.

$$(W\chi)(x^{(0)}) = \begin{cases} 2\pi\chi |x_3| [x_3^{-1} - (R^2 + x_3^2)^{-1/2}] & \text{for } x_3 \neq 0, \\ 0 & \text{for } x_3 = 0 \end{cases}$$

Hence, both when x is moving on to and moving off the surface, $(W\chi)(x)$ has a jump of absolute size $2\pi\chi$.

In the case of arbitrary surfaces and densities the picture is analogous, although the sizes of the jumps then depend on the points of the surface. More precisely, the limit values of the double layer potential on a smooth surface S are given by

$$\pm 2\pi\chi(x) + \int_S \chi(\xi) \frac{\partial}{\partial \mathbf{n}_\xi} \frac{1}{r} d_\xi S, \qquad x \in S.$$

Here one has to take the sign plus if the surface is approached from the side into which the normal $\mathbf{n}_\xi$ is directed; in the opposite case the sign minus has to be taken. Note that in the above-quoted formula x is assumed to be an inner point of the surface S. The integral appearing in this formula is called the *direct value of the double layer potential*.

The expression for the boundary values of a double layer potential given above allows us to reduce the *Dirichlet problem for the Laplace equation* to an integral equation. Let us, for example, consider the *internal Dirichlet problem*. This problem consists in finding a function u which satisfies Laplace's equation $\Delta u = 0$ in some domain bounded by a closed smooth surface S, on which u is required to be equal to a given function f. Seeking the solution in the form of a potential $W\chi$ with unknown density χ, which guarantees that Laplace's equation is satisfied, the boundary condition yields the following equation for χ:

$$-2\pi\chi(x) + \int_S \chi(\xi) \frac{\partial}{\partial \mathbf{n}_\xi} \frac{1}{r} d_\xi S = f(x), \qquad x \in S.$$

In an analogous manner the *external Dirichlet problem* can be reduced to an integral equation.

In order to reduce the *internal* or *external Neumann problems*, in which $\partial u / \partial \mathbf{n}$ is required to take prescribed values on S, to an integral equation, the solution is sought in the form of a simple layer potential. As we have pointed out above, its derivative in the direction of the normal has jumps on S, which thus leads to a situation that is analogous to the case of the double layer potential.

We defer a more detailed discussion of the results listed above to the next subsection. We here only wish to mention that the double layer potential whose density is identically equal to 1 (this is a so-called *Gaussian integral*) has a simple visual geometric interpretation. Namely, it is the *solid angle* at which the surface S is seen from the point x, this angle being counted negatively if $\cos(x - \xi, \mathbf{n}_\xi) \geq 0$ and being taken positively in the opposite case. Note that this at once explains why $(W\chi)(x)$ is zero if x lies outside the shpere S_0: in this case two spherical segments are seen under the same solid angle but from different sides.

We conclude these introductory remarks by noting that there is an important special case of potential fields: the so-called plane fields. Such fields are in particular set up by a mass distributed with a density independent of x_3 on an infinite cylindrical surface parallel to the x_3-axis (see, for instance, Kellogg [1929]). This sort of fields is usually studied with the help of logarithmic potentials in the plane. In that case S is a plane curve (the curve generating the cylinder) and the role of the function $(-4\pi r)^{-1}$ is played by the function $(2\pi)^{-1} \log[(x_1 - \xi_1)^2 + (x_2 - \xi_2)^2]^{1/2}$. Below we shall investigate plane potential fields simultaneously with the fields in spaces whose dimension is greater than two.

1.2. Properties of Potentials. Before proceeding further, we need some notations and definitions. We let Ω^+ denote a bounded simply-connected domain (region) in $\mathbb{R}^n$ ($n \geq 2$) whose *boundary S is of the class $C^{1,\alpha}$* ($0 < \alpha < 1$), i.e., in Cartesian coordinates, S can be locally given by functions the first derivatives of which satisfy a Hölder condition with the exponent α. Put $\Omega^- = \mathbb{R}^n \setminus \Omega^+$. We denote by $x = (x_1, \ldots, x_n)$ and $\xi = (\xi_1, \ldots, \xi_n)$ points in $\mathbb{R}^n$.

Definition. A function $u^{\pm}$ of the class $C^2(\Omega^{\pm})$ is said to be *harmonic* in $\Omega^{\pm}$ if it satisfies Laplace's equation,

$$\Delta u^{\pm} = 0 \qquad \text{in } \Omega^{\pm},$$

and, in the case of u^-, the following requirement on the behavior at infinity is met:

$$u^-(x) = \begin{cases} O(1) & \text{if } n = 2, \\ o(1) & \text{if } n \geq 3 \end{cases} \qquad \text{as } |x| \to \infty.$$

Here $\Delta = \sum_{i=1}^n \partial^2 / \partial x_i^2$ and $|x| = (\sum_{i=1}^n x_i^2)^{1/2}$.

In the case $n \geq 3$, the asymptotic behavior of a harmonic function u^- and its gradient ∇u^- as $|x| \to \infty$ is as follows:

$$u^-(x) = O(|x|^{-n+2}), \qquad |\nabla u^-(x)| = O(|x|^{-n+1}). \tag{1.1}$$

A *fundamental solution of the Laplace equation* is a distribution $\mathscr{E}(x, \xi)$ such that

$$\Delta_x \mathscr{E} = \delta(x - \xi) \qquad \text{in } \mathbb{R}^n,$$

where $\delta(x - \xi)$ is the Dirac measure concentrated at the point ξ. We henceforth suppose that $\mathscr{E}(x, \xi)$ is the fundamental solution specified by

$$\mathscr{E}(x, \xi) = \begin{cases} (2\pi)^{-1} \log r & \text{for } n = 2, \\ -[(n-2)\sigma_n r^{n-2}]^{-1} & \text{for } n \geq 3, \end{cases}$$

where $r = |x - \xi|$ and $\sigma_n = 2\pi^{n/2}/\Gamma(n/2)$ is the area of the unit sphere in $\mathbb{R}^n$.

We let $\mathbf{n}_x$ (resp. $\mathbf{n}_\xi$) refer to the normal vector of length 1 to the surface S at the point $x \in S$ (resp. $\xi \in S$) which is directed outwards, i.e. in to Ω^-. Since S is a surface of the class $C^{1,\alpha}$, we have, for $(x, \xi) \in S \times S$ with $x \neq \xi$, the estimates

$$(\partial/\partial \mathbf{n}_x)\mathscr{E}(x, \xi), (\partial/\partial \mathbf{n}_\xi)\mathscr{E}(x, \xi) = O(r^{-n+1+\alpha}). \tag{1.2}$$

Definition. We say that a function $v^{\pm} \in C^1(\Omega^{\pm})$ has a *regular normal derivative* $\partial v^{\pm}/\partial \mathbf{n}_{\pm}$ on S if, uniformly in $x \in S$,

$$\frac{\partial v^{\pm}}{\partial \mathbf{n}_x}(x \mp t\mathbf{n}_x) \to \frac{\partial v^{\pm}}{\partial \mathbf{n}_{\pm}}(x) \qquad \text{as } t \to +0.$$

In that case clearly $\partial v^{\pm}/\partial \mathbf{n}_{\pm} \in C(S)$.

If a function $u^{\pm}$ which is harmonic in $\Omega^{\pm}$ belongs to $C(\Omega^{\pm})$ and has a regular normal derivative, then it can be represented in the form

$$u^{\pm}(x) = \pm \int_S \left[u^{\pm}(\xi) \frac{\partial}{\partial \mathbf{n}_\xi} \mathscr{E}(x, \xi) - \mathscr{E}(x, \xi) \frac{\partial u^{\pm}}{\partial \mathbf{n}_\xi} \right] d_\xi S, \qquad x \in \Omega^{\pm}. \tag{1.3}$$

The integrals

$$(V\varrho)(x) = \int_S \varrho(\xi)\mathscr{E}(x, \xi)\, d_\xi S, \tag{1.4}$$

$$(W\chi)(x) = \int_S \chi(\xi)\frac{\partial}{\partial \mathbf{n}_\xi}\mathscr{E}(x, \xi)\, d_\xi S, \tag{1.5}$$

depending on the parameter $x \in \mathbb{R}^n \backslash S$ and being prompted by (1.3), are called the *simple layer* and *double layer potential with density* ϱ and χ, respectively. We shall suppose that ϱ and χ are in $C(S)$.

In case $n \geq 3$, the potentials $V\varrho$ and $W\chi$ are harmonic in $\Omega^\pm$. Moreover,

$$(V\varrho)(x) = O(|x|^{-n+2}), \qquad (W\chi)(x) = O(|x|^{-n+1}) \tag{1.6}$$

as $|x| \to \infty$. On condition that $n = 2$, these properties of $W\chi$ as well as the property of $V\varrho$ to be harmonic in Ω^+ remain valid. However, in general the potential $V\varrho$ is no longer harmonic in Ω^-, because it may increase logarithmically at infinity. Note that $V\varrho$ is bounded in Ω^- if $\int_S \varrho\, dS = 0$.

We now proceed to those properties of potentials which form the foundation for reducing boundary value problems to boundary integral equations (proofs are, for example, in the books Kellogg [1929], Günter [1953], Mikhlin [1959], [1977], Vladimirov [1976], Steklov [1983]).

By virtue of estimate (1.2), a double layer potential is well-defined at the points $x \in S$. In this case its *value* is called *direct* and denoted by $(W_0\chi)(x)$. The function $W_0\chi$ is continuous throughout S.

The potential

$$(W1)(x) = \int_S \frac{\partial}{\partial \mathbf{n}_\xi}\mathscr{E}(x, \xi)\, d_\xi S \qquad (\chi = 1 \text{ on } S)$$

is referred to as a *Gaussian integral*. It can be evaluated explicitly:

$$(W1)(x) = \begin{cases} 1 & \text{for } x \in \Omega^+, \\ 0 & \text{for } x \in \Omega^-, \\ 1/2 & \text{for } x \in S. \end{cases} \tag{1.7}$$

Using the latter equalities, one can show the following result on the boundary values of double layer potentials with arbitrary density (*Theorem on the jumps of the potential* $W\chi$).

Theorem 1. *There exist functions* $W_{\pm}\chi \in C(S)$ *such that* $(W\chi)(x) \to (W_{\pm}\chi)(x_0)$ *as* $x \in \Omega^\pm$ *approaches* $x_0 \in S$. *The convergence of the potential* $W\chi$ *to the boundary values* $W_{\pm}\chi$ *is uniform in* $x_0 \in S$ *and the following equalities hold:*

$$W_{\pm}\chi = \pm \tfrac{1}{2}\chi + W_0\chi. \tag{1.8}$$

Corollary 1. *We have*

$$W_+\chi - W_-\chi = \chi. \tag{1.9}$$

A property of the simple layer potential $V\varrho$ that is worth stating separately is its continuity in $\mathbb{R}^n$. Moreover, owing to (1.2), the integral

$$\frac{\partial(V\varrho)}{\partial \mathbf{n}_0}(x) = \int_S \varrho(\xi)\frac{\partial}{\partial \mathbf{n}_x}\mathscr{E}(x, \xi)\, d_\xi S$$

is well-defined for every $x \in S$. This integral is called the *direct value of the normal derivative of the simple layer potential*. The function $\partial(V\varrho)/\partial \mathbf{n}_0$ is continuous on S.

Theorem 2. *The potential $V\varrho$ possesses a regular normal derivative $\partial(V\varrho)/\partial \mathbf{n}_\pm$ on S, which can be given by the formula*

$$\frac{\partial(V\varrho)}{\partial \mathbf{n}_\pm} = \mp\frac{1}{2}\varrho + \frac{\partial(V\varrho)}{\partial \mathbf{n}_0}. \tag{1.10}$$

Corollary 2. *We have*

$$\partial(V\varrho)/\partial \mathbf{n}_- - \partial(V\varrho)/\partial \mathbf{n}_+ = \varrho. \tag{1.11}$$

1.3. The Dirichlet and Neumann Problems in $\Omega^\pm$ and the Integral Equations for Them. A harmonic function $u^\pm \in C(\Omega^\pm)$ such that

$$u^\pm = \varphi^\pm \qquad \text{on } S, \tag{1.12}$$

where $\varphi^\pm \in C(S)$ is a given function, is said to be a *solution of the Dirichlet problem* in $\Omega^\pm$. It turns out that the Dirichlet problem has no more than one solution.

In order to reduce the Dirichlet problem to an integral equation, we seek its solution in the form of a potential (1.5) with unknown density $\chi^\pm$. We infer from Theorem 1 that this potential is a solution of the problem if and only if $\chi^\pm$ satisfies the equation

$$\chi^\pm \pm T\chi^\pm = \pm 2\varphi^\pm, \tag{1.13}$$

where T is the integral operator with the kernel $2(\partial/\partial\mathbf{n}_\xi)\mathscr{E}(x, \xi)$ (see (1.8) and (1.12)).

A *solution of the Neumann problem* in the classical sense is a harmonic function $u^\pm \in C(\overline{\Omega}^\pm)$ satisfying the condition

$$\partial u^\pm/\partial \mathbf{n}_\pm = \psi^\pm \qquad \text{on } S, \tag{1.14}$$

where $\psi^\pm \in C(S)$ is a given function; $\partial u_\pm/\partial \mathbf{n}_\pm$ refers to the regular normal derivative. Any two solutions of the Neumann problem in Ω^+ differ only by an additive constant, and if $n = 2$, then the same is also true for any two solutions in Ω^-. In the case $n > 2$, the Neumann problem in Ω^- has at most one solution.

The condition

$$\int_S \psi^+\, dS = 0 \tag{1.15}$$

is *necessary* for the Neumann problem in Ω^+ to be solvable. If $n = 2$, the solvability of this problem in Ω^- necessitates (1.15) with ψ^+ replaced by ψ^-.

The reduction of the Neumann problem in $\Omega^{\pm}$ to an integral equation is accomplished by looking for a solution in the form of the potential (1.4) with unknown density $\varrho^{\pm}$. Theorem 2 implies that $V\varrho^{\pm}$ is a solution of the boundary value problem if and only if $\varrho^{\pm}$ fulfils the equation

$$\varrho^{\pm} \mp T^*\varrho^{\pm} = \mp 2\psi^{\pm}, \tag{1.16}$$

T^* being the integral operator with the kernel $2(\partial/\partial\mathbf{n}_x)\mathscr{E}(x, \xi)$ (see (1.10) and (1.14)).

The estimates (1.2) show that T and T^* are integral operators with a weak kernel. Therefore T and T^* are compact on $L_2(S)$, and because, moreover, the kernels of these operators are real-valued and result from each other by replacing x with ξ, they are adjoint to each other. Consequently, the Fredholm theory can be applied to the equations (1.13) and (1.16) considered in $L_2(S)$ (see Chap. 2 of S. Prössdorf's article contained in this volume). The following result is also a consequence of the circumstance that the kernels of the operators T and T^* have a weak singularity.

Theorem 3 (see, e.g., § 14 of Mikhlin [1959]). *If* $\varphi^{\pm}$, $\psi^{\pm} \in C(S)$, *then every solution* $\chi^{\pm}$, $\varrho^{\pm} \in L_2(S)$ *of* (1.13), (1.16) *is actually continuous on S.*

§ 2. Solvability of Boundary Integral Equations

2.1. The Integral Equations of the Internal Dirichlet and External Neumann Problems in the Case $n \geq 3$

Theorem 4. *If* $n \geq 3$, *then each of the equations*

$$\chi^+ + T\chi^+ = 2\varphi^+, \qquad \varrho^- + T^*\varrho^- = 2\psi^- \tag{1.17}$$

has a unique solution in $L_2(S)$ *for any right-hand side.*

Proof. By the Fredholm alternative, the theorem will be proved once we have shown that at least one of the homogeneous versions of the two equations (1.17) has only the trivial solution.

Let $\varrho_0^- \in L_2(S)$ be any solution of the equation

$$\varrho_0^- + T^*\varrho_0^- = 0. \tag{1.18}$$

Theorem 3 implies that actually $\varrho_0^- \in C(S)$. Consider the potential $V\varrho_0^-$. Due to (1.18) and (1.10), we have

$$\partial(V\varrho_0^-)/\partial\mathbf{n} = 0 \text{ on } S. \tag{1.19}$$

From the uniqueness theorem for the external Neumann problem we deduce that $V\varrho_0^- = 0$ on $\overline{\Omega^-}$. So since $V\varrho_0^-$ is harmonic in Ω^+ and continuous throughout $\mathbb{R}^n$, the uniqueness theorem for the internal Dirichlet problem gives that $V\varrho_0^- = 0$ on $\overline{\Omega^+}$. Hence $\partial(V\varrho_0^-)/\partial\mathbf{n}_+ = 0$ on S. Combining this with (1.19) and taking into account (1.11) we see that $\varrho_0^- = 0$ on S. $\square$

Thus, the equations (1.17) are uniquely solvable in $L_2(S)$ for any right-hand side. Since the solutions are continuous whenever the right-hand side is in $C(S)$, we obtain the following result.

Theorem 5. *If $n \geq 3$, then the Dirichlet and Neumann problems in Ω^+ and Ω^-, respectively, are solvable for arbitrary boundary data from $C(S)$. The solution of the Dirichlet (Neumann) problem is representable by the double (simple) layer potential whose density χ^+ (ϱ^-) satisfies the first (second) of the equations (1.17).*

2.2. The Integral Equations of the External Dirichlet and Internal Neumann Problems in the Case $n \geq 2$. Due to formula (1.7), the equation $\chi_0^- - T\chi_0^- = 0$ has the nontrivial solution $\chi_0^- = 1$ on S.

Lemma 1. *The solutions of the equation*

$$\varrho^+ - T^*\varrho^+ = 0 \tag{1.20}$$

span a one-dimensional subspace of $L^2(S)$.

Proof. We here content ourselves to the case $n \geq 3$. The changes that have to be introduced into the arguments for $n = 2$ will be given in 2.4.

Let ϱ_0^+ and ϱ_1^+ be any nontrivial solutions of equation (1.20). By Theorem 3, these solutions are continuous. Consider the potentials $V\varrho_i^+$ ($i = 0, 1$). From (1.20) and (1.10) we have

$$\partial(V\varrho_i^+)/\partial \mathbf{n}_+ = 0 \qquad \text{on } S. \tag{1.21}$$

The uniqueness theorem for the internal Neumann problem implies that $V\varrho_i^+$ takes on a constant value c_i throughout $\overline{\Omega^+}$. We claim that $c_i \neq 0$. To see this, assume the contrary, i.e. that $c_i = 0$. Then, by the harmonicity of $V\varrho_i^+$ in Ω^- and the continuity of $V\varrho_i^+$ in $\mathbb{R}^n$ and also by the uniqueness theorem for the external Dirichlet problem, $V\varrho_i^+$ must vanish identically in $\overline{\Omega^-}$. So $\partial(V\varrho_i^+)/\partial \mathbf{n}_- = 0$ on S, which in conjunction with (1.21) and (1.11) gives that $\varrho_i^+ = 0$ on S. This, however, is a contradiction, because ϱ_i^+ was assumed to be a nontrivial solution of equation (1.20).

Note that in passing we showed that if the potential (1.4) vanishes identically in Ω^+, then its density is identically zero on S. We remark that this is not true in the case $n = 2$, since $\int_{|\xi|=1} \mathscr{E}(x, \xi) \, d_\xi S = 0$ for $|x| \leq 1$.

Now put $\varrho_2^+ = c_1 \varrho_0^+ - c_0 \varrho_1^+$. It is obvious that ϱ_2^+ also fulfils equation (1.20). Because $V\varrho_2^+ = c_1 V\varrho_0^+ - c_0 V\varrho_1^+$, we obtain that $V\varrho_2^+ = 0$ on $\overline{\Omega^+}$. By what has been proved above, $\varrho_2^+ = 0$ on S. So it follows that $\varrho_1^+ = \varrho_0^+ c_1/c_0$, which completes the proof. $\square$

The solution ϱ_0^+ of (1.20) satisfying $\int_S \varrho_0^+ \, dS = 1$ is called the *density of the Robin potential*. From the proof of Lemma 1 we infer that $V\varrho_0^+$ assumes a constant value $C \neq 0$ in Ω^+; the number C^{-1} is referred to as the *capacity of $\overline{\Omega^+}$*.

Now the Fredholm theory gives the following result.

Theorem 6. *For the equation*

$$\varrho^+ - T^*\varrho^+ = -2\psi^+ \tag{1.22}$$

to be solvable it is necessary and sufficient that the orthogonality condition (1.15) *be satisfied.*

If the equation (1.22) has a solution, then the Neumann problem in Ω^+ is solvable also. Taking into consideration what was said in 1.3 we thus arrive at the following conclusion.

Theorem 7. *For $n \geq 2$, the internal Neumann problem is solvable if and only if condition* (1.15) *is fulfilled. In that case the solution can be represented as the sum of an arbitrary constant and the potential* (1.4), *the density ϱ being determined from equation* (1.22).

The next result, pertaining to the external Dirichlet problem, is a consequence of the Fredholm alternative and of Lemma 1.

Theorem 8. *A necessary and sufficient condition for the solvability of the equation*

$$\chi^- - T\chi^- = -2\varphi^- \tag{1.23}$$

is that

$$\int_S \varphi^- \varrho_0^+ \, dS = 0, \tag{1.24}$$

where ϱ_0^+ is any nontrivial solution of equation (1.20). *Any two solutions of equation* (1.23) *differ by an additive constant only.*

The external Dirichlet Problem is solvable whenever φ^- obeys (1.24), *and its solution is representable by the potential* (1.5) *whose density satisfies equation* (1.23).

2.3. The Modified Integral Equation of the External Dirichlet Problem. If condition (1.24) is violated, then equation (1.23) does not possess a solution and consequently, the external Dirichlet problem has no solution representable as a double layer potential. Note that this is no surprise, because the potential (1.5) decreases at infinity more rapidly than any harmonic function (recall (1.1) and (1.6)).

Without loss of generality assume the origin lies in the domain Ω^+. Since then the function $|x|^{-n+2}$ is harmonic in Ω^-, the solution of the external Dirichlet problem may be sought in the form

$$u^-(x) = \int_S \chi^-(\xi) \frac{\partial}{\partial \mathbf{n}_\xi} \mathcal{E}(x, \xi) \, d_\xi S + \frac{1}{|x|^{n-2}} \int_S \chi^-(\xi) \, d_\xi S. \tag{1.25}$$

By the theorem on the boundary values of the double layer potential, χ^- has to satisfy the equation

$$\chi^- - T_1 \chi^- = -2\varphi^-, \tag{1.26}$$

where T_1 is the integral operator with the kernel

$$2 \left[|x|^{-n+2} + (\partial/\partial \mathbf{n}_\xi) \mathcal{E}(x, \xi) \right].$$

Let $\chi_0^- \in L_2(S)$ be the solution of the homogeneous version of (1.26), that is, of equation (1.26) with $\varphi^- = 0$ on S. Since the kernel of the operator T_1 has a weak singularity, χ_0^- belongs to $C(S)$ (see Theorem 3).

The function u_0^- given by formula (1.25) with $\chi^- = \chi_0^-$ is harmonic in Ω^- and we have $u_0^- = 0$ on S. We so deduce from the uniqueness theorem for the external Dirichlet problem that $u_0^- = 0$ on $\overline{\Omega}^-$. Hence, if $x \in \overline{\Omega}^-$ then

$$|x|^{n-2} \int_S \chi_0^-(\xi)(\partial/\partial\mathbf{n}_\xi)\mathscr{E}(x, \xi)\, d_\xi S = -\int_S \chi_0^-\, dS.$$

Passing to the limit $|x| \to \infty$ in this equality and taking into consideration (1.6), we obtain that $\int_S \chi_0^-\, dS = 0$. Thus, χ_0^- satisfies the equation $\chi_0^- = T\chi_0^-$ and so, by what was shown in 2.2, χ_0^- is constant on S. As the mean value of χ_0^- is zero, χ_0^- itself vanishes identically on S.

The result just established, along with the Fredholm alternative, yields the following.

Theorem 9. *Equation* (1.26) *is uniquely solvable in* $L_2(S)$.

Theorems 9 and 3 combine to obtain the next result.

Theorem 10. *If* $n \geq 2$, *then the external Dirichlet problem has a solution for every boundary function* φ^-. *The solution can be represented in the form* (1.25), *where* χ^- *is defined by equation* (1.26).

2.4. Remarks on the Boundary Integral Equations for Boundary Value Problems in the Plane

1) In the case $n = 2$ the proof of Theorem 4 differs from the proof given in 2.1 only in that now all one can say is that $V\varrho_0^-$ is constant on $\mathbb{R}^2$ (while in the case $n \geq 3$ we had $V\varrho_0^- = 0$ on $\mathbb{R}^n$). However, in the case at hand formula (1.11) can be applied to deduce that $\varrho_0^- = 0$ on S. Thus, both Theorem 4 and the part of Theorem 5 concerning the Dirichlet problem remain valid for $n = 2$.

2) There are right-hand sides ψ^- such that the potential (1.4), built up from the solution of the integral equation for the external Neumann problem, is not bounded in Ω^-. We therefore have to find a condition ensuring that $V\varrho$ is a solution of the external Neumann problem.

Lemma 2. *Let* ϱ^- *be a solution of the equation* $\varrho^- + T^*\varrho^- = 2\psi^-$, *where* ψ^- *is subject to the condition*

$$\int_S \psi^-\, dS = 0. \tag{1.27}$$

Then the potential (1.4) *with the density* ϱ^- *is bounded in* Ω^-.

Proof. Integrating the equation for ϱ^- over S and taking into account (1.27) gives that

$$\int_S \varrho^-\, dS + 2\int_S \int_S \varrho^-(\xi)\frac{\partial}{\partial\mathbf{n}_x}\mathscr{E}(x, \xi)\, d_\xi S d_x S = 0.$$

By (1.7), the second term on the left of the equality obtained is equal to $\int_S \varrho^- \, dS$. Consequently, $\int_S \varrho^- \, dS = 0$. In Ω^-, the potential (1.4) can be written in the form

$$(V\varrho^-)(x) = (2\pi)^{-1} \int_S \varrho^-(\xi) \log \frac{r}{|x|} \, d_\xi S. \tag{1.28}$$

Here, as above, the origin is assumed to belong to Ω^+. So it is clear that (1.28) is bounded in Ω^-. $\square$

Since condition (1.27) is necessary for the external Neumann problem to have a solution, we arrive at the following conclusion.

Theorem 11. *The Neumann problem in $\Omega^- \subset \mathbb{R}^2$ is solvable if and only if (1.27) is fulfilled. The solution can be represented in the form of a sum of an arbitrary constant and the potential (1.4) whose density satisfies the equation $\varrho^- + T^*\varrho^- = 2\psi^-$.*

3) In the case $n = 2$, the proof of Lemma 1 is modified as follows.

We first show that the integral $\int_S \varrho_i^+ \, dS$ is nonzero. In the opposite case we would have that $(V\varrho_i^+)(\infty) = 0$. Furthermore, by the uniqueness theorem for the internal Neumann problem and equation (1.20), $V\varrho_i^+$ is constant in Ω^+. Hence, due to the uniqueness theorem for the external Dirichlet problem, $V\varrho_i^+$ is constant throughout $\mathbb{R}^2$. So formula (1.11) implies that $\varrho_i^+ = 0$ on S, which contradicts the assumption that ϱ_i^+ be a nontrivial solution of (1.20).

To show that ϱ_0^+ and ϱ_1^+ are linearly dependent, we remark that

$$\int_S \left\{ \varrho_0^+ \left(\int_S \varrho_1^+ \, dS \right) - \varrho_1^+ \left(\int_S \varrho_0^+ \, dS \right) \right\} dS = 0,$$

and since, by what was proved above, the integral of a nontrivial solution of equation (1.20) is nonzero, we obtain that $\varrho_0^1 (\int_S \varrho_1^1 \, dS) - \varrho_1^1 (\int_S \varrho_0^1 \, dS) = 0$ on S. This completes the proof.

§3. Spectral Properties of Boundary Operators

This section concerns the spectral properties of the operators T and T^* on $L_2(S)$. Here, as above, $T = 2W_0$, where W_0 is the direct value of the potential (1.5), and T^* is the adjoint of the operator T.

Definition. Any value of the parameter $\lambda \in \mathbb{C}$ for which the equation

$$\chi - \lambda T\chi = 0 \tag{1.29}$$

has a nontrivial solution is called a *characteristic value* of the operator T. If λ is a characteristic value of T, then each nonzero function χ satisfying (1.29) is referred to as an *eigenfunction* of the operator T for λ.

We know from Fredholm theory that λ is a characteristic value of T if and only if $\bar{\lambda}$ is a characteristic value of T^*. The multiplicities of λ and $\bar{\lambda}$ coincide.

The results of Sec. 2 reveal that $\lambda = 1$ is a characteristic value (cf. (1.7) and (1.29)), whereas $\lambda = -1$ is not a characteristic value (see 2.1). The multiplicity of $\lambda = 1$ equals 1, and $\chi = 1$ is an eigenfunction for $\lambda = 1$.

Theorem 12. *If λ is a characteristic value of the operator T, then* (i) $\lambda \in \mathbb{R}$ *and* (ii) $|\lambda| \geq 1$.

Proof. We confine ourselves to the case $n \geq 3$. By virtue of (1.8) and (1.10), we have

$$(1 - \lambda)W_+\chi - (1 + \lambda)W_-\chi = \chi - \lambda T\chi, \tag{1.30}$$

$$(1 - \bar{\lambda})\frac{\partial(V\varrho)}{\partial \mathbf{n}_-} - (1 + \bar{\lambda})\frac{\partial(V\varrho)}{\partial \mathbf{n}_+} = \varrho - \bar{\lambda}T^*\varrho. \tag{1.31}$$

Now let λ be a characteristic value of T and write $\lambda = \alpha + i\beta$ with $\alpha, \beta \in \mathbb{R}$. Then there exists a function $\varrho = \varrho_1 + i\varrho_2 \neq 0$ such that the right-hand side of (1.31) vanishes. The potentials $V\varrho_i$ are continuously differentiable functions in $\Omega^{\pm}$ (see Kellogg [1929]). Separating (1.31) into the real and imaginary parts gives

$$(1 - \alpha)\frac{\partial(V\varrho_1)}{\partial \mathbf{n}_-} - (1 + \alpha)\frac{\partial(V\varrho_1)}{\partial \mathbf{n}_+} - \beta\left[\frac{\partial(V\varrho_2)}{\partial \mathbf{n}_-} + \frac{\partial(V\varrho_2)}{\partial \mathbf{n}_+}\right] = 0, \tag{1.32}$$

$$(1 - \alpha)\frac{\partial(V\varrho_2)}{\partial \mathbf{n}_-} - (1 + \alpha)\frac{\partial(V\varrho_2)}{\partial \mathbf{n}_+} + \beta\left[\frac{\partial(V\varrho_1)}{\partial \mathbf{n}_-} + \frac{\partial(V\varrho_1)}{\partial \mathbf{n}_+}\right] = 0. \tag{1.33}$$

Multiply the first of these equalities by $V\varrho_2$, the second one by $V\varrho_1$, integrate over S, and then substract the first of the equalities obtained from the second one. Due to Green's formula and the harmonicity of the potentials $V\varrho_1$ and $V\varrho_2$, the terms containing α become zero, and what remains is

$$\beta[(I_1^+ + I_2^+) - (I_1^- + I_2^-)] = 0. \tag{1.34}$$

Here

$$I_i^{\pm} := \pm \int_S \frac{\partial(V\varrho_i)}{\partial \mathbf{n}_{\pm}} V\varrho_i dS = \int_{\Omega^{\pm}} |\nabla V\varrho_i|^2 dx \qquad (i = 1, 2).$$

Multiplying (1.32) and (1.33) by $V\varrho_1$ and $V\varrho_2$, respectively, adding and integrating the results over S leads to the equality

$$(1 + \alpha)(I_1^+ + I_2^+) + (1 - \alpha)(I_1^- + I_2^-) = 0. \tag{1.35}$$

The equations (1.34) and (1.35) may be viewed as a linear algebraic system for $I_1^+ + I_2^+$ and $I_1^- + I_2^-$. The determinant of the system is 2β. If $\beta \neq 0$, then $I_1^+ = I_2^+ = I_1^- = I_2^- = 0$. Hence $V\varrho_1$ and $V\varrho_2$ take on constant values in Ω^+ as well as in Ω^-, which fact, by formula (1.11), implies that $\varrho_1 = \varrho_2 = 0$ on S, contradicting our assumption that ϱ be nontrival. So it follows that $\beta = 0$, which proves (i).

We thus have $\lambda = \alpha$ and $\varrho = \varrho_1$. Therefore equality (1.35) assumes the form

$$(1 + \lambda)I_1^+ + (1 - \lambda)I_1^- = 0,$$

whence

$$\lambda = (I_1^- + I_1^+)/(I_1^- - I_1^+),$$

proving (ii). $\square$

Theorem 13. *Let λ_0 be a characteristic value of T (resp. T^*) and let χ_0 (resp. ϱ_0) be any eigenfunction for λ_0. Then χ_0 (resp. ϱ_0) has no associate function, i.e., there is no function χ_1 (resp. ϱ_1) such that*

$$\chi_1 - \lambda_0 T\chi_1 = T\chi_0 \qquad (resp.\ \varrho_1 - \lambda_0 T^*\varrho_1 = T^*\varrho_0).$$

Proof. As in the proof of the preceding theorem, we restrict ourselves to the case $n \geqq 3$. Assume the contrary, that is, let ϱ_0 have an associate function ϱ_1. From (1.31) and (1.10) we get

$$(1 - \lambda_0)\frac{\partial(V\varrho_0)}{\partial \mathbf{n}_-} - (1 + \lambda_0)\frac{\partial(V\varrho_0)}{\partial \mathbf{n}_+} = 0,$$

$$(1 - \lambda_0)\frac{\partial(V\varrho_1)}{\partial \mathbf{n}_-} - (1 + \lambda_0)\frac{\partial(V\varrho_1)}{\partial \mathbf{n}_+} = \frac{\partial(V\varrho_0)}{\partial \mathbf{n}_-} + \frac{\partial(V\varrho_0)}{\partial \mathbf{n}_+}.$$

Put

$$I_0^{\pm} = \pm \int_S \frac{\partial(V\varrho_0)}{\partial \mathbf{n}_{\pm}} V\varrho_0 \, dS = \int_{\Omega^{\pm}} |\nabla V\varrho_0|^2 \, dx$$

and proceed in a fashion similar to the one which gave (1.34) and (1.35) to obtain that $I_0^+ - I_0^- = 0$ and $(1 + \lambda_0)I_0^+ + (1 - \lambda_0)I_0^- = 0$. It follows that $I_0^+ = I_0^- = 0$ and hence, by (1.11), $\varrho_0 = 0$ on S, which, however, is impossible, because ϱ_0 is an eigenfunction. This contradiction completes the proof. $\square$

§4. Some More Methods for Reducing the Dirichlet and Neumann Problems to Boundary Equations

4.1. Direct Variants of the Integral Equations Method. Instead of representing the solution in the form of a potential whose density is determined from an integral equation, one may use other methods to reduce boundary value problems to integral equations on S. When the solutions of the integral equations are explicitly expressed in terms of the solution of the boundary value problems, we speak of *direct variants of boundary equations*.

We shall make use of formula (1.3), with $x \in \Omega^{\pm}$. Letting x approach the surface S and taking into account the continuity of $u^{\pm}$ and of the simple layer potential as well as the theorem on the jumps of the double layer potential, we obtain that

$$\frac{1}{2}u^{\pm}(x) = \pm \int_S \left[u^{\pm}(\xi)\frac{\partial}{\partial \mathbf{n}_{\xi}}\mathscr{E}(x, \xi) - \mathscr{E}(x, \xi)\frac{\partial u^{\pm}}{\partial \mathbf{n}_{\xi}} \right] d_{\xi}S, \qquad x \in S. \quad (1.37)$$

When dealing with the Neumann problem in $\Omega^{\pm}$, the functions $\partial u^{\pm}/\partial n$ are given on S, and so equation (1.37) may be interpreted as an integral equation for the boundary values of the functions $u^{\pm}$:

$$u^{\pm} \mp Tu^{\pm} = \mp 2V_0\varphi^{\pm}. \qquad (1.38)$$

In 2.1 and 2.4 we showed that the equation for u^- is uniquely solvable for every right-hand side.

According to 2.2, the equation (1.38) for u^+ has a solution if and only if the orthogonality condition

$$\int_S \varrho_0^+ V_0\varphi^+ \, dS = 0$$

is satisfied, where ϱ_0^+ denotes a nonzero solution of equation (1.20). Since the operator V_0 is symmetric and the function $V\varrho_0^+$ is constant on $\overline{\Omega}^+$ (see the proof of Lemma 1), the latter equality is equivalent to equality (1.15), which, in turn, is a necessary and sufficient condition for the solvability of the internal Neumann problem.

Now consider the Dirichlet problem in $\Omega^{\pm}$. In this case inserting the given boundary values $u^{\pm}$ into (1.37) produces an integral equation of the first kind for the values of $\partial u^{\pm}/\partial n$ on S. The kernel of the integral operator generated by this equation of the first kind is just the fundamental solution $\mathscr{E}(x, \xi)$. To prove that an integral equation of the first kind with such a kernel is solvable will be the main concern of the following Subsection 4.2.

Before concluding the present subsection, let us mention still another method of reducing the Dirichlet problem to a boundary integral equation. Suppose S is of the class $C^{2,\alpha}$ and $\varphi^{\pm} \in C^{1,\alpha}(S)$. It is well-known that then $u^{\pm} \in C^{1,\alpha}(\overline{\Omega})$ (see Miranda [1955]). So we may differentiate the equality (1.3) to obtain that

$$\frac{1}{2}\frac{\partial u^{\pm}}{\partial n_x} = \mp \int_S \frac{\partial u^{\pm}}{\partial n_\xi}\frac{\partial}{\partial n_x}\mathscr{E}(x, \xi)\, d_\xi S \pm \frac{\partial}{\partial n_x}\int_S \varphi^{\pm}\frac{\partial}{\partial n_\xi}\mathscr{E}(x, \xi)\, d_\xi S, \qquad x \in S, \tag{1.39}$$

which may be regarded as an equation for $\partial u^{\pm}/\partial n$ on S. To investigate equation (1.39) one can resort to the results of 2.1–2.4.

4.2. The Equation of the First Kind for the Dirichlet Problem.

Our subject here is equations of the form

$$\int_S \varrho(\xi)\mathscr{E}(x, \xi)\, d_\xi S = f(x), \qquad x \in S. \tag{1.40}$$

In 4.1 we pointed out that equations of this type arise from the direct variant of the boundary equations method for the Dirichlet problem (see(1.37)). Notice that equation (1.40) also emerges from seeking the solution of the Dirichlet problem in $\Omega^{\pm}$ via the potential ansatz $V\varrho$, in which case $f = \varphi^{\pm}$. For the simplicity's sake, we shall suppose that $n \geq 3$ and that S is of the class $C^{2,\alpha}$.

Lemma 3. *If* $f \in C^2(S)$, *then equation* (1.40) *possesses a solution* $\varrho \in C(S)$.

Proof. Consider the Dirichlet problem

$$\Delta v = 0 \quad \text{in } \Omega^+, \qquad v = f \quad \text{on } S,$$

which has a unique solution $v \in C^1(\overline{\Omega^+})$. For $x \in S$, put $g(x) = (\partial v/\partial \mathbf{n})(x)$. It is clear that v solves the Neumann problem

$$\Delta v = 0 \quad \text{in } \Omega^+, \qquad \partial v/\partial \mathbf{n} = g \quad \text{on } S.$$

From the results of 2.2 we infer that $v = V\varrho_* + C$, with some constant C and a certain function $\varrho_* \in C(S)$. Furthermore, we also can write $1 = V(K\varrho_0^+)$, where ϱ_0^+ is a nontrivial solution of equation (1.20) and K denotes some constant. So $v = V(\varrho_* + CK\varrho_0^+)$ and thus, $\varrho = \varrho_* + CK\varrho_0^+$. $\square$

The operator V_0 induced by the left-hand side of (1.40) is well-known to be bounded, symmetric, and compact on $L_2(S)$. Moreover, zero is not among its eigenvalues. *Indeed*, if $V_0\varrho_0 = 0$, then, for every function $f \in C^2(S)$, we have

$$\int_S \varrho_0 f dS = \int_S \varrho_0 V_0 \varrho dS = \int_S \varrho V_0 \varrho_0 dS = 0$$

(see the proof of Lemma 1), where ϱ is a solution of (1.40), whose existence is guaranteed by Lemma 3. Consequently, $\varrho_0 = 0$ on S, $\square$, and it also follows that the operator V_0^{-1} exists and that, by Lemma 3, V_0^{-1} maps $C^2(S)$ into $C(S)$. The following theorem shows that even much more is true.

Theorem 14. *The operator* V_0^{-1} *extends to a continuous operator of* $L_2(S)$ *onto* $H^{-1}(S)$. (*Here and in what follows* $H^{-r}(S), r > 0$, *refers to the space of generalized functions which is the dual of* $H^r(S) = W_2^r(S)$; *see Hörmander* [1963, Chap. 2]).

The *proof* of this result can, for instance, be based on the fact that V_0 is a pseudodifferential operator for which the principal part of the symbol coincides with the principal part of the symbol of the operator $(-\delta)^{-1/2}$, where δ is the Laplace-Beltrami operator on S. $\square$ Assuming that the surface S is sufficiently smooth, one can show that V_0^{-1} maps $H^r(S)$ onto $H^{r-1}(S)$ for every real number r.

In the same way as equation (1.40) can be used to solve the Dirichlet problem by means of the simple layer potential, the double layer potential can be applied to solve the Neumann problem. An example of such a kind was considered by Giroire and Nedelec [1978], who studied the external Neumann problem in $\mathbb{R}^3$. They established the following integral identity for the density χ of the potential (1.5):

$$\int_S \int_S [\chi(\xi) - \chi(x)][\tau(\xi) - \tau(x)] \frac{\partial^2}{\partial \mathbf{n}_x \partial \mathbf{n}_\xi} \mathscr{E}(x, \xi) \, d_x S d_\xi S = -2 \int_S \psi \tau dS.$$

$$(1.41)$$

Here $\psi \in H^{-1/2}(S)$, while χ and τ belong to the quotient space $H^{1/2}(S)/\mathbb{R}^1$, τ being an arbitrary element.

The bilinear form on the left of (1.41) is symmetric and positive-definite on the space $H^{1/2}(S)/\mathbb{R}^1$, which guarantees that there is a unique element χ of this space satisfying the identity (1.41) for arbitrary τ.

Chapter 2
Integral Equations for the Equations of Lamé and Stokes

Boundary integral equations are a powerful tool for investigating and solving the boundary value problems of continuum mechanics. This is demonstrated in the present chapter for the equations of Lamé and Stokes. The boundary value problems for these equations are reduced to systems of integral equations for which results analogous to those established in Chap. 1 hold. Other examples of the application of boundary integral equations to continuum mechanics will be given in Chap. 3.

The first section concentrates on certain integral equations, primarily singular ones. These equations arise from the first and second fundamental boundary value problems for the three-dimensional Lamé equations of the linear homogeneous isotropic theory of elasticity. We consider the problem of regularizing the singular integral equations obtained, i.e., the problem of converting them into equations tractable by the Fredholm theory.

In the last subsection of Sec. 1 we discuss the systems of integral equations for the mixed problem of elasticity.

What will not be dealt with in Sec. 1 is the application of complex function theory to the integral equations of plane elasticity. This question will be touched upon in Chap. 3. There we shall consider integral equations emerging from dynamical problems.

Sec. 2 concerns results pertaining to the boundary integral equations for the Stokes equations. The latter equations arise from linearizing the Navier-Stokes equations. Exactly as in the theory of harmonic potentials, in this case the kernels of the integral operators prevailing have a weak singularity.

§1. The Boundary Integral Equations Method for the Lamé Equations

1.1. The Fundamental Principles of the Theory of Equilibrium of the Isotropic Solid Body. A detailed treatment of the elements of continuum mechanics is of course impossible within the scope of this survey article, and so we refer e.g. to the book Duvaut, Lions [1972] for this topic.

An elastic body undergoing a deformation is usually described by the displacement vector $\mathbf{u}(x) = (u_1(x), u_2(x), u_3(x))$ and the stress tensor $\sigma_{ij}(x)$ $(i, j = 1, 2, 3)$.

Here $x \in \mathbb{R}^3$, and the vector $\mathbf{u}(x)$ characterizes the terminal position of the point of the body whose initial position is given by the coordinates x_1, x_2, x_3.

The momentum conservation law for the elastic body implies that in the equilibrium case the following equations hold:

$$\sum_{j=1}^{3} \frac{\partial \sigma_{ij}}{\partial x_j} + F_i = 0; \qquad i = 1, 2, 3;$$

here F_i are the components of the vector describing the body force per unit volume induced by external forces. Another consequence of the momentum conservation law is that the stress tensor is symmetric:

$$\sigma_{ij} = \sigma_{ji}; \qquad i, j = 1, 2, 3.$$

To describe equilibrium of deformed solid bodies we need, in addition to the dynamical equations given above, also the state equations, which relate the stress tensor to the kinematical characteristics of the deformation. The latter characteristics are in elasticity theory recorded by the strain tensor

$$e_{ij} = \tfrac{1}{2}\left(\frac{\partial u_i}{\partial x_j} + \frac{\partial u_j}{\partial x_i}\right); \qquad i, j = 1, 2, 3.$$

The state equations commonly used in the theory of elasticity are *Hooke's law*, which states that there is a linear relation between the components of the strain and stress tensors. In the case of a homogeneous isotropic body, Hooke's law assumes the form

$$\sigma_{ij} = \lambda \delta_i^j \sum_{k=1}^{3} e_{kk} + 2\mu e_{ij}; \qquad i, j = 1, 2, 3,$$

where δ_i^j is the Kronecker delta, and λ and μ are physical constants of the medium, called the Lamé constants. They are subject to the restrictions $\mu > 0$ and $\lambda > -2\mu/3$. Note that, by the definition of the strain tensor, Hooke's law may also be expressed in the form

$$\sigma_{ij} = \lambda \delta_i^j \operatorname{div} \mathbf{u} + \mu\left(\frac{\partial u_i}{\partial x_j} + \frac{\partial u_j}{\partial x_i}\right); \qquad i, j = 1, 2, 3.$$

Lamé's equations result from substituting the latter expresssion into the dynamical equations and they are usually written in the form

$$\mu \Delta \mathbf{u} + (\lambda + \mu) \nabla \operatorname{div} \mathbf{u} + F = 0.$$

In what follows we shall apply the potential method to the *homogeneous Lamé equations*

$$\mu \Delta \mathbf{u} + (\lambda + \mu) \nabla \operatorname{div} \mathbf{u} = 0. \tag{2.1}$$

This elliptic system of equations is an analogue of the Laplace equation and it can be tackled by a boundary integral equations method, which will now be discussed according to the scheme pursued in Chap. 1.

1.2. The Fundamental Solution of Lamé's Equations. The matrix whose entries are

$$\Gamma_{i,j}(x, \xi) = -\frac{\lambda + \mu}{8\pi\mu(\lambda + 2\mu)}\left[\frac{\lambda + 3\mu}{\lambda + \mu}\frac{\delta_i^j}{r} + \frac{(x_i - \xi_i)(x_j - \xi_j)}{r^3}\right]; \qquad i, j = 1, 2, 3$$

may serve as a *fundamental solution of the Lamé equations*. Letting the operator defined by the left-hand side of (2.1) act on the kth column of that matrix gives $\delta(x - \xi)\mathbf{e}^k$, where $\mathbf{e}^k$ is the unit vector parallel to the kth axis of the Cartesian system at hand and $\delta(x - \xi)$ denotes the Dirac measure concentrated at ξ. The fundamental solution introduced here is referred to as the *Kelvin-Somigliana matrix*.

Let Ω^+ be a simply connected and bounded domain in $\mathbb{R}^3$ the boundary $\partial\Omega^+ = S$ of which belongs to the class $C^{1,\alpha}$ $(0 < \alpha < 1)$. Put $\Omega^- = \mathbb{R}^3 \setminus \overline{\Omega^+}$. The differential operator acting by the rule

$$\mathcal{T}_\varkappa(\partial/\partial x, \mathbf{n}_x)\mathbf{u} = (\mu + \varkappa)\partial\mathbf{u}/\partial\mathbf{n}_x + (\lambda + \mu - \varkappa)\mathbf{n}_x \operatorname{div} \mathbf{u} + \varkappa\mathbf{n}_x \times \operatorname{rot} \mathbf{u}, \quad \varkappa > 0, \tag{2.2}$$

is called the *generalized stress operator*. Its entries are given by

$$(\mathcal{T}_\varkappa)_{i,j} = \mu\delta_i^j\frac{\partial}{\partial\mathbf{n}_x} + (\lambda + \mu)n_i(x)\frac{\partial}{\partial x_j} + \varkappa\left(n_j(x)\frac{\partial}{\partial x_i} - n_i(x)\frac{\partial}{\partial x_j}\right),$$

where $n_j(x) = \cos(\mathbf{n}_x, x_j)$.

If $\varkappa = \mu$, this operator is simply termed the *stress operator*, since in this case $\mathcal{T}_\mu(\partial/\partial x, \mathbf{n}_x)\mathbf{u}$ is nothing but the stress vector in the direction $\mathbf{n}_x$. If $\varkappa = \mu(\lambda + \mu)$ $(\lambda + 3\mu)^{-1}$, the operator (2.2) is referred to as the *pseudo-stress-operator* and denoted by $\mathcal{N}(\partial/\partial x, \mathbf{n}_x)$.

A solution $\mathbf{u}^\pm = (u_1^\pm, u_2^\pm, u_3^\pm)$ of the Lamé equations is said to be *regular* in $\Omega^\pm$ if $u_k^\pm \in C^2(\Omega^\pm) \cap C^1(\overline{\Omega^\pm})$ and, in Ω^-, the following estimates as $|x| \to \infty$ are valid:

$$|\mathbf{u}^-(x)| = O(|x|^{-1}); \qquad |\nabla u_k^-| = O(|x|^{-2}), \qquad k = 1, 2, 3. \tag{2.3}$$

A solution of (2.1) regular in $\Omega^\pm$ can be represented in the form

$$\mathbf{u}^\pm(x) \pm \int_S \{[\mathcal{T}_\varkappa(\partial/\partial\xi, \mathbf{n}_\xi)\Gamma(x, \xi)]'\mathbf{u}^\pm(\xi)$$

$$- \Gamma(x, \xi)(\mathcal{T}_\varkappa(\partial/\partial\xi, \mathbf{n}_\xi)\mathbf{u}^\pm)(\xi)\}d_\xi S, \qquad x \in \Omega^\pm \tag{2.4}$$

(compare this with formula (1.3)). Here the prime denotes the operation of taking the transposed matrix, and the integral is understood in the principal value sense.

1.3. The Potentials of Elasticity Theory and Their Properties. As in the theory of harmonic potentials, we again have *simple* and *double layer potentials*. What is new now is the possibility of introducing a family $\{W^{(\varkappa)}\chi\}$ of double layer potentials which depends on the parameter $\varkappa$.

The potential $W^{(x)}\chi$ is defined as

$$(W^{(x)}\chi)(x) = \int_S [\mathcal{T}_x(\partial/\partial\xi, \mathbf{n}_\xi)\Gamma(x, \xi)]'\chi(\xi)d_\xi S, \qquad x \in \Omega^\pm, \qquad (2.5)$$

or, in vectorial notation,

$$(W^{(x)}\chi)(x) = \frac{\varkappa(\lambda + 3\mu) - \mu(\lambda + \mu)}{8\pi\mu(\lambda + 2\mu)} \int_S \frac{(\mathbf{r} \times \mathbf{n}_\xi) \times \chi(\xi)}{r^3} d_\xi S$$

$$+ \frac{\mu(\lambda + 3\mu) - \varkappa(\lambda + \mu)}{8\pi\mu(\lambda + 2\mu)} \int_S \frac{(\mathbf{r} \cdot \mathbf{n}_\xi)\chi(\xi)}{r^3} d_\xi S$$

$$+ \frac{3(\lambda + \mu)(\mu + \varkappa)}{8\pi\mu(\lambda + 2\mu)} \int_S \frac{(\mathbf{r} \cdot \mathbf{n}_\xi)(\mathbf{r} \cdot \chi)\mathbf{r}}{r^5} d_\xi S, \qquad (2.6)$$

where $\mathbf{r}$ is the vector starting in x and terminating in ξ and $\chi(\xi) = (\chi_1(\xi), \chi_2(\xi), \chi_3(\xi))$ is a density built up by functions belonging to $C^{0,\beta}(S)$, $0 < \beta < \alpha$.

The *direct value* $(W_0^{(x)}\chi)(x)$, $x \in S$, *of the double layer potential* can be shown to exist. To evaluate the direct value by formula (2.5), the integral has to be understood in the principal value sense. The same also applies to the first integral on the right of (2.6). The only exception from what was said occurs for $\varkappa = \mu(\lambda + \mu)(\lambda + 3\mu)^{-1}$, in which case $W_0^{(x)}\chi$ is an ordinary improper integral which converges whenever $\chi_k \in C(S)$ ($k = 1, 2, 3$), since the surface S is of the class $C^{1,\alpha}$.

Theorem 1. *If the components χ_1, χ_2, χ_3, of the density are from $C^{0,\beta}(S)$, then the components of the potential $W^{(x)}\chi$, continued onto $\overline{\Omega^\pm}$ by*

$$W_\pm^{(x)}\chi = \pm\tfrac{1}{2}\chi + W_0^{(x)}\chi, \qquad (2.7)$$

belong to $C^{0,\beta}(\overline{\Omega^\pm})$.

The vector function

$$(V\varrho)(x) = \int_S \Gamma(x, \xi)\varrho(\xi)d_\xi S \qquad (2.8)$$

is referred to as the *simple layer potential with the density* ϱ.

Theorem 2. *If $\varrho_1, \varrho_2, \varrho_3$ belong to $C(S)$, then the potential $V\varrho$ is continuous in $\mathbb{R}^3$. If moreover, $\varrho_1, \varrho_2, \varrho_3 \in C^{0,\beta}(S)$ $(0 < \beta < \alpha)$, then the first derivatives of the components of the potential (2.8), extended to functions on $\overline{\Omega^\pm}$ by taking on S the boundary values*

$$\left[\frac{\partial(V\varrho)_k}{\partial x_j}\right]_\pm = \mp\frac{1}{2}\left\{\mu^{-1}\varrho_k - \frac{\lambda + \mu}{\mu(\lambda + 2\mu)}n_k(\mathbf{n} \cdot \varrho)\right\}n_j + \left[\frac{\partial(V\varrho)_k}{\partial x_j}\right]_0; \qquad k = 1, 2, 3,$$

(2.9)

belong to $C^{0,\beta}(\overline{\Omega^\pm})$; the direct value $[\partial(V\varrho)_k/\partial x_j]_0$ in (2.9) is understood in the principal value sense.

Formula (2.9) implies that the generalized stress of the potential $V\varrho$ behaves analogously to the normal derivative of the harmonic simple layer potential (for which see Sec. 1.2 of Chap. 1). Viz, we have

$$[\mathscr{T}_x(\partial/\partial x, \mathbf{n}_x)(V\varrho)]_\pm = \mp\tfrac{1}{2}\varrho + [\mathscr{T}_x(\partial/\partial x, \mathbf{n}_x)(V\varrho)]_0. \qquad (2.10)$$

Here the direct value of the generalized stress of the potential (2.8) is a singular integral operator for all $\varkappa$, except for $\varkappa = \mu(\lambda + \mu)(\lambda + 3\mu)^{-1}$. In the latter case it is an operator with a weak singularity.

The operators $W_0^{(\varkappa)}$ and $[\mathscr{T}_x V]_0$ are adjoint to each other.

To conclude this subsection, we remark that the potentials (2.5) and (2.8) satisfy the equations (2.1) in $\Omega^\pm$. The potential $V\varrho$ admits an estimate of the form (2.3), while the potential $W^{(\varkappa)}\chi$ decreases at infinity one order faster.

1.4. The Boundary Integral Equations of the Fundamental Problems of Elasticity Theory.

By a *solution of the first fundamental boundary value problem of elasticity theory* we mean a solution $\mathbf{u}^\pm$ of the Lamé equations in $\Omega^\pm$ satisfying the boundary condition

$$\mathbf{u}^\pm = \boldsymbol{\varphi}^\pm \qquad \text{on } S, \qquad (2.11)$$

where $\varphi_i^\pm \in C^{0,\beta}(S)$, $0 < \beta < \alpha$ $(i = 1, 2, 3)$. We also require that in Ω^- the estimates (2.3) hold as $|x| \to \infty$.

To reduce this problem to a system of boundary integral equations, we look for a solution of the problem in the form of the potential $W^{(\varkappa)}\chi^\pm$ with unknown density $\chi^\pm$. Then by formula (2.7), we have the equation

$$\chi^\pm \pm 2W_0^{(\varkappa)}\chi^\pm = \pm 2\boldsymbol{\varphi}^\pm. \qquad (2.12)$$

In order to proceed in analogy to the theory of harmonic potentials (Sections 1 and 2 of Chapter 1), we consider a family of boundary value problems depending on a parameter $\varkappa$ which are reduced to integral equations adjoint to the equation (2.12). These problems consist in finding solutions $\mathbf{u}^\pm$ of the Lamé equations in $\Omega^\pm$ such that

$$\mathscr{T}_x(\partial/\partial x, \mathbf{n}_x)\mathbf{u}^\pm = \boldsymbol{\psi}^\pm \qquad \text{on } S, \qquad (2.13)$$

where $\psi_i^\pm \in C^{0,\beta}(S)$ $(i = 1, 2, 3)$. In the case of Ω^- we also require that (2.3) hold.

The condition (2.13) admits a physical interpretation if $\varkappa = \mu$. In this case the problem is called the *second fundamental problem of elasticity theory*. For arbitrary $\varkappa$, we shall speak of the *generalized second fundamental problem*. If the solution is sought in the form $V\varrho^\pm$, then, by (2.10), the unknown density $\varrho^\pm$ must satisfy the equation

$$\varrho^\pm \pm 2[\mathscr{T}_x(\partial/\partial x, \mathbf{n}_x)(V\varrho^\pm)]_0 = \pm 2\boldsymbol{\psi}^\pm. \qquad (2.14)$$

The Fredholm theory can be immediately applied to the equations (2.12) and (2.14) only if $\varkappa = \mu(\lambda + \mu)(\lambda + 3\mu)^{-1}$. For all other values of $\varkappa$, neither the operator $W_0^{(\varkappa)}$ nor its adjoint are compact. Nevertheless, the Fredholm theorems extend to the singular integral equations (2.12) and (2.14). This is a consequence of the following result.

Theorem 3. *The equations* (2.12) *and* (2.14) *possess two-sided regularizers.*

(For the definition of a regularizer see Sec. 3 of Chap. 1 of S. Prössdorf's article in this volume).

Proof (Maz'ya, Sapozhnikova [1964]). We denote by $\mathscr{W}$ the symbol matrix of the operator $W_0^{(\varkappa)}$ (see Sec. 6 of Chap. 4 of S. Prössdorf's article). The computations performed in §45 of the book Mikhlin [1962] show that

$$\mathscr{W}(\xi) = \frac{ic_\varkappa}{2}\begin{bmatrix} 0 & 0 & \cos\varphi \\ 0 & 0 & \sin\varphi \\ -\cos\varphi & -\sin\varphi & 0 \end{bmatrix},$$

where ξ lies on the sphere and has the coordinates ϑ, φ ($0 \le \vartheta \le \pi, 0 \le \varphi \le 2\pi$); the constant $c_\varkappa$ is given by

$$c_\varkappa = [\varkappa(\lambda + 3\mu) - \mu(\lambda + \mu)][2\mu(\lambda + 2\mu)]^{-1}.$$

Because $((2/c_\varkappa)\mathscr{W})^3 = (2/c_\varkappa)\mathscr{W}$, we have

$$(\mathscr{I} - 2\mathscr{W})^{-1} = \mathscr{I} + \frac{2}{1 - c_\varkappa^2}\mathscr{W} + \frac{4}{1 - c_\varkappa^2}\mathscr{W}^2,$$

$\mathscr{I}$ being the identity matrix, which implies that the operator

$$I + \frac{2}{1 - c_\varkappa^2}W_0^{(\varkappa)} + \frac{4}{1 - c_\varkappa^2}(W_0^{(\varkappa)})^2$$

is a regularizer of the lower sign version of equation (2.12) (now, of course, I is the identity operator) and that the regularized operator is of the form

$$I + \frac{c_\varkappa^3}{1 - c_\varkappa^2}\left[\frac{2}{c_\varkappa}W_0^{(\varkappa)} - \left(\frac{2}{c_\varkappa}W_0^{(\varkappa)}\right)^3\right].$$

Since $\mathscr{W}^* = -\mathscr{W}$, a regularizer of the upper sign version of equation (2.14) is provided by the operator

$$I - \frac{2}{1 - c_\varkappa^2}W_0^{(\varkappa)} + \frac{4}{1 - c_\varkappa^2}(W_0^{(\varkappa)})^2.$$

The remaining pair of equations in (2.12) and (2.14) can be regularized analogously. □

Theorem 3 justifies the application of the Fredholm theorems to the equations (2.12) and (2.14). To gain deeper insight into the solvability properties of these equations, we can make use of the scheme employed in Sec. 2 of Chap. 1 to analyze the integral equations of the theory of harmonic functions and we so come to similar results in the present situation.

In particular, the homogeneous equation $\chi^+ + 2W_0^{(\varkappa)}\chi^+ = 0$ and its adjoint have only the trivial solution. A consequence of this fact is as follows.

Theorem 4. *The equation* $\chi^+ + 2W_0^{(\varkappa)}\chi^+ = 2\varphi^+$ *and its adjoint are uniquely solvable.*

Note that this theorem enables us to find the solutions of the external second and internal first fundamental problems for Lamé's equations in the form of appropriate potentials (see Kupradze, Gegelia, Basheleĭshvili, Burchuladze [1976]).

The homogeneous equation

$$\chi_0^- - 2W_0^{(\mu)}\chi_0^- = 0 \tag{2.15}$$

has six linearly independent solutions:

$$\chi_{0k}^- = e^k, \qquad k = 1, 2, 3; \tag{2.16}$$

$$\chi_{04}^- = (0, x_3, -x_2), \qquad \chi_{05}^- = (-x_3, 0, x_1), \qquad \chi_{06}^- = (x_2, -x_1, 0).$$

Theorem 5. *For the equation*

$$\varrho^- - 2[\mathcal{T}_\mu(\partial/\partial x, \mathbf{n}_x)(V\varrho^-)]_0 = -2\psi^-$$

to be solvable it is necessary and sufficient that

$$\int_S \psi^- dS = 0, \qquad \int_S \mathbf{x} \times \psi^- dS = 0. \tag{2.17}$$

As the conditions (2.17) are also necessary for the second fundamental problem in Ω^+ to possess a solution, we arrive at the following conclusion.

Theorem 6. *If the conditions (2.17) are fulfilled, then the second fundamental problem for the equations (2.1) has a solution, which is unique up to a rigid displacement and which can be represented by the simple layer potential whose density satisfies the integral equation occuring in Theorem 5.*

From Fredholm theory we know that the homogeneous equation adjoint to (2.15) also possesses six linearly independent solutions ϱ_{0k}^+ ($k = 1, \ldots, 6$). By means of these solutions, the integral equation of the external first fundamental problem (see (2.12)) can be altered so as to become solvable for every right-hand side (cf. Sec. 2.3 of Chap. 1).

The solution of the first fundamental problem in Ω^- is sought in the form

$$\mathbf{u}^- = W^{(\mu)}\chi^- + \sum_{k=1}^6 c_k V\varrho_{0k}^+. \tag{2.18}$$

Herin the density χ^- as well as the constants are unknown (cf. Sec. 1.6 of Chap. 1). Note that χ^- is given by the equation

$$\chi^- - 2W_0^{(\mu)}\chi^- = -2\left(\varphi^- + \sum_{k=1}^6 c_k V_0\varrho_{0k}^+\right) \tag{2.19}$$

(see (2.7)). The constants c_k on the right of (2.19) must be previously determined from the equations

$$\sum_{k=1}^6 c_k \int_S \varrho_{0i}^+ \cdot V_0\varrho_{0k}^+ dS = -\int_S \varphi^- \cdot \varrho_{0i}^+ dS. \tag{2.20}$$

The equations (2.20) form a uniquely solvable linear algebraic system for the constants c_k. At the same time the equalities (2.20) tell us that the right-hand side of equation (2.19) is orthogonal to the linearly independent solutions ϱ_{0k} of the conjugate homogeneous equation. Thus, we have the following.

Theorem 6'. *If* $(c_1, \ldots, c_6)$ *is the solution of the uniquely solvable linear algebraic system* (2.20), *then equation* (2.19) *has a solution for every function* $\boldsymbol{\varphi}^-$. *Its solution* χ^- *is unique up to an arbitrary linear combination of the functions* χ_{01}^-, $\ldots, \chi_{06}^-$.

Here is a consequence of the preceding theorem.

Theorem 7. *The first fundamental problem of elasticity theory is solvable in* Ω^-. *The solution can be given by* (2.18), χ^- *and* $(c_1, \ldots, c_6)$ *being defined as in Theorem 6'.*

We conclude this section with a theorem on the spectral properties of the operator $2W_0^{(\mu)}$ which is analogous to the corresponding result of harmonic potential theory (Sec. 3 of Chap. 1).

Theorem 8. *All characteristic values of the operator* $2W_0^{(\mu)}$ *and its adjoint are real und have modulus at least one. There do not exist associate functions.*
    The number $\lambda = -1$ *is no characteristic value for* $2W_0^{(\mu)}$ *and its adjoint. The number* $\lambda = +1$ *is a characteristic value for these operators and it owns six linearly independent eigenfunctions.* (See Sec. 4 of Chap. 6 of the book Kupradze et al. [1976].)

1.5. The System of Integral Equations for the Mixed Problem of Elasticity Theory. Let S_1 and S_2 be disjoint open subsets of the surface S such that $\overline{S_1} \cup \overline{S_2} = S$. The *mixed problem* consists in finding a solution $\mathbf{u}^\pm$ of the Lamé equations in $\Omega^\pm$ such that

$$\mathbf{u}^\pm = \boldsymbol{\varphi}^\pm \qquad \text{on } S_1,$$
$$\mathscr{T}_\mu(\partial/\partial x, \mathbf{n}_x)\mathbf{u}^\pm = \boldsymbol{\psi}^\pm \qquad \text{on } S_2. \tag{2.21}$$

Let $V_k\varrho$ ad $W_k\chi$ denote the potentials defined by the formulas (2.5) and (2.8) with S replaced by S_k ($k = 1, 2$). We seek the solution $\mathbf{u}^\pm$ in the form $W_1\chi^\pm + V_2\varrho^\pm$, where the unknown densities $\chi^\pm$ and $\varrho^\pm$ defined on S_1 and S_2, respectively, are, by (2.21), to be determined as solutions of the equations

$$\pm\tfrac{1}{2}\chi^\pm + W_{10}\chi^\pm + V_{20}\varrho^\pm = \boldsymbol{\varphi}^\pm \qquad \text{on } S_1,$$
$$\mp\tfrac{1}{2}\varrho^\pm + [\mathscr{T}_\mu(W_1\chi^\pm)]_0 + [\mathscr{T}_\mu(V\varrho^\pm)]_0 = \boldsymbol{\psi}^\pm \qquad \text{on } S_2. \tag{2.22}$$

This system was studied by the author in the case of surfaces which are allowed to have edges (see Maz'ya [1986] and also Sec. 1.5 of Chap. 5).

To find a solution $\mathbf{u}^\pm$ of the mixed problem one can also resort to the representation $W_2\chi^\pm + V_1\varrho^\pm$, which leads to the following system of equations for $\chi^\pm$ and $\varrho^\pm$ (these functions are given on S_2 and S_1, respectively):

$$W_{20}\chi^\pm + V_{10}\varrho^\pm = \varphi^\pm \qquad \text{on } S_1,$$

$$[\mathscr{T}_\mu(W_2\chi^\pm)]_0 + [\mathscr{T}_\mu(V_1\varrho^\pm)]_0 = \psi^\pm \qquad \text{on } S_2. \qquad (2.23)$$

Note that a system of the form (2.23) also arises from using the direct method of solving the mixed problem. In this case the unknowns are no longer densities of potentials and the right-hand sides have a form different from the one prevailing here (cf. Sec. 4.1 of Chap. 1). The solvability of the system like (2.23) obtained by employing the direct method follows from the existence of solutions of the corresponding boundary value problems of elasticity theory (see Duvaut, Lions [1972]). The uniqueness of the solution of (2.23) was in the case $S_1 \neq \varnothing$ proved by Ugodchikov and Khutoryanskiĭ [1979].

We remark that system (2.23) takes advantage of the symmetry of the operator generated by the left-hand side. Several other properties of this system are discussed in the book Ugodchikov, Khutoryanskiĭ [1986].

With the help of the direct variant of the boundary integral equations method the mixed problem of elasticity can also be reduced to the system

$$\mp\mathbf{g}^\pm/2 + W_{10}\mathbf{g}^\pm \mp V_{20}\mathbf{h}^\pm = \mathbf{f}_1^\pm \qquad \text{on } S_1,$$

$$\pm\mathbf{h}^\pm/2 + [\mathscr{T}_\mu(V_2\mathbf{h}^\pm)]_0 \mp [\mathscr{T}_\mu(W_1\mathbf{g}^\pm)]_0 = \mathbf{f}_2^\pm \qquad \text{on } S_2. \qquad (2.24)$$

Here $\mathbf{g}^\pm$ and $\mathbf{h}^\pm$ are unknown functions, while $\mathbf{f}_1^\pm$ and $\mathbf{f}_2^\pm$ are given functions defined on S_1 and S_2, respectively. The existence of solutions of the system (2.24) again follows from the solvability of the corresponding boundary value problems (Duvaut, Lions [1972]). Ugodchikov and Khutoryanskiĭ [1979] showed that the solution $(\mathbf{g}^+, \mathbf{h}^+)$ is unique the case $S_1 \neq \varnothing$ and that the solution $(\mathbf{g}^-, \mathbf{h}^-)$ is unique provided $S_2 \neq \varnothing$.

Khutoryanskiĭ [1981] proved that all the eigenvalues λ of a certain extension of the operator defined by the system (2.24) lie in the disk $|\lambda| \leq 1/2$.

We finally remark that a detailed discussion of the boundary integral equations method for Lamé's equations can be found in the book Kupradze, Gegelia, Basheleĭshvili, Burchuladze [1976], where this method is also applied to a variety of other problems of the mechanics of deformable solid bodies. An account of the further development of this method is the monograph Burchuladze, Gegelia [1985], which also contains a historical survey on the topic. In this connection we also refer to §45 of the book Mikhlin [1962] as well as the books Parton, Perlin [1977] and Ugoduchikov, Khutoryanskiĭ [1986]. An ample bibliography of applications of integral equations to elasticity theory is in Burchuladze, Gegelia [1985] and Kupradze et al. [1976].

§2. The Boundary Integral Equations Method for the Stokes Equations

2.1. The Stokes Equations as a Model for Describing the Slow Steady Flow of a Viscous Fluid. The motion of a viscous fluid can be characterized by the velocity vector $\mathbf{v}(x, t) = (v_1(x, t), v_2(x, t), v_3(x, t))$ and the stress tensor $\sigma_{ij}(x, t)$ $(i, j = 1, 2, 3)$.

Here $x \in \mathbb{R}^3$, and x_1, x_2, x_3 are the coordinates of the point at which the vector $\mathbf{v}$ and the tensor σ_{ij} is attached.

The mass conservation law for an incompressible fluid appears as the *continuity equation*

$$\operatorname{div} \mathbf{v} = 0.$$

Vectors satisfying this equation are called *solenoidal*.

The momentum conservation law assumes the form

$$\varrho \frac{dv_i}{dt} = \sum_{j=1}^{3} \frac{\partial \sigma_{ij}}{\partial x_j} + F_i; \qquad i = 1, 2, 3.$$

Here ϱ denotes the density of the fluid, F_1, F_2, F_3 are the components of the volume forces acting in the fluid, and d/dt refers to the complete derivative, which is defined as $\partial/\partial t + \mathbf{v} \cdot \nabla$. As in the case of a deformable solid body, the stress tensor is symmetric:

$$\sigma_{ij} = \sigma_{ji}; \qquad i, j = 1, 2, 3.$$

The *Stokes law* says that the state equation for a viscous fluid, which establishes a connection between the stress tensor on the one hand and the pressure p and the rate-of-deformation tensor $(\partial v_i/\partial x_i + \partial v_j/\partial x_i)/2$ on the other, is of the form

$$\sigma_{ij} = -p\delta_i^j + \varrho v \left(\frac{\partial v_i}{\partial x_j} + \frac{\partial v_j}{\partial x_i} \right); \qquad i, j = 1, 2, 3,$$

v being the coefficient of kinematic viscosity.

On substituting the latter equalities into the momentum conservation law and taking into account the continuity equation we arrive at the *Navier-Stokes equations*

$$\frac{\partial \mathbf{v}}{\partial t} + (\mathbf{v} \cdot \nabla)\mathbf{v} = -\varrho^{-1}\nabla p + v\Delta \mathbf{v} + \mathbf{F}.$$

The first term on the left disappears in the case of a steady flow. When dealing with slow flows (Reynolds number $\ll 1$), we may *linearize* the Navier-Stokes equations, that is, we may neglect the nonlinear term, which is small compared with the remaining terms. What results is the *Stokes equations*, describing the slow steady flow of a viscous fluid:

$$v\Delta \mathbf{v} = \varrho^{-1}\nabla p + \mathbf{F} = 0,$$

$$\operatorname{div} \mathbf{v} = 0.$$

Without loss of generality suppose $\varrho = 1$. We shall focus our attention on the application of the method of potentials to the *homogeneous Stokes equations*

$$v\Delta \mathbf{v} - \nabla p = 0,$$
$$\operatorname{div} \mathbf{v} = 0. \tag{2.25}$$

This is a system of four equations with the four unknowns v_1, v_2, v_3 and p. The pressure p is determined up to an additive constant, which will be tacitly assumed henceforth.

The results following below go back to Lichtenstein [1928] and Odqvist [1930]. For a more detailled discussion of these results we refer to the book by O.A. Ladyzhenskaya [1970]. Several concrete questions arising in connection with the practice of applying boundary integral equations to the solution of boundary value problems for the Stokes equations are dealt with in the book by S.M. Belonosov and K.E. Chernous [1985], which also contains a bibliography.

2.2. The Fundamental Solution of the Stokes Equations. If a square matrix $U(x, \xi) = [u_i^j(x, \xi)]$ built up by three velocity vectors and a pressure vector $Q(x, \xi) = (q^1(x, \xi), q^2(x, \xi), q^3(x, \xi))$ corresponding to $U(x, \xi)$ satisfy the equations

$$v\Delta_x \mathbf{u}^j - V_x q^j = \delta(x - \xi)\mathbf{e}^j,$$

$$\text{div } \mathbf{u}^j = 0,$$

then $U(x, \xi)$ and $Q(x, \xi)$ are referred to as *fundamental solutions of the Stokes system*. Here $\mathbf{e}^j$ is the unit vector parallel to the jth axis of the Cartesian system chosen and $\delta(x - \xi)$ is Dirac measure concentrated at ξ.

A fundamental solution whose components tend to zero as $|x|$ goes to infinity is given by

$$u_i^j = -\frac{1}{8\pi v}\left[\frac{\delta_i^j}{r} + \frac{(x_i - \xi_i)(x_j - \xi_j)}{r^3}\right],$$

$$q^j = -\frac{x_j - \xi_j}{4\pi r^3}; \qquad i, j = 1, 2, 3, \tag{2.26}$$

δ_i^j being the Kronecker delta.

Notice that the expression in the square brackets in (2.26) is obtained from the analogous term in the Kelvin-Somigliana tensor by letting $(\lambda + 3\mu)/(\lambda + \mu) = 1$ in the latter term.

Now let Ω^+ be a simply connected and bounded region in $\mathbb{R}^3$. We shall assume that the boundary S of Ω^+ belongs to the class $C^{1,\alpha}$, where $0 < \alpha < 1$ (cf. 1.2). Set $\Omega^- = \mathbb{R}^3 \setminus \Omega^+$. Let finally $\mathcal{S}(\partial/\partial x, \mathbf{n}_x)$ and $\mathcal{S}'(\partial/\partial x, \mathbf{n}_x)$ be the matrix operators acting on the pair $(\mathbf{v}, p)$ by the rule

$$\mathcal{S}(\partial/\partial x, \mathbf{n}_x)\mathbf{v} = -p\mathbf{n}_x + v(2\,\partial\mathbf{v}/\partial\mathbf{n}_x + \mathbf{n}_x \times \text{rot } \mathbf{v}),$$

$$\mathcal{S}'(\partial/\partial x, \mathbf{n}_x)\mathbf{v} = p\mathbf{n}_x + v(2\,\partial\mathbf{v}/\partial\mathbf{n}_x + \mathbf{n}_x \times \text{rot } \mathbf{v}).$$

Any solution $(\mathbf{v}^\pm, p^\pm)$ of Stokes' equations in $\Omega^\pm$ admits a representation in the following form (cf. (2.4)):

$$\mathbf{v}^\pm(x) = \pm\int_S \{[\mathcal{S}'(\partial/\partial\xi, \mathbf{n}_\xi)U(x, \xi)]\mathbf{v}^\pm(\xi) - U(x, \xi)\mathcal{S}(\partial/\partial\xi, \mathbf{n}_\xi)\mathbf{v}^\pm\}d_\xi S,$$

$$p^\pm(x) = \mp\int_S [Q(x, \xi)\mathcal{S}(\partial/\partial\xi, \mathbf{n}_\xi)\mathbf{v}^\pm - 2v\mathbf{v}^\pm(\xi)\partial Q/\partial\mathbf{n}_\xi]d_\xi S. \tag{2.27}$$

In Ω^-, the formulas (2.27) hold under the additional assumption that

$$|\mathbf{v}^-(x)| = O(|x|^{-1}),$$

$$|\nabla v_k(x)|, |p(x)| = O(|x|^{-2}), \qquad k = 1, 2, 3, \tag{2.28}$$

as $|x| \to \infty$. The representations (2.27) are analogues of formula (1.3) for harmonic functions.

2.3. Hydrodynamic Potentials and Their Properties. *Hydrodynamic potentials are pairs $(V, \mathscr{V})\varrho$ and $(W, \mathscr{W})\chi$ whose components are integrals depending on a parameter* $x \in \mathbb{R}^3 \backslash S$. *The simple layer potential is defined by*

$$(V\varrho)(x) = \int_S U(x, \xi)\varrho(\xi)d_\xi S,$$

$$(\mathscr{V}\varrho)(x) = \int_S \mathbf{Q}(x, \xi) \cdot \varrho(\xi)d_\xi S,$$

while the *double layer potential* is of the form

$$(W\chi)(x) = \int_S [\mathscr{S}'(\partial/\partial\xi, \mathbf{n}_\xi)U(x, \xi)]\chi(x, \xi)d_\xi S,$$

$$(\mathscr{W}\chi)(x) = 2v \int_S \chi(\xi)(\partial\mathbf{Q}/\partial\mathbf{n}_\xi)(x, \xi)d_\xi S.$$

The components ϱ_k and χ_k $(k = 1, 2, 3)$ of the densities are assumed to be continuous functions.

The potentials $(V, \mathscr{V})\varrho$ and $(W, \mathscr{W})\chi$ satisfy the equations (2.25) in $\Omega^\pm$. Furthermore, as $|x| \to \infty$ we have the estimates

$$(V\varrho)(x) = O(|x|^{-1});$$

$$(\mathscr{V}\varrho)(x), (W\chi)(x), (\mathscr{W}\chi)(x) = O(|x|^{-2}). \tag{2.29}$$

Writing the potential $W\chi$ more explicitly in the form

$$W\chi = -\frac{3}{4\pi}\int_S \frac{(\mathbf{r} \cdot \mathbf{n}_\xi)(\mathbf{r} \cdot \chi)\mathbf{r}}{r^5}d_\xi S, \tag{2.30}$$

notation being as in (2.6), reveals that, for surfaces $S \in C^{1,\alpha}$, the kernel of the operator W admits the estimate $O(r^{-2+\alpha})$ on $S \times S$. This, in turn, yields the existence of the direct value $W_0\chi$ on S.

For double layer potentials with constant density χ analogues of the formulas (1.7) are valid. In the general case we have the following theorem on the boundary values.

Theorem 9. *There exist continuous vector functions $W_{\pm}\chi$ on S such that $(W\chi)(x) \to (W_{\pm}\chi)(x_0)$ as $x \in \Omega^\pm$ approaches $x_0 \in S$. Moreover,*

$$W_{\pm}\chi = \pm\tfrac{1}{2}\chi + W_0\chi. \tag{2.31}$$

The vector function $V\varrho$ is continuous in $\mathbb{R}^3$ and the expression $\mathscr{S}(\partial/\partial x, \mathbf{n}_x)$ $(V\varrho)$ owns properties which are analogous to the ones of the normal derivate of the harmonic simple layer potential (see Sec. 1.2 of Chap. 1).

Theorem 10. *As the point $y \in \Omega^\pm$ approaches a point $x \in S$ along the normal $\mathbf{n}_x$, the values of $\mathscr{S}(\partial/\partial y, \mathbf{n}_x)$ $(V\varrho)$ tend to a limit, which is denoted by $[\mathscr{S}(\partial/\partial x, \mathbf{n}_x)$ $(V\varrho)]_\pm$ and can be given by*

$$[\mathscr{S}(\partial/\partial x, \mathbf{n}_x)\,(V\varrho)]_\pm = \mp \tfrac{1}{2}\varrho - [\mathscr{S}(\partial/\partial x, \mathbf{n}_x)\,(V\varrho)]_0. \tag{2.32}$$

The last term in (2.32) (the direct value of the function $\mathscr{S}(\partial/\partial x, \mathbf{n}_x)\,(V\varrho)$) exists since S is of the class $C^{1,\alpha}$ (cf. formula (2.10)).

2.4. Boundary Value Problems for the Stokes Equations and Their Solution by Means of Boundary Integral Equations.

We shall consider only two boundary value problems for the equations (2.25) in $\Omega^\pm$. The first of them, the *Dirichlet problem*, consists in finding a pair $(\mathbf{v}^\pm, p^\pm)$ whose components are continuous in $\Omega^\pm$ and satisfy the equations (2.25) and the condition

$$\mathbf{v}^\pm = \boldsymbol{\varphi}^\pm \quad \text{on } S; \qquad \varphi_i^\pm \in C(S), \qquad i = 1, 2, 3. \tag{2.33}$$

In the Ω^- case it is also required that

$$|\mathbf{v}^-(x)| = O(|x|^{-1}), \qquad p^-(x) = O(|x|^{-2}) \tag{2.34}$$

as $|x| \to \infty$.

Because the vector $\mathbf{v}^+$ in demand is solenoidal, the solvability of the Dirichlet problem in Ω^+ necessitates the equality

$$\int_S \boldsymbol{\varphi}^+ \cdot \mathbf{n}\,ds = 0. \tag{2.35}$$

To reduce the Dirichlet problem in Ω^+ to an integral equation on S, we look for its solution in the form of a double layer potential with the unknown density χ^+. This density must fulfil the equation

$$\chi^+ + 2W_0\chi^+ = 2\boldsymbol{\varphi}^+ \tag{2.36}$$

(see (2.31)). In order to apply Fredholm theory to equation (2.36), we have to consider the second boundary value problem in Ω^- and its boundary integral equation (cf. Sec. 2.1 of Chap. 1).

The second boundary value problem in $\Omega^\pm$ requires to find a pair $(\mathbf{v}^\pm, p)$ satisfying the equations (2.25) and the condition

$$\mathscr{S}(\partial/\partial x, \mathbf{n}_x)\mathbf{v}^\pm = \boldsymbol{\psi}^\pm \qquad \text{on } S. \tag{2.37}$$

In Ω^-, the problem also involves the condition (2.34).

Representing the solution of the second boundary value problem in Ω^- by a simple layer potential with the unknown density ϱ^- gives the equation

$$\varrho^- - 2[\mathscr{S}(\partial/\partial x, \mathbf{n}_x)\,(V\varrho^-)]_0 = 2\boldsymbol{\psi}^-. \tag{2.38}$$

Note that the equations (2.36) and (2.38) are adjoint to each other.

Theorem 11. *The only nontrivial solution of the homogeneous version of equation (2.38) is* **n**. *Condition (2.35) is necessary and sufficient for equation (2.36) to possess a solution. The Dirichlet problem for the Stokes equations in Ω^+ is solvable for each φ^+ subject to condition (2.35). Its solution is representable by the potential $(W, \mathcal{W})\chi^+$ whose density χ^+ satisfies equation (2.36). The Dirichlet problem in Ω^+ is uniquely solvable.*

The following result should be compared with what was said in Sec. 1.4.

Lemma 1. *The homogeneous integral equation*

$$-\chi_0^- + 2W_0\chi_0^- = 0 \tag{2.39}$$

has six linearly independent solutions, which can be given by the formulas (2.15) (recall that the origin is always assumed to lie in Ω^+).

Hence, by standard Fredholm theory, the equation adjoint to (2.39) also possesses six linearly independent solutions, ϱ_{0k}^+ ($k = 1, \dots, 6$). The conclusion is that, as in the case of Lamé's equations, the equation for the external Dirichlet problem for the Stokes equations must be modified so that the new equation be solvable for arbitrary boundary data.

By literally repeating the scheme presented before Theorem 5 we arrive at the following result.

Theorem 12. *If $(c_1, \dots, c_6)$ is the solution of the uniquely solvable system*

$$\sum_{k=1}^{6} c_k \int_S \varrho_{0i}^+ V_0 \rho_{0k}^+ dS = -\int_S \varphi^- \varrho_{0i}^+ dS,$$

then the equation

$$\chi^- - 2W_0\chi^- = -2\left(\varphi^- + \sum_{k=1}^{6} c_k V_0 \varrho_{0k}^+\right)$$

has a solution χ^- for every function φ^-. The function χ^- is uniquely determined up to an arbitrary linear combination of the solutions $\chi_{01}^-, \dots, \chi_{06}^-$ of the homogeneous version of (2.39).

Duplicating Theorem 7 for the Lamé equations we obtain a theorem on the solvability of the Dirichlet problem in Ω^- for the Stokes equations.

We conclude by remarking that the method of boundary integral equations is applicable to the Stokes equations in the plane case as well. In this case one has to use the fundamental solution given by

$$u_i^j = \frac{1}{4\pi\nu}\left[\delta_i^j \log r - \frac{(x_i - \xi_i)(x_j - \xi_j)}{r}\right];$$

$$q^j = -\frac{1}{2\pi}\frac{\partial(\log r)}{\partial x_j}.$$

Chapter 3
Some More Applications of the Boundary Integral Equations Method

The two preceding chapters concentrated on applications of the boundary integral equations method to the Dirichlet and Neumann problems for the Laplace equation and to the first and second boundary value problems of elasticity and hydromechanics of viscous fluids. The examples we have considered are not only the typical but also the chronologically first applications of the method of boundary integral equations. Since its pioneering time this method has found many applications not only to boundary value problems but also to initial-boundary value problems of mathematical physics. The aim of the present chapter is to give a more or less complete picture of the variety of those problems. For lack of space we cannot embark on all the various directions into which the boundary integral equations method has been developed and so several of its applications are merely listed in the concluding remarks to this chapter. With different degree of minuteness, the present chapter deals with the following topics: the singular integral equations of the oblique derivative problem for the Laplace equation (Sec. 1); systems of singular integral equations connected with boundary value problems for the biharmonic operator (Sec. 2); the integral equations of time-dependent boundary value problems for equations of the theory of heat conduction and wave propagation, of dynamical elasticity theory, and also of the quasi-static and dynamical theories of visco-elasticity (Sec. 3).

§ 1. Integral Equations for the Oblique Derivative Problem

We confine ourselves to the oblique derivative problem for the Laplace equation. Let Ω^+ be a simply connected bounded region in $\mathbb{R}^n$ ($n \geq 2$) whose boundary S belongs to the class $C^{1,\alpha}$ ($0 < \alpha < 1$). The *oblique derivative problem* consists in determining a harmonic function $u^+ \in C^1(\Omega^+)$ satisfying the boundary condition

$$\partial u^+/\partial \mathbf{l} = \psi^+ \qquad \text{on } S, \tag{3.1}$$

where ψ^+ is a given function in $C(S)$ and $\mathbf{l}$ denotes any field of unit vectors on S. This problem first appeared in Poincaré's investigations devoted to the theory of tides (see Poincaré [1910]), and therefore some authors call it the *Poincaré problem*. Note that it was just this problem which led Poincaré to introducing singular integral equations over an arbitrary contour into mathematics.

We first cite some results on the plane case. If $n = 2$, condition (3.1) can be rewritten as

$$\cos \lambda u^+_{x_1} + \sin \lambda u^+_{x_2} = \psi^+, \tag{3.2}$$

where $\lambda(s)$ is the angle made by the vector $\mathbf{l}_s$ occuring in (3.1) and the positively oriented x_1-axis; s denotes arc length on S.

Seeking a function u^+ harmonic in Ω^+ amounts to finding an analytic function Φ of the complex variable $z = x_1 + ix_2$ whose real part is u^+. We assume again that the origin is located in Ω^+. So the function Φ can be conveniently sought in the form

$$\Phi(z) = \int_S \varrho(\zeta)\log\frac{\zeta - z}{\zeta}\,d_\zeta s,$$

ϱ being an unknown real-valued density satisfying a Hölder condition. With this ansatz, equation (3.2) leads to the following singular integral equation (see Muskhelishvili [1968]):

$$\text{Re}\{-\pi i(\overline{dz/ds})\}\varrho(z) - \int_S \varrho(\zeta)\text{Re}\frac{e^{i(\zeta-z)}}{\zeta - z}\,d_\zeta s = \psi^+(z); \qquad z, \zeta \in S. \qquad (3.3)$$

Theorem 1 (Muskhelishvili [1968]). *The index $\varkappa$ of equation (3.3) is finite and given by*

$$\varkappa = 2 - \pi^{-1}[\lambda(s)]_S,$$

where $[\lambda]_S$ denotes the increment of the angle λ as the result of a counter-clockwise circuit of the contour S; $\varkappa$ is an even integer.

In case $\varkappa \leq 0$, equation (3.3) with $\psi^+ = 0$ on S possesses one nontrivial solution, and for the solvability of equation (3.3) it is necessary and sufficient that its right-hand side satisfy $1 - \varkappa$ orthogonality conditions.

If $\varkappa > 0$, then the homogeneous version of equation (3.3) has $\varkappa$ linearly independent solutions and the inhomogeneous equation is solvable for arbitrary right-hand sides.

What Theorem 1 says about the integral equation (3.3) also applies to the oblique derivative problem. Notice that if $\varkappa \leq 0$, then the only solution of the homogeneous boundary value problem is the functions u^+ taking on constant values throughout $\overline{\Omega^+}$.

In higher dimensions ($n \geq 3$) a boundary integral equation for the oblique derivative problem was established and studied by Giraud [1934]. The solution of the problem is sought as a simple layer potential (see (1.4)) and for the unknown density one obtains the singular integral equation

$$-\frac{\cos(\mathbf{v}, \mathbf{l})}{2}\varrho(x) + \int_S \varrho(y)\frac{\partial}{\partial\mathbf{l}_x}\mathscr{E}(x, y)d_y S = \psi^+(x), \qquad x \in S, \qquad (3.4)$$

where $\mathbf{v}$ is the normal at $x \in S$ directed towards the outside of Ω^+.

By working out his theory of singular integral equations with nonvanishing symbol, S.G. Mikhlin was able to simplify the arguments used by Giraud to investigate equation (3.4). The symbol $A(x, \xi)$ of the singular operator generated by the left-hand side of (3.4) is

$$1/2(-\cos(\mathbf{v}, \mathbf{l}) + i \cos(\boldsymbol{\xi}, \mathbf{l}))$$

(see Sec. 3 of Chap. 4 of S. Prössdorf's article contained in this volume). This symbol does not degenerate in the case $n = 2$. If $n > 2$, it has zeros if and only if $\cos(\mathbf{v}, \mathbf{l}) = 0$ at some points on the surface S. In the latter case one speaks of the degenerate oblique derivative problem.

The theory of multidimensional singular integral equations yields the following result (see Mikhlin [1962]).

Theorem 2. *If the vector field* l *is nowhere tangent to the surface S, then equation* (3.4) *is normally solvable and its index equals zero.*

By virtue of the theorem on the sign of the oblique derivative at the extremal points of a harmonic function, the only solution of the homogeneous equation (3.4) is the density of the Robin potential (see Sec. 2 of Chap. 1). This combined with Theorem 2 shows that for the non-degenerate oblique derivative problem to have a solution it is necessary and sufficient that the function ψ^+ satisfy one orthogonality condition. The solution of the problem is then uniquely determined up to an arbitrary additive constant.

There is a number of works devoted to the degenerate oblique derivative problem (see, for example, Hörmander [1966], Malyutov [1969], Egorov, Kondrat'ev [1969], Maz'ya [1972], Egorov [1984], Maz'ya and Paneyakh [1974]). The most complete results were obtained for the case where the field l is tangent to S on some $(n - 2)$-dimensional submanifold S_0, at the same time being nontangential to S_0. The components of S_0 can so be divided into three classes: so-called "entrance" points (of l into Ω^+), "exit" points, and "status quo" points. With regard to the solvability of the problem, these three situations manifest themselves in the following manner. The entrance components yield an overdetermination of the problem, the exit components involve an under-determination, and the components of S_0 along which the field l remains on the same side of S do no affect the solvability of the problem.

The restriction to the case that there be no points on S_0 at which the field is tangent to S_0 was removed in the paper Maz'ya [1972], where a theorem on the unique solvability of the oblique derivative problem in a generalized sense was established. The latter theorem includes the situation where on the exit set the boundary condition $u = 0$ is given and also the possibility that the solution is discontinuous at the entrance points. Both sorts of points are permitted to belong to one and the same component of the manifold S_0.

A part of the results on the degenerate oblique derivative problem quoted above was obtained in the context of the general theory of pseudodifferential operators on manifolds and another part of them was established by resorting to the methods of the theory of elliptic differential equations of the second order (maximum principle, a-priori estimates, etc.). We remark that the boundary integral equation (3.4) has not yet been studied in detail in the case of sufficiently general degeneracies of the symbol.

§2. Integral Equations of Boundary Value Problems for the Biharmonic Equation

2.1. Systems of Boundary Integral Equations for the Biharmonic Equation.
Ya. B. Lopatinskiĭ [1953] developed a method of reducing boundary value problems for elliptic systems in a bounded domain to regular integral equations. To exemplify his method, he considered the first boundary value problem for the biharmonic equation. The potentials and boundary integral equations for the biharmonic equation are thoroughly studied in the works Panich [1960], [1961], [1962], [1966].

Let Ω^+ be a simply connected bounded region in $\mathbb{R}^n$ ($n = 2, 3$) with a boundary of the class $C^{1,\alpha}$ ($0 < \alpha < 1$). Set $\Omega^- = R^n \backslash \Omega^+$.

A function $u^\pm$ in $C^4(\Omega^\pm)$ is said to be *biharmonic* in $\Omega^\pm$ if it satisfies the equation

$$\Delta^2 u^\pm = 0 \quad \text{in } \Omega^\pm, \tag{3.5}$$

where $\Delta^2 = \sum_{j=1}^n \partial^4/\partial x_j^4 + 2\sum_{j=2, i<j}^n \partial^4/\partial x_i^2\, \partial x_j^2$.

A *fundamental solution of equation* (3.5) is any distribution $\mathscr{E}(x, \xi)$ such that

$$\Delta_x^2 \mathscr{E} = \delta(x - \xi) \quad \text{in } \mathbb{R}^n,$$

$\delta(x - \xi)$ being the Dirac measure concentrated at the point ξ. Usually one works with the fundamental solution given by

$$E(x, \xi) = \begin{cases} -r/(8\pi) & \text{if } n = 3, \\ r^2 \log r/(8\pi) & \text{if } n = 2 \end{cases}$$

(see §2 of Chap. 12 of the book Sobolev [1974]). Here $r = |x - \xi|$. Biharmonic functions admit a representation similar to formula (1.3) for harmonic functions (see §1 of Chap. 12 of Sobolev [1974]). Furthermore, any biharmonic function u can be expressed in terms of two harmonic functions v_1 and v_2, viz

$$u(x) = |x|v_1(x) + |x|^2 v_2(x) + u(0).$$

A large number of properties of biharmonic functions extends to polyharmonic functions, those which satisfy the equation $\Delta^m u = 0\, (m \geq 2)$ (see Sobolev [1974]).

For $n = 2, 3$ one can introduce the *biharmonic potentials* by

$$(V\varrho)(x) = \frac{-1}{2(n-1)\pi} \int_S \frac{(\mathbf{r}, \mathbf{n}_\xi)^2}{r^n} \varrho(\xi)\, d_\xi S, \tag{3.6}$$

$$(W\chi)(x) = \frac{(-1)^n n}{2(n-1)\pi} \int_S \frac{(\mathbf{r}, \mathbf{n}_\xi)^3}{r^{n+2}} \chi(\xi)\, d_\xi S, \tag{3.7}$$

where $\mathbf{r} = \xi - x$. The densities ϱ and χ are assumed to be continuous. Note that the potentials $V\varrho$ and $W\chi$ are biharmonic functions in $\Omega^\pm$. Furthermore, we have as $|x| \to \infty$

$$(V\varrho)(x) = O(|x|^{-n+2}); \qquad (W\chi)(x) = O(|x|^{-n+1}). \tag{3.8}$$

Since, by (1.2), $r^{-1}(\mathbf{r}, \mathbf{n}_\xi) = O(r^\alpha)$, the potential (3.7) possesses a continuous direct value $W_0\chi$ on S.

Theorem 3. *As $x \in \Omega^\pm$ tends to $x_0 \in S$, the boundary value $(W_\pm\chi)(x_0)$ of $(W\chi)(x)$ exists and one has*

$$W_\pm\chi = \pm\tfrac{1}{2}\chi + W_0\chi. \tag{3.9}$$

The following theorem shows that the normal derivative of the potential (3.7) has a jump on S.

Theorem 4. *If S belongs to the class $C^{2,\alpha}$ and χ is in $C^1(S)$, then the limit values $\partial(W\chi)/\partial\mathbf{n}_\pm$ of the normal derivative of the potential (3.7) exist and we have*

$$\frac{\partial(W\chi)}{\partial\mathbf{n}_\pm} = \mp\, 2^{n-2}\varkappa\chi + \frac{\partial(W\chi)}{\partial\mathbf{n}_0}, \tag{3.10}$$

where $\varkappa$ is the mean curvature of S and $\partial(W\chi)/\partial\mathbf{n}_0$ denotes the direct value of the normal derivative.

Theorem 5. *The potential (3.6) is continuous in $\mathbb{R}^n$ and the boundary values of its regular normal derivative are given by the formula*

$$\frac{\partial(V\varrho)}{\partial\mathbf{n}_\pm} = \mp\frac{1}{2}\varrho + \frac{\partial(V\varrho)}{\partial\mathbf{n}_0}, \tag{3.11}$$

the second term on the right being the direct value of the normal derivative.

A solution of the first boundary value problem for the biharmonic equation is any function $u^\pm$ given in $\Omega^\pm$ for which

$$\begin{aligned} \Delta^2 u^\pm &= 0 \quad \text{in } \Omega^\pm, \\ u^\pm = \varphi^\pm \quad \text{and} \quad \partial u^\pm/\partial\mathbf{n} &= \psi^\pm \quad \text{on } S. \end{aligned} \tag{3.12}$$

Looking for $u^\pm$ in the form $V\varrho^\pm + W\chi^\pm$, we obtain, due to (3.9)–(3.11), the following system for $\varrho^\pm$ and $\chi^\pm$:

$$\pm\chi^\pm + 2W_0\chi^\pm + 2V_0\varrho^\pm = 2\varphi^\pm,$$

$$\mp\varrho^\pm \mp 2^{n-1}\varkappa\chi^\pm + 2\frac{\partial(W\chi^\pm)}{\partial\mathbf{n}_0} + 2\frac{\partial(V\varrho^\pm)}{\partial\mathbf{n}_0} = 2\psi^\pm.$$

In the case $n = 2$, Panich [1961] proposed employing other potentials in order to reduce problem (3.12). Let

$$(W^{(1)}\chi)(x) = \frac{1}{\pi}\int_S \chi(\xi)\mathscr{Y}(x, \xi)\, d_\xi S,$$

where

$$\mathscr{Y}(x, \xi) = \frac{-(\mathbf{r}, \mathbf{n}_\xi)^3}{r^4} + \varkappa(\xi)\frac{(\mathbf{r}, \mathbf{n}_\xi)^2}{r^2} + \frac{1}{4}\frac{\partial}{\partial s_\xi}\left\{\varkappa(\xi)\frac{\partial}{\partial s_\xi}[r^2\log r]\right\}.$$

Here $\partial/\partial s_\xi$ denotes differentiation along the arc S. Seeking the solution of the first boundary value problem in Ω^+ in the form $u^+ = V^{(1)}\varrho^+ + W^{(1)}\chi^+$, where $V^{(1)}\varrho^+$ is the potential with the kernel $-(2\pi)^{-1}\log r$, we conclude that the densities ϱ^+ and χ^+ must satisfy the system

$$\chi^+ + 2W_0^{(1)}\chi^+ + 2V_0^{(1)}\varrho^+ = \varphi^+,$$

$$\varrho^+ + 2\frac{\partial(W^{(1)}\chi^+)}{\partial\mathbf{n}_0} + 2\frac{\partial(V^{(1)}\varrho^+)}{\partial\mathbf{n}_0} = \psi^+.$$

(3.13)

This system is not equivalent to the internal problem (3.12), since the latter problem is uniquely solvable whereas the solutions of the homogeneous system (3.13) span a one-dimensional space. The solvability of the system (3.13) entails an orthogonality condition for the vector (φ^+, ψ^+). What eventually results is the following.

Theorem 6. *Let S be of the class $C^{5,\alpha}$ $(0 < \alpha < 1)$ and suppose $\varphi^+ \in C^{3,\alpha}(S)$ and $\psi^+ \in C^{2,\alpha}(S)$. Seek the solution of problem (3.12) in Ω^+ in the form*

$$u^+ = W^{(1)}\chi^+ + V^{(1)}\varrho^+ + Av^+,$$

where $v^+ \in C^1(\overline{\Omega^+})$ is harmonic in Ω^+ and A is an unknown constant. Then the system (3.13) has a solution (χ^+, ϱ^+) for any A determined by the afore-mentioned orthogonality condition.

In order to solve problem (5.12) in Ω^-, Panich [1961] utilized a technique akin to the one applied in Sec. 2.3 of Chap. 1 to the external Dirichlet problem for the Laplace equation.

Boundary value problems of a more general form for the biharmonic equation in the plane are studied by means of potentials in the article Panich [1966]. Note that there polyharmonic equations are considered as well.

In conclusion we remark that the system of integral equations arising from the first internal boundary value problem for the biharmonic equation was investigated in Cohen, Gosselin [1983] in the case where S is a C^1 curve in the plane and the boundary data are subject to the conditions that $\varphi^+ \in W_p^1(S)$ and $\psi^+ \in L_p(S)$. The matrix operator generated by the direct values of the potentials of Agmon [1957] turns out to be compact on the direct product $W_p^1(S) \times L_p(S)$ (compare this with the results of Fabes, Jodeit, and Riviere that will be quoted in Sec. 3 of Chap. 4). The system of integral equations and thus also the boundary value problem are uniquely solvable.

2.2. Methods of Complex Function Theory and Boundary Integral Equations for Plane Problems of Elasticity Theory. Plane problems of elasticity can be reduced to finding the so-called *Airy function W*, which is a solution of the biharmonic equation (see Novatskiĭ [1975]). The function W in turn can be expressed by two complex-valued analytic functions φ and ψ in the form

$$W(x, y) = \text{Re}\{\overline{z}\varphi(z) + \chi(z)\},$$

(3.14)

where $z = x + iy$ and $d\chi/dz = \psi$ (see Muskhelishvili [1966]). The latter formula is now named after Goursat, who established it in 1898, and it enables us to convert the fundamental boundary value problems of plane elasticity theory into problems of finding the two analytic functions φ and $\psi = d\chi/dz$.

The first problem, which concerns the case where the displacements on the contour are given, reduces itself to the determination of two function φ and ψ analytic in Ω^+ satisfying the boundary condition

$$\varkappa\varphi(\zeta) - \zeta\overline{\varphi'(\zeta)} - \overline{\psi(\zeta)} = g(\zeta), \qquad \zeta \in S. \tag{3.15}$$

Here $\varkappa = (\lambda + 3\mu)(\lambda + \mu)^{-1}$, $g = 2\mu(u + iv)$, λ and μ are the Lamé constants, and u, v denote the displacements.

In the case of the second problem of elasticity theory the boundary condition also assumes the form (3.15), but now $\varkappa = -1$ and the function g is a certain curvilinear integral of the stresses on S.

N.I. Muskhelishvili [1966] showed that the analytic functions φ and ψ satisfying condition (3.15) can be obtained by solving an integral equation for the boundary values of the function φ. N.I. Muskhelishvili's equation is as follows:

$$-\varkappa\overline{\varphi(t)} - \frac{\varkappa}{2\pi i}\int_S \overline{\varphi(\zeta)}d\log\frac{\overline{\zeta} - \overline{t}}{\zeta - t} - \frac{1}{2\pi i}\int_S \varphi(\zeta)d\frac{\overline{\zeta} - \overline{t}}{\zeta - t} = A(t); \qquad \zeta, t \in S, \tag{3.16}$$

where $A(t) = -(1/2)g(t) + (2\pi i)^{-1}\int_S g(\zeta)(\zeta - t)^{-1}d\zeta$, the integral understood in the Cauchy principal value sense. The derivation of this equation relies upon the Cauchy theorem and the Sokhotskiĭ-Plemelj formulas.

Equation (3.16) can be rewritten as

$$\varkappa\overline{\varphi(t)} - \frac{\varkappa}{\pi}\int_S \overline{\varphi(\zeta)}\frac{\partial\vartheta}{\partial s}d_\zeta S - \frac{1}{\pi}\int_S \varphi(\zeta)e^{-2i\vartheta}\frac{\partial\vartheta}{\partial s}d_\zeta S = -A(t). \tag{3.17}$$

Here $\zeta - t = re^{i\vartheta}$ and $\partial\vartheta/\partial s = -\cos(\mathbf{r}, \mathbf{n}_\zeta)/r$. If the curve S belongs to the class $C^{1,\alpha}$ ($0 < \alpha < 1$), the real and imaginary parts of equation (3.17) form a system for which the Fredholm alternative is in force.

In case $\varkappa = 1$, the homogeneous version of (3.17) has the nontrivial solution $\varphi_0(t) = i\alpha t + \beta$ ($\alpha \in \mathbb{R}, \beta \in \mathbb{C}$). This leads to the following result.

Theorem 7. *If $\varkappa = 1$, then for equation (3.17) to be solvable it is necessary and sufficient that*

$$\mathrm{Re}\int_S \overline{g}dS = 0, \tag{3.18}$$

g being the right-hand side of (3.15). The solution of equation (3.17) admits analytic continuation into the domain Ω^+.

We remark that condition (3.18) is also entailed by the solvability of the second boundary value problem of elasticity theory.

Theorem 8. *If $\varkappa > 1$, equation (3.17) is uniquely solvable for every right-hand side.*

Note that the equations (3.16) and (3.17) can be also employed in the case of multiply connected domains.

To determine the functions φ and ψ analytic in Ω^+ one may also resort to the *Sherman-Lauricella integral equation*. In order to derive this equation represent the analytic functions φ and ψ in the form

$$\varphi(z) = \frac{1}{2\pi i} \int_S \frac{w(\zeta)}{\zeta - z} d\zeta,$$

$$\psi(z) = \frac{1}{2\pi i} \int_S \frac{\overline{w(\zeta)}}{\zeta - z} d\zeta - \frac{1}{2\pi i} \int_S \frac{\overline{\zeta} w(\zeta)}{(\zeta - z)^2} d\zeta + \frac{1}{2\pi i} \int_S \frac{w(\zeta)}{\zeta - z} d\overline{\zeta},$$

$$(3.19)$$

with an unknown function $w(\zeta)$ defined on S. Suppose (3.18) is fulfilled. From the boundary condition (3.15) with $\varkappa = -1$ we then get the following equation for w:

$$w(t) + \frac{1}{2\pi i} \int_S w(\zeta) d \log \frac{\zeta - t}{\overline{\zeta} - \overline{t}} - \frac{1}{2\pi i} \int_S \overline{w(\zeta)} d \frac{\zeta - t}{\overline{\zeta} - \overline{t}} = g(t). \qquad (3.20)$$

This equation is referred to as *Lauricella's equation*. It is solvable if and only if (3.18) holds.

To obtain an equation that is solvable for every right-hand side, D.I. Sherman replaced (3.20) by the equation

$$w(t) + \frac{1}{\pi} \int_S w d\vartheta - \frac{1}{\pi} \int_S \overline{w(\zeta)} e^{2i\vartheta} d\vartheta$$

$$+ \left(\frac{1}{t - a} - \frac{1}{\overline{t} - \overline{a}} + \frac{(t - a)}{(\overline{t} - \overline{a})^2} \right) \frac{1}{\pi i} \operatorname{Re} \int_S \frac{w(\zeta) d\zeta}{(\zeta - a)^2} = f(t). \qquad (3.21)$$

Here $\zeta - t = r e^{i\vartheta}$ and a is an inner point of Ω^+. Equation (3.21) has a solution for arbitrarily chosen right-hand side, and provided condition (3.18) is fulfilled, the solution of (3.21) satisfies equation (3.20).

In a similar fashion N.I. Muskhelishvili's equation with $\varkappa = 1$ can be replaced by an equation solvable for every right-hand side.

Let us finally remark that the system emerging from separation of (3.21) into the real and imaginary parts has an adjoint system of integral equations that makes a good physical sense (see Morozov, Paukshto [1986]). The latter system appears when we use the direct variant of the boundary equations method (see Sec. 4.1 of Chap. 1) for the plane problem of elasticity theory in the formulation of B.E. Pobedrya [1980].

§3. The Solution of Nonstationary Boundary Value Problems by the Boundary Integral Equations Method

The method of boundary integral equations applies to non-stationary boundary value problems as well. In the first subsection we discuss the integral equations arising from initial-boundary value problems for equations of the parabolic type. Subsection 3.2 deals with the application of the boundary integral equations method to initial-boundary value problems for the wave equation. The concluding two subsections are devoted to the time-dependent boundary integral equations for nonstationary problems of elasticity and visco-elasticity.

3.1. Heat Potentials. Let Ω^+ be a simply connected and bounded region in $\mathbb{R}^n$ ($n \geq 2$) whose boundary S is in the class $C^{1,\alpha}$ ($0 < \alpha < 1$), denote $\mathbb{R}^n \backslash \overline{\Omega^+}$ by Ω^-, and put $Q_T^{\pm} = \Omega^{\pm} \times (0, T)$ and $S_T = S \times (0, T)$.

Our subject here is the equation of heat conduction

$$u_t - a^2 \Delta u = 0, \qquad a > 0, \tag{3.22}$$

Δ being the Laplace operator. A *fundamental solution of equation* (3.22) *is given by*

$$\mathscr{E}(x, t; \xi, \tau) = \frac{\Theta(t - \tau)}{[2a\sqrt{\pi(t - \tau)}]^n} \exp\{-r^2/(4a^2 t)\}$$

(see Vladimirov [1976]). Here $r = |x - \xi|$ and $\Theta(t)$ is the Heaviside function. It is easily seen that

$$\frac{\partial \mathscr{E}}{\partial t} - a^2 \Delta_x \mathscr{E} = \delta(x - \xi, t - \tau).$$

Let $\varrho, \chi \in C(\overline{S}_T)$. The integrals

$$(V\varrho)(x, t) = \int_0^t d\tau \int_S \varrho(\xi, \tau) \mathscr{E}(x, t; \xi, \tau) \, d_\xi S,$$

$$(W\chi)(x, t) = \int_0^t d\tau \int_S \chi(\xi, \tau) \frac{\partial}{\partial \mathbf{n}_\xi} \mathscr{E}(x, t; \xi, \tau) \, d_\xi S,$$

depending on the parameters (x, t), are referred to as the *simple* and *double layer heat potential*, respectively (cf. (1.4) and (1.5)). These integrals satisfy equation (3.22) in $Q_T^{\pm}$ for every $T > 0$. Clearly, $(V\varrho)(x, 0) = (W\chi)(x, 0) = 0$ on $\Omega^{\pm}$.

Theorem 9. *The limit values* $W_{\pm}\chi$ *on* S_T *of the potential* $W\chi$ *from out of* $Q_T^{\pm}$ *do exist. They are linked with the direct value* $W_0\chi$ *by the equalities (cf. (1.8))*

$$W_{\pm}\chi = \mp \tfrac{1}{2}\chi + W_0\chi. \tag{3.23}$$

Theorem 10. *The regular normal derivative of the potential* $V\varrho$ *exists and equals*

$$\frac{\partial(V\varrho)}{\partial \mathbf{n}_{\pm}} = \pm \frac{1}{2}\varrho + \frac{\partial(V\varrho)}{\partial \mathbf{n}_0} \tag{3.24}$$

(cf. (1.10)), $\partial(V\varrho)/\partial \mathbf{n}_0$ *denoting the direct value of the normal derivative.*

The simple layer potential is continuous in x as well as in t on $\mathbb{R}^n \times [0, T]$.

Definition. A function $u^{\pm}$ is said to be a classical solution of the *first mixed problem for equation* (3.22) in $Q_T^{\pm}$ if all its derivates occuring in this equation exist and are continuous and if the function itself belongs to $C(\overline{Q_T^{\pm}})$ and satisfies the initial condition

$$u^{\pm} = 0 \qquad \text{for } t = 0 \text{ on } \overline{\Omega}^{\pm} \tag{3.25}$$

as well as the boundary condition

$$u^{\pm} = \varphi^{\pm} \qquad \text{on } \overline{S}_T. \tag{3.26}$$

Here $\varphi^{\pm} \in C(\overline{S}_T)$ is a given function such that $\varphi^{\pm} = 0$ on S for $t = 0$.

In order to reduce this problem to a boundary integral equation, its solution may be sought as a potential $W\chi^{\pm}$ with an unknown density $\chi^{\pm}$. Due to formula (3.23), the density must fulfil the equation

$$\chi^{\pm} \mp T\chi^{\pm} = \mp 2\varphi^{\pm}, \qquad T = 2W_0. \tag{3.27}$$

The *second mixed problem for equation* (3.22) differs from the first one only in that now instead of (3.26) a Neumann boundary condition on S is given. Looking for the solution in the form $V\varrho^{\pm}$ produces an equation for the unknown density $\varrho^{\pm}$ which turns out to be the adjoint of (3.27).

Theorem 11. *The integral equations for the first and second mixed problems for equation* (3.22) *are Volterra equations whose kernels have a weak singularity (see Sec. 1 of Chap. 2 of S. Prössdorf's article in this volume). Their solutions are continuous throughout S_T whenever their right-hand sides are so. Moreover, the solutions can be constructed by means of the method of successive approximation and are thus available in the form of uniformly convergent series.*

If S is a contour with continuous curvature, the nth term of the Neumann series does not exceed

$$(Ct)^{n/2} \max |\varphi^{\pm}| / \Gamma(n/2 + 1).$$

The constant C depends on S (see Müntz [1934]). For arbitrary convex (but not necessarily smooth) contours S the convergence of Neumann's series is proved in Mikhlin [1949].

The material of the present subsection is classical (see, for example, Mikhlin [1949] or Müntz [1934]). The results cited above were generalized into several directions.

So the Laplace operator in equation (3.22) may be replaced by a general elliptic operator of the second order with smooth coefficients (see the book Ladyzhenskaya, Solonnikov, Ural'tseva [1967]).

The mixed problems for equation (3.22) were studied in cylinders $Q_T^{\pm}$ the base $\Omega^{\pm}$ of which is permitted to have a nonsmooth boundary S. Kresin [1981] showed that if S possesses nonintersecting edges, then the spectral radius of the operator T equals $(\pi - \varphi)/\pi$, where φ is the least of the angles contained by the tangential planes at the points of the edges. In the same paper, he also proved that if there are finitely many circular conic points O_i ($1 \leq i \leq N$) on the surface, then the spectral radius is $\cos(\psi/2)$, where $\psi = \min\{\varphi_1, \ldots, \varphi_N, 2\pi - \varphi_1, \ldots, 2\pi - \varphi_N\}$ and φ_i is the solid angle made by the vertex of O_i (measured from the inside of Ω^+). Moreover, Kresin [1981] determined the essential norm of the operator T and considered the matrix case.

The first initial-boundary value problem with homogeneous initial conditions for the heat equation in the quarter-space was investigated by Fabes, Jodeit, Lewis [1977]. They showed that in the case $p > 3/2$ the corresponding boundary integral equation is solvable in L_p whenever the data are taken from L_p and that in the case $1 < p < 3/2$ the space of the solutions of the homogeneous equation is infinite-dimensional.

The articles Dont [1972], [1975a], [1975b], [1976], [1981], [1983] and Veselý [1973], [1975] deal with heat potentials and boundary value problems in domains satisfying a condition analogous to condition (A) of Sec. 2 of Chap. 4.

Heat potentials for noncylindrical domains are the subject of the paper Kamynin [1965]. Finally the works Mikhailov [1960], Solonnikov [1965], and Eĭdel'man [1964] concern generalizations of the method of potentials to systems of parabolic equations.

3.2. Wave Potentials. The first hints to the possibility of applying potentials to the wave equation are contained in the work of Volterra and Hadamard (see the book Hadamard [1932]). The same ideas were wielded by Müntz [1932] to treat the dynamical plane problem of elasticity, which can be reduced to the problem of finding scalar- and vector-valued solutions of the wave equation which are, in a sense, connected with the boundary conditions. *Wave potentials* (in the multi-dimensional case) were thoroughly studied by S.G. Mikhlin and V.D. Sapozhnikova [1977], and all what will follow in this subsection is taken from their paper we have just cited.

Consider the *wave equation*

$$u_{tt} - \Delta u = 0 \qquad \text{in } Q_\infty^\pm \tag{3.28}$$

with the homogeneous initial conditions

$$u = u_t = 0 \qquad \text{for } t = 0 \text{ in } \Omega^\pm. \tag{3.29}$$

Put

$$\mathscr{E}(x - \xi, t - \tau) = \frac{1}{(n-2)!\sigma_n} \frac{\partial^{n-1}}{\partial t^{n-1}} q\left(\frac{t - \tau}{r}\right),$$

where

$$q(\lambda) = \begin{cases} \displaystyle\int_1^\lambda (z^2 - 1)^{(n-3)/2} dz & \text{for } \lambda \geqq 1, \\ 0 & \text{for } \lambda < 1. \end{cases}$$

The distribution $\mathscr{E}$ is a *fundamental solution of equation* (3.28).

Denote by $K_{x,t}$ the characteristic cone with the vertex at (x, t) and set $S_{x,t} = K_{x,t} \cap S_\infty$. Any function $u \in C^2(\overline{Q_\infty^\pm})$ satisfying (3.28) and (3.29) admits the representation

$$u(x, t) = \frac{1}{(n-2)!\sigma_n} \frac{\partial^{n-1}}{\partial t^{n-1}} \int_{S_{x,t}} \left[q\left(\frac{t - \tau}{r}\right) \frac{\partial u}{\partial \mathbf{n}_\xi} - u \frac{\partial}{\partial \mathbf{n}_\xi} q\left(\frac{t - \tau}{r}\right) \right] d_\xi S \, d\tau.$$

The integrals

$$(V\varrho)(x, t) = \frac{1}{(n-2)!\sigma_n} \frac{\partial^{n-1}}{\partial t^{n-1}} \int_{S_{x,t}} \varrho(\xi, \tau) q\left(\frac{t-\tau}{r}\right) d_\xi S \, d\tau,$$

$$(W\chi)(x, t) = \frac{1}{(n-2)!\sigma_n} \frac{\partial^{n-1}}{\partial t^{n-1}} \int_{S_{x,t}} \chi(\xi, \tau) \frac{\partial}{\partial n_\xi} q\left(\frac{t-\tau}{r}\right) d_\xi S \, d\tau$$

are called the *simple* and *double layer wave potentials*, respectively.

Theorem 12 (Mikhlin, Sapozhnikova [1977]). *The following conditions ensure the validity of the formulas* (3.23) *and* (3.24):

(a) $n = 2$; $\varrho \in C^{(0,1)}(\bar{S}_\infty)$ *and* $\varrho(x, 0) = 0$; $\chi \in C^{(0,2)}(\bar{S}_\infty)$ *and* $\chi(x, 0) = \chi_t(x, 0) = 0$;

(b) $n = 2k, k > 1$ *or* $n = 2k + 1, k = 0, 1, \ldots$; $\varrho \in C^{(0,k+1)}(\bar{S}_\infty)$, $\partial^j \varrho / \partial t^j = 0$ *for* $j = 0, \ldots, k$; $\chi \in C^{(0,k+2)}(\bar{S}_\infty)$, $\partial^j \chi / \partial t^j = 0$ *for* $j = 0, \ldots, k + 1$.

The simple and double layer wave potentials satisfy equation (3.28). Furthermore, if certain conditions analogous to but stronger than (a) and (b) are satisfied then (3.29) holds (see Mikhlin, Sapozhnikova [1977] for the precise statement).

If a function u fulfils equation (3.28) and the initial conditions (3.29) and, in addition, satisfies a Dirichlet condition on $\bar{S}_\infty$ (see (3.26)), it is called a solution of the *first mixed problem for the wave equation*. Replacing the Dirichlet by Neumann gives the *second problem* on $\bar{S}_\infty$. Seeking the solutions of these problems in the form of a double and simple layer potential, respectively, leads to the following equations for the unknown densities $\chi^\pm$ and $\varrho^\pm$:

$$\chi^\pm \mp 2W_0\chi^\pm = \mp 2\varphi^\pm,$$

$$\varrho^\pm \mp 2\frac{\partial(V\varrho^\pm)}{\partial n_0} = \pm 2\psi^\pm.$$

In case n is an odd integer, take the Laplace transform with respect to t to obtain

$$\tilde{\chi}^\pm(x, \eta) \pm \frac{2}{(n-2)!\sigma_n} \int_S \frac{1}{r^{n-3}} \frac{\partial}{\partial n_\xi} \frac{1}{r} K(r, \eta) \tilde{\chi}^\pm(\xi, \eta) d_\xi S = \pm 2\tilde{\varphi}^\pm(x, \eta),$$

$$(3.30)$$

$$\tilde{\varrho}^\pm(x, \eta) \pm \frac{2}{(n-2)!\sigma_n} \int_S \frac{1}{r^{n-3}} \frac{\partial}{\partial n_x} \frac{1}{r} K(r, \eta) \tilde{\varrho}^\pm(\xi, \eta) d_\xi S = \pm 2\tilde{\psi}^\pm(x, \eta).$$

$$(3.31)$$

Here, of course, $\tilde{\chi}^\pm, \tilde{\varrho}^\pm, \tilde{\varphi}^\pm, \tilde{\psi}^\pm$ are the Laplace transforms of the corresponding functions; $K(r, \eta)$ is given by

$$K(r, \eta) = \left(\frac{n-3}{2}\right)! \sum_{(n-3)/2 \leq k \leq n-2} \frac{e^{-\eta r}}{\eta^{k-n+2}} \binom{k}{\frac{n-3}{2}}$$

$$\times \frac{k+1}{2} \frac{n-3}{2} \frac{n-5}{2} \ldots (n-k-1)(2r)^{n-k-2}.$$

Theorem 13. *The equations* (3.30) *and* (3.31) *are solvable for every* $\varphi^{\pm}$, $\psi^{\pm} \in$ $L_2(S_\infty)$ *if only the surface S belongs to the class* $C^{[n/2]+5}$.

The conclusion is that there exist solutions of the first and second mixed problems representable by double and simple layer potentials, respectively. In the case where n is even, one may operate with a lifting argument, that is, one may apply the result for $n + 1$ dimensions to the situation where both the solution and the data are independent of one and the same spatial variable (see Mikhlin, Sapozhnikova [1977] for details).

Theorems on the solvability of integral equations of retarded potential theory in certain pairs of Sobolev spaces are proved in Bamberger and Ha Duong [1986].

3.3. Potentials of Nonstationary Problems of Elasticity Theory. Boundary integral equations methods have also found application to *dynamical elasticity theory*, whose subject is initial-boundary value problems for the equation

$$A(\partial/\partial x)\mathbf{u}(x, t) - \varrho\frac{\partial^2\mathbf{u}(x, t)}{\partial t^2} = 0. \tag{3.32}$$

Here ϱ is the density of the material and A denotes the operator generated by the left-hand side of (2.1) in the case of a homogeneous isotropic body or an anisotropic medium.

First of all we need a representation of the displacement vector analogous to formula (2.4). Such sort of representations were constructed in a series of works (De Hoop [1958], Novatskiĭ [1975], and others) under the assumption that a fundamental solution $\mathscr{E}(x, y, t)$ is available. In the case of a homogeneous isotropic medium a fundamental solution was already found by Stokes (1849). A fundamental solution for the case of a three-dimensional homogeneous anisotropic medium was given by D.G. Natroshvili [1979] in the form of a triple integral and by N.M. Khutoryanskiĭ [1985] as a line integral (resembling the Herglotz-Petrovskiĭ formula, for which see Gelfand, Shilov [1958]). Note that in all these situations $\mathscr{E}(x, y, t)$ is matrix-valued.

The simple and double layer potentials are now defined by

$$(V\mathbf{\varrho})(x, t) = \int_S \mathscr{E}(x, y, t) * \mathbf{\varrho}(\xi, t)\, d_\xi S,$$

$$(W\mathbf{\chi})(x, t) = \int_S [\mathscr{T}(\partial/\partial\xi, \mathbf{n}_\xi)\mathscr{E}(x, \xi, t)]' * \mathbf{\chi}(\xi, t)\, d_\xi S,$$

where $\mathscr{T}$ is the stress operator and the asterisk refers to convolution with respect to time:

$$f * g = \int_{\mathbb{R}^1} f(t - \tau)g(\tau)\, d\tau.$$

Several types of integral equations emerging from the initial-boundary value problems of dynamical elasticity are discussed in the books Kupradze et al.

[1976] and Ugodchikov, Khutoryanskiĭ [1986]. For example, when considering the second initial-boundary value problem with homogeneous initial data, the time-dependent boundary integral equation obtained by a limit passage in the representation formula for the displacements $\mathbf{u}(x, t)$ takes the form

$$\tfrac{1}{2}\mathbf{u}(x, t) + \int_S [\mathcal{T}(\partial/\partial \xi, \mathbf{n}_\xi)\mathscr{E}(x, \xi, t)]' * \mathbf{u}(\xi, t)\, d_\xi S$$

$$= \int_S \mathscr{E}(x, \xi, t) * \mathbf{p}(\xi, t)\, d_\xi S, \quad x \in S,$$

$\mathbf{p}$ being the vector of surface forces. The existence of solutions of the latter equation follows from the theorem on the solvability of the second boundary value problem of dynamical elasticity theory (see Kupradze et al. [1976] and Duvaut, Lions [1972]).

The integral equations for mixed boundary value problems bring out analogies to the equations (2.22)–(2.24) of statics. The analogue of the boundary equation (2.23) was studied in Khutoryanskiĭ [1986b], where the corresponding operator was proved to be positive-definite.

3.4. Time-Dependent Boundary Integral Equations in Visco-Elasticity. Another field of application of boundary equations is *quasi-static visco-elasticity*. Here, unlike the theory of elasticity where Hooke's law is acting, the stresses at a given moment depend on the entire pre-history of the deformation. More precisely, the connection between stress and strain in a homogeneous body occupying a domain $\Omega^+ \subset \mathbb{R}^3$ is given by

$$\sigma_{ij}(x, t) = G_{ijkl}(t, \cdot) ** \varepsilon_{kl}(x, \cdot); \qquad x \in \Omega^+; i, j = 1, 2, 3,$$

where the double asterisk denotes *Stieltjes convolution*:

$$\int_{-\infty}^t G_{ijkl}(t, \tau) d_\tau \varepsilon_{kl}(x, \tau).$$

The $G_{ijkl}(t, \tau)$ are the so-called *relaxation functions*.

There are at least two ways of reducing the problems of visco-elasticity to integral equations: the Laplace transform may be applied either to equations with a parameter (see Shippy [1963]) or immediately to the time-dependent boundary equations (see Khutoryanskiĭ [1979]). Let us dwell on the second approach.

The matrix-valued fundamental solution $\mathscr{E}(x, t) = [\mathscr{E}_{ij}]_{i,j=1}^3$ corresponding to a unit force acting at the moment τ satisfies the equation

$$-\int_{-\infty}^t G_{ijml}(t, \lambda) d_\lambda \mathscr{E}_{mk,lj}(x, \lambda, \tau) = \delta_{ik}\delta(x)\Theta(t - \tau),$$

where $\delta(x)$ and $\Theta(t)$ are the Dirac and Heaviside functions.

By means of the matrix $\mathscr{E}$ the displacements can be expressed via the following formula (see Ugodchikov, Khutoryanskiĭ [1986] and Khutoryanskiĭ [1983]):

$$u(x, t) = \int_S \mathscr{E}(x - \xi, t, \cdot) ** \mathbf{p}(\xi, \cdot) \, d_\xi S - \int_S T(x, \xi, t, \cdot) ** \mathbf{u}(\xi, \cdot) \, d_\xi S,$$

$$(3.33)$$

$\mathbf{p}$ being surface force and T being the 3 by 3 matrix whose entries are

$$T_{ij}(x, \xi, t, \tau) = n_k(\xi) \frac{\partial}{\partial \xi_l} \mathscr{E}_{im}(x - \xi, t, \cdot) ** G_{mljk}(\cdot, \tau). \qquad (3.34)$$

Notice that, in contrast to elasticity theory, the quantities (3.34) are in general not the stresses at the point ξ. This happens only for stable media, in which case the relaxation functions merely depend on the difference $t - \tau$.

In the case of an anisotropic medium a matrix-valued fundamental solution $\mathscr{E}(x, t, \tau)$ can be explicitly given in terms of an integral over the circle (which tallies with elasticity):

$$\mathscr{E}(x, t, \tau) = \frac{1}{8\pi^2 |x|} \int_{\{\xi \in \mathbb{R}^3 : |\xi| = 1, \xi \cdot x = 0\}} C^{-1}(\xi, t, \tau) \, d_\xi S.$$

Here $C^{-1}(\xi, t, \tau)$ is the matrix determined by the Volterra equation

$$C^{-1}(\xi, t, \cdot) ** C(\xi, \cdot, \tau) = \Theta(t - \tau) I,$$

where

$$I = [\delta_{ij}], \qquad C(\xi, t, \tau) = [G_{ijkl}(t, \tau) \xi_j \xi_l], \qquad i, j, k, l = 1, 2, 3.$$

If the body is in undeformed state until the moment $t = 0$ and the stresses are given on the boundary, then (3.34) yields the equation $u + 2W_0 u = f$, where

$$(W_0 u)(x, t) = \int_S \int_0^t T(x, \xi, t, \tau) d_\tau u(\xi, \tau) \, d_\xi S.$$

The properties of the operator W_0 (including its spectral properties) are studied in Khutoryanskiĭ [1983]; also see Ugodchikov, Khutoryanskiĭ [1986].

In the case of stable media, the time-dependent integral equations for the nonstationary dynamical problems of visco-elasticity match those of dynamical elasticity (see Khutoryanskiĭ [1979]). The only difference comes from the fundamental solution utilized. Any endeavor to obtain it in an explicit form leads to the problem of finding the inverse Laplace transform of the function $s \mapsto \exp(-|x|s(s\tilde{J}\varrho)^{1/2})$, where $\tilde{J}$ is the Laplace transform of the creep function and ϱ the density of the medium. For some special models of the viscoelastic medium this problem was solved in Achenbach, Chao [1962], Khutoryanskiĭ, Igumnov [1982], Igumnov, Khutoryanskiĭ [1983].

3.5. Concluding Remarks. So far we have examined only a few examples of the application of the boundary integral equations method to problems of mathematical physics. In point of fact, the range of the applicability of this method is considerably wider. It is merely lack of space which prevents us from dwelling upon a series of other important directions into which the boundary equations method has been developed. So let us at least mention some of these directions.

1°. First and foremost we emphasize that the boundary integral equations method bears universal character and is, in essence, applicable to all the linear differential equations whose fundamental solutions are available. Methods of reducing general elliptic boundary value problems to pseudodifferential equations on the boundary were worked out by Ya.B. Lopatinskiĭ [1953], A.P. Calderón [1963], R.T. Seeley [1966], L. Hörmander [1966], M.I. Vishik and G.I. Eskin (see Eskin [1973]), to mention only some of the principal figures.

2°. The method of boundary integral equations has won great popularity in solving external boundary problems for the Helmholtz and Maxwell equations (see, e.g., Romanov [1980], Kleinman [1978], Kleinman, Wendland [1977], Angell, Kleinman [1982]). The manner in which the corresponding boundary equations are investigated is, in principle, the same as that for the Laplace equation (see Chap. 1). However, for the values of the wave number belonging to the discrete spectrum of the corresponding internal boundary value problem, the integral equations are not solvable for every right-hand side, although the external boundary value problems with radiation possess a unique solution. There exist several ways to gain boundary integral equations which are solvable for every right-hand side; one has, for instance, the so-called modified Green's function, null-field equations, or reproducing kernel methods (see Ahner [1986]).

3°. In the linear theory of surface waves boundary integral equations are applied to both stationary and nonstationary problems (see, for example, Vainberg, Maz'ya [1973], Maz'ya [1977], Angell, Hsiao, Kleinman [1986], Kuznetsov, Maz'ya [1988]). The basic Green functions used to derive the integral equations are contained in Stoker [1957] and Wehausen, Laitone [1960]. To investigate the boundary equations of the stationary problems one has to overcome the same difficulties as for the Helmholtz equation (see John [1950], Martin [1984], Kuznetsov, Maz'ya [1988]). Moreover, a series of problems concerning the unique solvability of the boundary value problems under consideration are still open (see Vainberg, Maz'ya [1973], Maz'ya [1977], Kuznetsov, Maz'ya [1988]).

4°. In addition to the applications of the boundary integral equations method to continuum mechanics presented above, this method has found broad use in the theory of elastic vibrations, couple stress elasticity, fracture mechanics, thermo-elasticity, etc. For these topics we refer to the books Kupradze et al. [1976], Parton, Perlin [1977], Burchuladze, Gegelia [1985], Ugodchikov, Khutoryanskiĭ [1986].

5°. An important aspect of the method of boundary integral equations is techniques for their numerical solution (in particular the *boundary element method*). This subject has its own set of problems and is now an independent field of active research. An account of the intense development of numerical analysis for boundary integral equations is given in the monographs Kantorovich, Krylov [1962], Brebbia, Walker [1980], Banerjee, Butterfield [1981], Crouch, Starfield [1983] and in the collections of papers "*Boundary-integral equation method: computational applications in applied mechanics*", Amer. Soc. Mech. Eng.: New York 1975 and "*Contributions of mathematical analysis to the numerical solution*

of partial differential equations", Proc. Centre Math. Anal. Austral. Nat. Univ.,
Vol. 7: Canberra 1984. Also see Wendland [1981], [1982].

The history of the boundary element method is reflected in the books Banerjee,
Butterfield [1981] and Crouch, Starfield [1983]. The first of these two books also
provides a classification of the different variants of this method. It should be
pointed out that, in comparison with other algorithms for the numerical solution
of boundary value problems, the boundary element method has a series of
advantages. One of its benefits is that it lowers the dimension of the problem by
a unit, another favour results from its convenience for solving external boundary
value problems.

We finally remark that V.S. Ryaben'kii [1987] elaborated a potential method
for systems of difference equations with constant coefficients, which leads to
discrete analogues of boundary integral equations.

Chapter 4
The Integral Equations of Potential Theory in
the Spaces C and L_p

§ 1. The Fredholm Radius of Boundary Integral Operators

1.1 Introduction. The theory of boundary integral equations exhibited in the
preceding chapters was based on the consistent assumption that the boundary
is sufficiently smooth. However, the needs of practical applications as well as the
intrinsic logic of the theory itself dictate the consideration of more general curves
and surfaces.

The pioneering work in this direction was done by Korn [1901–1902],
Zaremba [1904], and Carleman [1916], who utilized the methods of potential
theory to study boundary value problems for the Laplace equation in the case
where the boundary contains a finite number of corner points. Radon [1919b]
generalized the results of Korn and Zaremba to plane contours with "bounded
rotation" and no cusp points. He considered the corresponding integral operators in the space C of continuous functions. The new issue consists in the
circumstance that boundary integral operators on irregular curves or surfaces
loose the property of being compact, although they remain bounded operators
for a large class of boundaries. Radon conquered this difficulty by invoking an
approach to the theory of linear equations in functional spaces which had been
worked out by him earlier and which was founded on the notion of the Fredholm
radius.

1.2. Radon's Theory and Its Outgrowth. Let L be a linear bounded operator
acting on a Banach space $\mathscr{B}$. We denote by L^* the adjoint operator of L, which
is defined on the dual space $\mathscr{B}^*$ of $\mathscr{B}$.

The *Fredholm radius* $R(L)$ of the operator L (introduced by Radon [1919a]) is the radius of the largest disk centered at the origin of the complex λ-plane such that in its interior the Fredholm theorems are true for the equations

$$u - \lambda L u = f, \qquad v - \bar{\lambda} L^* v = g$$

$(u, f \in \mathcal{B}, v, g \in \mathcal{B}^*)$. For the Fredholm theorems see Sec. 5 of Chap. 1 of the article by S. Prössdorf contained in this volume. Given a compact operator K on $\mathcal{B}$, let $c_L(K)$ denote the radius of the largest disk in the λ-plane in which the series

$$I + \lambda(L - K) + \lambda^2(L - K)^2 + \cdots \tag{4.1}$$

converges uniformly. Following Radon [1919a], the number $\sup c_L(K)$, the supremum over all compact operators K on $\mathcal{B}$, is called the *continuity degree of the operator* L.

Theorem 1 (Radon [1919a]). *The Fredholm radius of an operator coincides with its continuity degree.*

The results of Radon layed the foundation of an important part of functional analysis (see the works by S.M. Nikol'skiĭ [1943], F.B. Atkinson [1951], I. Gohberg and M.G. Krein [1957]).

Definition. The quantity

$$|L| = \inf \| L - K \|_{\mathcal{B}},$$

the infimum taken over all compact operators K on $\mathcal{B}$, is referred to as the *essential norm* of the operator L.

The essential norm first appeared in Radon [1919b], where it was employed in order to gain an estimate of $R(L)$ from below: Since the operator series (4.1) converges for $|\lambda| \leq \| L - K \|_{\mathcal{B}}^{-1}$, one has

$$R(L) \geq |L|^{-1}.$$

The latter inequality in turn can be used to derive sufficient conditions ensuring that the Fredholm theorems hold for concrete integral equations or systems of such equations. This was just what Radon was aiming at when applying the above inequality to the operator $T = 2W_0$, where W_0 is the direct value of the logarithmic double layer potential (see Sec. 1.2 of Chap. 1).

Now let S be a rectifiable curve and let s denote arc length on S $(0 \leq s \leq l)$. If the angle $\vartheta(s)$ made by the positively oriented tangent and the abscissa is a function of bounded variation on $[0, 1]$, the curve S is said to be a *curve with bounded rotation* (or simply a *Radon curve*).

The following theorem provides the value of the Fredholm radius of the operator $T = 2W_0$.

Theorem 2 (Radon [1919b]). *If S is a closed curve with bounded rotation and the operator T is considered as acting on $C(S)$, we have the equality*

$$R(T) = |T|^{-1} = \pi/\alpha, \tag{4.2}$$

where

$$\alpha = \sup_{0 \leq s \leq l} |\vartheta(s+0) - \vartheta(s-0)|.$$

Note that the direction of the tangent to S is permitted to have jumps and that α is, up to the sign, nothing else than the greatest of these jumps. From (4.2) we infer that if $\alpha < \pi$, that is, if there are no *cusp points* (*peaks*) on S, then $R(T) > 1$ and the Fredholm theorems apply to the integral equations for the Dirichlet and Neumann problems. Starting with this observation and developing some further (nontrivial) machinery, Radon proved theorems on the solvability of the above integral equations, thus extending the classical theory of logarithmic potentials to the case of curves with bounded rotation and no cusps.

It is interesting to note that three years earlier Carleman [1916] in his dissertation showed, in principle, that if S is the union of finite number N of closed arcs $p_j p_{j+1}$ ($p_{N+1} = p_1$) of the class C^2, then the essential norm of T regarded as an operator on the space $C(S)$ with the norm

$$\sup_{j} \sup_{x \in S} |x - p_j|^{\varkappa_j} |u(x)|, \quad 0 \leq \varkappa_j \leq 1,$$

admits the estimate

$$|T| \leq \sup_{j} \frac{\sin(|\pi - \alpha_j|\varkappa_j)}{\sin(\pi \varkappa_j)},$$

where α_j is the angle between the half-tangents to S at p_j. In the case $\varkappa_j = 0$, Carleman's estimate coincides with the one of Radon.

Carleman [1916] also established an analogous result for two-dimensional surfaces comprised by two surfaces S_1 and S_2 of the class C^2 having a common border E, which is assumed to be a twice continuously differentiable curve. He proved in particular that for the essential norm of T on the space $C(S)$ the estimate

$$|T| \leq \sup_{p \in E} |1 - \alpha(p)/\pi|$$

holds, $\alpha(p)$ being the least of the two angles made by the tangential planes to S_1 and S_2 at the point $p \in E$.

A long time after Radon's paper [1919b] it reigned the attitude (even the conviction) towards that he generalized the theory of potentials in spaces of continuous functions "to its natural limit" (Riesz, Sz.-Nagy [1952]). However, in the sixties it was revealed by the works Král [1964], [1965], [1966], Burago, Maz'ya [1967], Burago, Maz'ya, Sapozhnikova [1962], [1966] that Radon's theory can be extended to a larger class of curves; this class may be described in terms of the variation of the angle at which the subsets of the curve are seen from an arbitrary point of it. More about these results will be said in Sec. 2 in the context of the multidimensional case.

Another development line of the theory of boundary integral equations on nonsmooth curves traces back to Ya.B. Lopatinskiĭ [1963], who resorted to results of I. Gohberg and M.G. Krein [1957] on integral equations over the half-line with kernels depending of the difference of the arguments. Lopatinskiǐ studied systems of equations of the form $\chi - T\chi = f$ on piecewise smooth contours without cusp points, where

$$(T\chi)(s) = \int_S K\left(s, \sigma, v_s, v_\sigma, \frac{z(s) - z(\sigma)}{|z(s) - z(\sigma)|}\right) \frac{\chi(\sigma)d\sigma}{|z(s) - z(\sigma)|}$$

and the matrix K admits the estimate

$$K(s, \sigma, \xi, \eta, \zeta) = O(|(\xi, \zeta)| + |(\eta, \zeta)|).$$

The operator T is considered in the space $L_1(\varrho^{-\varepsilon}, S)$ whose norm is given by

$$\|u; L_1(\varrho^{-\varepsilon}, S)\| = \int_S |u(\sigma)| \frac{d\sigma}{\varrho(\sigma)^\varepsilon},$$

$\varrho(\sigma)$ being the distance between the point σ and the nearest corner point of S.

Lopatinskiǐ's approach can be outlined as follows. In the first place it is shown that T coincides up to a compact operator with an integral operator with positively homogeneous kernel of the degree -1 on the legs of an angle. The explicit solution of the equations generated by the latter ("model") operator is then accomplished by means of the Mellin transform. The index formula established in Gohberg, Krein [1957] finally yields an index formula for the operator $I - T$.

Thus, instead of estimating the essential norm of T from below, as it was done by Radon, Lopatinskiǐ constructed a regularizer for $I - T$ (see Sec. 3 of Chap. 1 of S. Prössdorf's article in this volume). Proceeding in exactly the same way, B.V. Bazaliǐ and V.Yu. Shelepov [1980] succeeded in describing the Noether and Fredholm domains (see Sec. 4 of Chap. 1 of the article by S. Prössdorf) of the operator T on $C(S)$, thus gaining deeper insight into the nature of Theorem 2. Here is the precise result.

Theorem 3 (Bazaliǐ, Shelepov [1980]). *If S is a closed Radon curve without cusp points, then the operator $I + \lambda T$ is Noetherian on $C(S)$ if and only if λ does not belong to the real half-lines*

$$\left(-\infty, -\min_{s \in S} \frac{\pi}{|\pi - \alpha_s|}\right], \left[\min_{s \in S} \frac{\pi}{|\pi - \alpha_s|}, \infty\right),$$

where α_s is the internal angle (with respect to the domain Ω^+) made by S at s. In case this condition is satisfied, the index of the operator $I + \lambda T$ equals zero.

In Shelepov [1983], Theorem 3 was extended to the operator defined by

$$(T\chi)(s) = \int_S K\left(\frac{z(s) - z(\sigma)}{|z(s) - z(\sigma)|}\right) \chi(\sigma) \frac{(v(\sigma), z(s) - z(\sigma))}{|z(s) - z(\sigma)|^2} d\sigma$$

on the space of continuous vector-valued functions with m components. Here K is an $m \times m$ matrix whose entries are continuous "even" functions on the unit circle such that the average (the integral) of K over the unit circle is a non-singular matrix. By v we mean the unit normal to S directed inside.

The following theorem on the system of integral equations of plane elasticity represents a special case of the main result of Shelepov [1983].

Theorem 4 (Shelepov [1983]). *Let S be a closed Radon curve having no cusp points. Then for the operator $I - \lambda T$, answering for the first and second fundamental boundary value problems of plane elasticity,*

$$K_{ij}(\mathbf{e}) = -\frac{1}{\pi}[(1 - \varkappa)\delta_{ij} - \varkappa(\mathbf{e}, \mathbf{e}_j)(\mathbf{e}, \mathbf{e}_j)]; \qquad i, j = 1, 2$$

($\mathbf{e}_j$ being the jth basis vector), to be a Noether operator on $C(S)$ it is necessary and sufficient that λ do not belong to any of the real half-lines

$$\left(-\infty, -\min_{s \in S} \frac{\pi}{|\pi - \alpha_s| + |\varkappa|\sin|\pi - \alpha_s|}\right], \left[\min_{s \in S} \frac{\pi}{|\pi - \alpha_s| + |\varkappa|\sin|\pi - \alpha_s|}, \infty\right),$$

where $\varkappa = -1$ for the first and $\varkappa = (\lambda_0 + \mu_0)(\lambda_0 + 3\mu_0)^{-1}$ for the second fundamental boundary value problem (λ_0, μ_0 being the Lamé constants). In particular

$$R(T) = \min_{s \in S} \frac{\pi}{|\pi - \alpha_s| + |\varkappa|\sin|\pi - \alpha_s|}.$$

It is worth mentioning that, in contrast to the situation considered by Radon, the essential norm of the operator T prevailing in the previous theorem is strictly less than $R(T)^{-1}$. This is a consequence of the following theorem established by G. I. Kresin and the author [1979], [1981].

Theorem 5. *Let Ω^+ be a bounded domain the plane which is $C^{1,\gamma}$-diffeomorphic to some polygon with N vertices ($\gamma > 0$) and let α_j denote the least of the two angles contained by the two half-tangents to the boundary of Ω^+ at the corner point p_j ($1 \leq j \leq N$). Then the essential norm of the operator T figuring in Theorem 4 is given by*

$$|T| = \frac{2}{\pi}(1 + \varkappa) E\left(\frac{\pi - \alpha_{\min}}{2}, \frac{2\sqrt{\varkappa}}{1 + \varkappa}\right),$$

where E refers to the elliptic integral of the second kind and $\alpha_{\min} = \min\{\alpha_j: 1 \leq j \leq N\}$. In particular, if T is the operator corresponding to the Stokes equations ($\varkappa = 1$), then

$$|T| = \frac{4}{\pi}\cos\frac{\alpha_{\min}}{2}.$$

The next result on the logarithmic double layer potential on the space $L_p(S)$ ($1 < p < \infty$) is closely related to Theorem 3 (see Danilyuk, Shelepov [1967] and Shelepov [1969]).

Theorem 6 (Shelepov [1969]). *Let S be a closed Radon curve free of peaks and let T be the doubled direct value of the logarithmic double layer potential. Then if only $p \neq 1 + |\pi - \alpha_k|/\pi$, the operator $I + T$ is Noetherian on $L_p(S)$ and its index equals*

$$\varkappa_p = \sum_{k=1}^{\infty} \frac{1}{2}(1 - \text{sign}(p - 1 - |\pi - \alpha_k|/\pi)).$$

The Fredholm radius of T on $L_p(S)$ $(1 < p < \infty)$ is equal to

$$\min_{k \geq 1} \frac{\sin(\pi/p)}{\sin(|\pi - \alpha_k|/p)}.$$

In case S is a Radon curve without cusps and if $2 \leqq p < \infty$, then the Fredholm theorems are applicable to the operators on $L_p(S)$ induced by the integral equations of Sherman-Lauricella, Muskhelishvili, and Fredholm arising from plane elasticity (see Sec. 2.2 of Chap. 3).

In recent papers of R. Duduchava (see Duduchava [1984, 1986]) a general class of singular integral operators on piecewise smooth contours S without cusps was investigated. This class includes the Sherman-Laurichella and the Muskhelishvili equations. Solutions were considered in the weighted Sobolev spaces $W_q^1(S, a)$ with the norm

$$\|\phi\|_{W_q^1(S, a)} = \left(\int_S a(t)(|\phi'|^q + |\phi|^q)|dt| \right)^{1/q},$$

where $a(t) = \Pi_{j=1}^n |t - p_j|^{\gamma_j}$, $1 < q < \infty$, $-1 < \gamma_j < q - 1$. Necessary and sufficient conditions for the operators to be Noetherian as well as an index formula were found and the equivalence of the integral equations and the corresponding boundary value problems of planar elasticity was proved.

In connection with the development line we are discussing, we wish to direct the reader's attention to the monograph by I. I. Danilyuk [1975], which is devoted to the theory of integral equations for plane boundary value problems. We emphasize that this monograph focuses on the investigation of problems with discontinuous coefficients in domains bounded by nonsmooth contours (including Radon curves).

The same circle of ideas is also the dominant theme of the paper Fabes, Jodeit, Lewis [1977], dealing with the solvability in $L_p(S)$ of the integral equations for the internal and external Dirichlet problems for the Laplace equation in certain special domains (quarter-plane, circular cone, quarter-space, and bounded plane domains with an angle of size $\pi/2$ on the boundary). For instance in the quarter-plane case, where the equation is solved by using the Mellin transform, Fabes, Jodeit, and Lewis show that the integral equation has a solution in $L_p(S)$ if only $p \neq 3/2$, although the value $p = 3/2$ plays no exceptional role when representing the solution of the Dirichlet problem by the Poisson integral. On the other hand, however, for $1 < p < 3/2$ the solution of the Dirichlet problem in the complement of the quarter-plane is representable by a double layer potential but not by the Poisson integral. Fabes, Jodeit, and Lewis also point out that the integral equation of the internal Dirichlet problem in the cone $\{(x, y, z): z^2 > x^2 + y^2,$ $z > 0\}$ is uniquely solvable for all $p \in (1, \infty)$ except for a sequence $\{p_k\}_{k=1}^{\infty}$ of values p_k belonging to the open interval $(1, 2)$. They prove that in the quarter-plane case the integral equation has a unique solution in $L_p(S)$ whenever $p > 3/2$ (without explicitly determining this solution) and that the homogeneous equation has infinitely many linearly independent solutions in $L_p(S)$ provided $1 < p < 3/2$. Because the operator is a convolution with a kernel that is non-negative in either of the variables, its L_p-norm is available and can be shown to be less than 1 in case $p > 3/2$.

§2. Multidimensional Potential Theory in the Space $C(S)$

2.1. The Class of Surfaces. A series of works having their beginning in 1962 are devoted to the theory of boundary integral equations in the spaces C and C^* over irregular curves and surfaces of a fairly wide class: Burago, Maz'ya, Sapozhnikova [1962], [1966], Burago, Maz'ya [1967], Král [1964], [1965], [1966], [1980]. It was a remark made in the famous course of functional analysis by F. Riesz and B. Sz.-Nagy [1952], namely that "in the case of the spatial problem an analogue of curves with bounded rotation has not yet been found", which served as the starting-point of the note Burago, Maz'ya, Sapozhnikova [1962]. It turned out that a "proper" generalization of Radon's result to higher dimensions can be achieved in terms of a certain function $\omega(\xi, B)$ replacing in a sense the solid angle at which the set B is seen from the point ξ (the precise definition of $\omega(\xi, B)$ will be given in 2.2). The basic results, i.e. theorems on the solvability of the integral equations of the fundamental boundary value problems, were obtained for domains subject to the two conditions

$$\sup\{\text{var } \omega(\xi, S \setminus \xi) : \xi \in S\} < \infty, \tag{A}$$

$$\limsup_{r \to 0}\{\text{var } \omega(\xi, S \cap B_r(\xi)) : \xi \in S\} < \sigma_n/2. \tag{B}$$

Here S is the boundary of the domain Ω^+, var denotes the variation of the charge, $B_r(\xi)$ is the ball with center ξ and radius r, and σ_n is the area of the unit sphere.

We wish to point out that *condition* (A) is a corner-stone of the entire theory, whereas *condition* (B) is solely needed to prove the Fredholm alternative.

Here are simple examples illustrating the meaning of conditions (A) and (B).

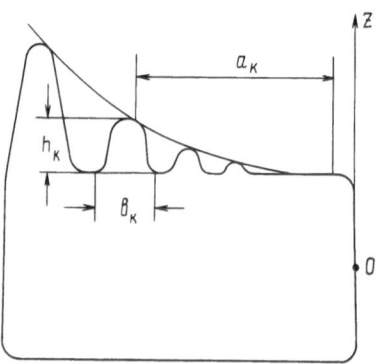

Fig. 3

Example 1. Let S be the curve drawn in Fig. 3, where $a_k = k2^{-k}, b_k = 2^{-k}$, $h_k = k^{-1}2^{-k}$. It is not difficult to see that S satisfies the conditions (A) and (B). However, the total variation of the rotation of this curve is infinite. Also notice that the contour S is smooth but that it belongs to none of the classes $C^{1,\gamma}$ ($\gamma > 0$). Letting $a_k = k^{-1}$, $b_k = k^{-2}$, $h_k = k^{-3}$ we obtain a curve of the class $C^{1,1/2}$ whose total variation of the rotation is infinite.

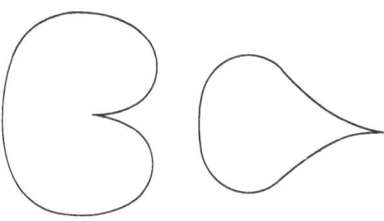

Fig. 4

Example 2. The curves represented by Fig. 4 fulfil condition (A) but do not satisfy condition (B). Note that this property is also shared by the curve of Fig. 3 in the case where $a_k = k^{-1}$, $b_k = k^{-4}$, $h_k = k^{-3}$.

Recently Král [1985] constructed an example of a set $\Omega^+ \subset \mathbb{R}^2$ for which the inequality in (B) becomes an equality and which has positive two-dimensional Hausdorff measure. He also showed that if $n = 2$ then condition (B) entails that for each point $x \in S$ there exist two disjoint open Lipschitz arcs $C_1, C_2 \subset S$ with the joint endpoint x such that $C_1 \cup \{x\} \cup C_2$ is a neighborhood of x on S. Furthermore, the one-sided tangents to C_1 and C_2 at x (understood in a well-defined sense) are shown to exist and to span a nonzero angle (i.e. x is not a cusp).

It is easily seen that a domain in $\mathbb{R}^n$ bounded by a surface of the class C^1 satisfies the condition (A) and (B) whenever

$$\int_0^{\cdot} \varrho^{-1}\omega(\varrho)d\varrho < \infty,$$

where $\omega(\varrho)$ is the continuity modulus of the normal to S.

Concluding the discussion about the several sorts of restrictions imposed on the surface S we remark that already Radon [1919b] proved the validity of condition (A) for arbitrary curves with bounded rotation. In this connection it is interesting to notice that in $\mathbb{R}^n$ ($n > 2$) there exists a surface S of the class C^1 such that $S \setminus \{0\} \in C^\infty$ and which possesses an integrable Gaussian curvature but does not satisfy condition (A). An example of such a surface is the surface obtained by rotating the curve of Fig. 3 (with $a_k = 2^{1-k}$, $b_k = 2^{-k}$, $h_k = k^{-1}2^{-k}$) about the Oz-axis (see Burago, Maz'ya [1967]).

2.2. Potentials and Integral Equations. A major part of what follows is taken from the paper by Yu. D. Burago and the author [1967]. An exposition in the same spirit may also be found in the book Král [1980]. We first provide some information about the perimeter of a set (see, for example, Federer [1969]).

Definition. Let $B \subset \mathbb{R}^n$ be a Borel set and suppose the characteristic function χ_B belongs to the space $BV(\mathbb{R}^n)$ of all locally integrable functions on $\mathbb{R}^n$ whose gradients (in the distributional sense) are finite vector-valued charges on $\mathbb{R}^n$. The total variation of the charge $\nabla\chi_B$ is called the *perimeter of the set B* (*in the sense of Caccioppoli and De Giorgi*) and denoted by $P(B)$.

In the following Ω^+ will always be a fixed Borel subset of $\mathbb{R}^n$ with a compact closure $\overline{\Omega}^+$ and a finite perimeter $P(\Omega^+)$; the boundary of Ω^+ will be consistently denoted by S.

Definition. A unit vector $\mathbf{n}_x$ is referred to as the (outer) normal (in the sense of Federer) to the set Ω^+ at the point x if

$$\lim_{r \to 0} r^{-n} \operatorname{mes}_n(\{y: y \in \Omega^+ \cap B_r(x), (y - x)\mathbf{n}_x > 0\}) = 0,$$

$$\lim_{r \to 0} r^{-n} \operatorname{mes}_n(\{y: y \in B_r(x)\backslash\Omega^+, (y - x)\mathbf{n}_x < 0\}) = 0.$$

The set of all points $x \in S$ at which a normal to Ω^+ exists is termed the *reduced boundary* of Ω^+ and designated by $\partial^*\Omega^+$.

The following important result often proves useful.

Theorem 7 (Federer [1958]). *The reduced boundary $\partial^*\Omega^+$ is measurable with respect to H_{n-1} and var $\nabla\chi_{\Omega^+}$. One has var $\nabla\chi_{\Omega^+}(\mathbb{R}^n\backslash\partial^*\Omega^+) = 0$, and if B is any subset of $\partial^*\Omega^+$ then*

$$\nabla\chi_{\Omega^+}(B) = - \int_B \mathbf{n}_x H_{n-1}(dx). \tag{4.3}$$

Here H_{n-1} denotes $(n-1)$-dimensional Hausdorff measure.

Definition. The *solid angle* at which a set $B \cap S$ is seen from a point $\xi \notin \overline{B}$ is the set function

$$\omega(\xi, B) = \begin{cases} -\displaystyle\int_B \nabla \log|x - \xi|^{-1}\nabla\chi_{\Omega^+}(dx) & \text{for } n = 2, \\[2mm] \dfrac{1}{2-n}\displaystyle\int_B \nabla|x - \xi|^{2-n}\nabla\chi_{\Omega^+}(dx) & \text{for } n > 2. \end{cases}$$

In view of (4.3) the latter integrals may also be taken over $\partial^*\Omega^+ \cap B$, and $\nabla\chi_{\Omega^+}(dx)$ may be replaced by $-\mathbf{n}_x H_{n-1}(dx)$.

Under the assumption that

$$\operatorname{var} \omega(\xi, \cdot)(B) \leqq c(\xi) < \infty$$

for $\bar{B} \subset \mathbb{R}^n \backslash \xi$, the definition of $\omega(\xi, B)$ extends to all Borel sets B in the following way:

$$\omega(\xi, B) = \omega(\xi, B \backslash \xi) + \omega(\xi, \xi),$$

where $\omega(\xi, \xi) := \sigma_n/2 - \omega(\xi, \mathbb{R}^n \backslash \xi)$ if $\xi \in S$ and $\omega(\xi, \xi) := 0$ if $\xi \notin S$.

Definition. The *harmonic double layer potential* with the continuous density χ is the function defined for $x \notin S$ by

$$(W\chi)(x) = \frac{1}{\sigma_n} \int_S \chi(\xi) \omega(x, d\xi).$$

Theorem 8. *Let $P(\Omega^+) < \infty$ and $S = \partial(\Omega^-)$. Then in order that for every continuous function χ the limit values $W_{\pm}\chi$ of the potential $W\chi$ exist from within and without Ω^+ it is necessary and sufficient that*

$$\sup\{\operatorname{var} \omega(\xi, \cdot)(S) : \xi \in \mathbb{R}^n \backslash S\} < \infty.$$

We have

$$W_+ \chi = \tfrac{1}{2}(\chi + T\chi), \tag{4.4}$$

$$W_- \chi = \tfrac{1}{2}(-\chi + T\chi), \tag{4.5}$$

and the doubled direct value of the double layer potential,

$$(T\chi)(x) = \frac{2}{\sigma_n} \int_S \chi(\xi) \omega(x, d\xi),$$

generates a continuous operator on $C(S)$.

The question on the existence of the nontangential limits (cf. § 3 for the definition) of the double layer potential was considered in Král [1970] and Dont [1972]. The boundary behavior of the double layer potential is also the subject of Král [1985].

Owing to the equalities (4.4) and (4.5), continuous solutions of the internal and external Dirichlet problems with prescribed boundary functions φ^+ and φ^- in $C(S)$ may be determined by means of the integral equations

$$\chi + T\chi = 2\varphi_+, \tag{4.6}$$

$$-\chi + T\chi = 2\varphi_-. \tag{4.7}$$

To pose, in some sense, the Neumann problem for surfaces satisfying condition (A) we need the notion of the *boundary flow*.

Definition. A function $u \in C^1(\operatorname{int} \Omega^+)$ is said to possess an *inner boundary flow* if

(i) for each infinitely differentiable function φ on $\mathbb{R}^n$ with compact support and for each sequence of sets $\Omega_m \subset \operatorname{int} \Omega^+$ with smooth boundaries which converge to Ω^+ there exists the limit

$$l_u(\varphi) = \lim_{m \to \infty} \int_{\partial \Omega_m} \varphi(x) \frac{\partial u}{\partial v_x} H_{n-1}(dx),$$

where v_x is the outer normal to $\partial\Omega_m$ and H_{n-1} denotes $(n-1)$-dimensional Hausdorff measure;

(ii) the functional l_u is bounded in the norm of $C(S)$.

Definition. Let the requirements (i), (ii) of the previous definition be met. The finite charge Σ^+ on S generating the extension of the functional l_u to all of $C(S)$,

$$l_u(\varphi) = \int_S \varphi(x)\,\Sigma^+(dx),$$

is called the *inner boundary flow* of the function u.

The *outer boundary flow* Σ^- of a function $u \in C^1(\Omega^-)$ is defined similarly.

In our understanding of the internal and external Neumann problems the charges Σ^+ and Σ^- play the part of the normal derivatives.

Definition. Given a finite charge ϱ on $S \subset \mathbb{R}^n$ $(n \geq 3)$, the *simple layer potential with the charge ϱ* is defined by

$$(V\varrho)(x) = \frac{1}{\sigma_n} \int_S r^{2-n}\varrho(d\xi), \qquad x \notin S.$$

The function $V\varrho$ turns out to be harmonic in $\mathbb{R}^n \backslash S$.

Theorem 9. *Suppose* $\mathrm{mes}_n(S) = 0$, $S = \partial(\Omega^-)$, *and condition* (A) *is fulfilled. Then the potential $V\varrho$ possesses inner and outer boundary flows, which equal*

$$-\frac{1}{2}\varrho(B) + \frac{1}{\sigma_n}\int_S \omega(x, B)\varrho(dx),$$

$$\frac{1}{2}\varrho(B) + \frac{1}{\sigma_n}\int_S \omega(x, B)\varrho(dx),$$

B being an arbitrary Borel subset of S.

The condition (A) is not only sufficient but also necessary for the boundary flows of an arbitrary $V\varrho$ to exist.

Thus, looking for the solution of the Neumann problem in the form of a simple layer potential leads to the equations

$$-\varrho + T^*\varrho = 2\Sigma^+, \tag{4.8}$$

$$\varrho + T^*\varrho = 2\Sigma^-. \tag{4.9}$$

Here T^* is the operator adjoint of T, acting on the dual space $C^*(S)$ of $C(S)$.

2.3 The Essential Norm of the Operator T. The applicability of the Fredholm theory to the integral equations derived in 2.2 is based on inequality (B) and the following representation of the essential norm of the operator T on the space $C(S)$.

Theorem 10 (Král [1965], [1980], Burago, Maz'ya [1967]). *If $S = \partial(\Omega^-)$ and condition (A) is satisfied, then*

$$|T| = \frac{2}{\sigma_n} \limsup_{r \to 0} \{\mathrm{var}\,\omega(\xi, S \cap B_r(\xi)) : \xi \in S\}.$$

Consequently, condition (B) is equivalent to the inequality $|T| < 1$. In case the region Ω^+ is convex, the formula for $|T|$ can be simplified as follows.

Theorem 11 (Netuka [1975]). *If Ω^+ is a convex region in $\mathbb{R}^n$, we have*

$$|T| = \sup\{|1 - 2d(\xi)| : \xi \in S\},$$

where $d(\xi)$ denotes the volume density of Ω^+ at ξ.

From this result we infer almost at once that $|T| < 1$ whenever Ω^+ is convex. On the other hand, it is easy to see that the latter inequality, or condition (B) tandamount to it, is not true even for sufficiently simple polyhedra, and this is the sore point of the theory. Angell, Kleinman, Král [1986] and Král, Wendland [1986] succeeded in compelling the inequality $|T| < 1$ to hold by replacing the usual norm in $C(S)$ with an equivalent weighted norm. The polyhedral surfaces considered in Král, Wendland [1986] are constituted by a finite number of rectangles parallel to the coordinate planes.

It is not out of the question that condition (B) merely emerges from our technique of proving the Fredholm theorems, viz, the employment of the essential norm. This point of view is supported by the results of Maz'ya [1985], Grachev, Maz'ya [1989], more about which will be said in Chapter 5.

The problem of the solvability of the integral equations of the theory of harmonic potentials is deferred to the next subsection. We now give some information about the essential norm of certain matrix-valued operators arising from the investigation of the boundary values of potentials of the double layer type

$$(W\chi)(x) = \frac{1}{2} \int_S K(\mathbf{e}_{x\xi})\chi(\xi)\omega(x, d\xi), \qquad x \notin S.$$

Here K is an "even" $(m \times m)$-matrix-valued function on the unit sphere $\mathbb{S}^{n-1}$ of $\mathbb{R}^n$, $\mathbf{e}_{x\xi}$ denotes $(\xi - x)|\xi - x|^{-1}$, and χ is an element of the space $C_m(S)$ of all continuous vector-valued functions on S with m components; the norm in $C_m(S)$ is defined by $\|\chi\| = \sup\{|\chi(x)| : x \in S\}$.

Theorem 12 (Kresin, Maz'ya [1979], [1981]). *Assume $S = \partial(\Omega^-)$ and condition (A) is satisfied. Then*

$$|T| = \lim_{\delta \to 0} \sup_{\xi \in \partial^*\Omega^+} \sup_{\mathbf{z} \in \mathbb{S}^{m-1}} \int_{S \cap B_\delta(\xi)} |K^*(\mathbf{e}_{\xi x})\mathbf{z}||\omega|(\xi, dx),$$

where $T\chi$ is the doubled direct value of the potential $W\chi$ viewed as an operator on $C_m(S)$, K^* refers to the adjoint matrix, and $\partial^*\Omega^+$ is the reduced boundary of Ω^+.

In particular, if Ω^+ is a bounded convex region in $\mathbb{R}^n$, we have

$$|T| = \sup_{\xi \in S \setminus \partial^*\Omega^+} \sup_{\mathbf{z} \in \mathbb{S}^{m-1}} \left[\frac{1}{2} \int_{\mathbb{S}^{n-1}} |K^*(\sigma)\mathbf{z}| d\sigma - \int_{\mathscr{E}_\xi} |K^*(\sigma)\mathbf{z}| d\sigma \right],$$

where $\mathscr{E}_\xi$ is the topological limit of the projections on the sphere $\partial B_1(\xi)$ of the set $\Omega^+ \cap \partial B_\delta(\xi)$ as $\delta \to 0$.

The potential W may be the two- or three-dimensional double layer potential of elasticity. In the plane case the last formula for the essential norm implies Theorem 4 stated in Sec. 1. For $n = 3$, let $K_\varkappa(\mathbf{e})$ denote the matrix whose entries are

$$-\frac{1}{4\pi}[(1 - \varkappa)\delta_{ij} + 3\varkappa(\mathbf{e}, \mathbf{e}_i)(\mathbf{e}, \mathbf{e}_j)] \qquad (i, j = 1, 2, 3).$$

Putting $\varkappa = (\lambda + \mu)(\lambda + 3\mu)^{-1}$ and $\varkappa = 1$, we obtain the matrix occuring in the boundary integral operator of the Dirichlet problem for the Lamé and Stokes equations, respectively (see Secs. 1 and 2 of Chap. 2).

In the case where conical singularities prevail on S, we have the following result on the operator $T(\varkappa)$ on $C_3(S)$ induced by the matrix $K_\varkappa(\mathbf{e})$.

Theorem 13 (Kresin, Maz'ya [1979], [1981]). *Let Ω^+ be a bounded domain in $\mathbb{R}^3$ on the boundary of which there are points $\xi_1, \xi_2, \ldots, \xi_N$ such that $S \backslash U_{\xi_i}$ is a manifold of the class $C^{1,\gamma}$ $(\gamma > 0)$. Suppose, for each $i = 1, 2, \ldots, N$, there exists a $C^{1,\gamma}$-diffeomorphism of the intersection of Ω^+ and some neighborhood of ξ_i onto a circular cone with the angle α_i at the vertex; the Jacobi matrix of this diffeomorphism is required to be the identity matrix at ξ_i. Finally let $\alpha_{min} = \min\{\alpha_1, \ldots, \alpha_N, 2\pi - \alpha_N\}$. Then $|T_3(\varkappa)|$ equals*

$$\frac{|1 - \varkappa|}{\pi}\left[\sqrt{1 + \Gamma^2\sin^2\frac{\alpha_{min}}{2}}\, E\left(\frac{\Gamma\sin\dfrac{\alpha_{min}}{2}}{\sqrt{1 + \Gamma^2\sin^2\dfrac{\alpha_{min}}{2}}}\right)\cos\frac{\alpha_{min}}{2}\right.$$

$$\left. + \int_0^{\pi/2}\frac{1 + \Gamma^2\cos^2\vartheta}{\Gamma\cos\vartheta}\arcsin\frac{\Gamma\cos\vartheta\cos\dfrac{\alpha_{min}}{2}}{\sqrt{1 + \Gamma^2\cos^2\vartheta}}\, d\vartheta\right],$$

where $\Gamma = |1 - \varkappa|^{-1}\sqrt{3\varkappa(\varkappa + 2)}$ and E is the complete elliptic integral of the second kind. For the operator of the Stokes equations this can be simplified to

$$|T(1)| = (2\pi)^{-1}(\pi - \alpha_{min} + \sin\alpha_{min}).$$

In the following theorem Ω^+ is a bounded region in $\mathbb{R}^3$ on the boundary of which a closed subset M is distinguished. It is assumed that $S \backslash M$ is a two-dimensional $C^{1,\gamma}$-submanifold of $\mathbb{R}^3$ and that each point of M has a neighborhood whose intersection with Ω^+ is $C^{1,\gamma}$-diffeomorphic to some dihedral angle. So at each point $\zeta \in M$ two tangential planes are well-defined. The least of the two angles contained by the tangential planes at ζ is denoted by $\alpha(\zeta)$ and, finally, α_{min} is defined as $\min\{\alpha(\zeta): \zeta \in M\}$.

Theorem 14 (Kresin, Maz'ya [1979], [1981]). *Under the afore-mentioned assumptions,*

$$|T(\varkappa)| = \frac{|1 - \varkappa|}{2}\left[\frac{\pi - \alpha_{min}}{2} + \int_{\alpha_{min}/2}^{\pi/2}\frac{1 + \Gamma^2\sin^2\vartheta}{\Gamma\sin\vartheta}\arccos(1 + \Gamma^2\sin^2\vartheta)^{-1/2}\, d\vartheta\right],$$

where Γ is as in Theorem 13. In particular, $|T(1)| = (3/2)\cos(\alpha_{min}/2)$.

2.4. Solvability of the Integral Equations of Harmonic Potential Theory. We now return to the Dirichlet and Neumann problems for the Laplace equation in a domain Ω^+ subject to the conditions (A) and (B). Our exposition again follows the paper Burago, Maz'ya [1967] (also see Král [1980]).

Once the Fredholm theorems for the operator T have been proved, the solvability of equations (4.6) and (4.9) can be studied proceeding along the lines of Radon's approach to curves with bounded rotation.

The following theorem is the basic result on the solvability of the integral equations of the theory of harmonic potentials. We assume that the sets int Ω^+ and $\mathbb{R}^n \backslash \overline{\Omega^+}$ are connected; the only reason for this assumption is to simplify the wording of the theorem.

Theorem 15. Let Ω^+ be a Borel subset of $\mathbb{R}^n$ with a compact closure and a finite perimeter. Also suppose that Ω^+ satisfies condition (B) (note that, as shown in Burago, Maz'ya [1967], these hypotheses automatically imply that $S = \partial(\Omega^-)$ and $\text{mes}_n(S) = 0$).

Then the following assertions are true.

1) The integral equation (4.6) of the internal Dirichlet problem has a unique solution in $C(S)$ for every continuous right-hand side φ_+.

2) The integral equation (4.9) of the external Neumann problem is uniquely solvable for every finite charge Σ^-.

The classical results on the integral equations of the external Dirichlet and the internal Neumann problems may be rephrased in an analogous fashion.

Notice that Theorem 15 does not guarantee uniqueness in the case of the Neumann problem. What the theorem states is solely that there is a unique solution representable as a simple layer potential. Theorem 16 below, whose proof relies upon Theorem 8, concerns the uniqueness question for the Neumann problem in a certain class of harmonic functions.

Definition. Suppose that $S = \partial(\Omega^-)$. A finite charge ϱ is said to belong to the class C_V if the simple layer potential $V\varrho$ generated by ϱ possesses finite and equal limits on S from inside and outside of S.

Theorem 16. Let Ω^+ be a Borel subset of $\mathbb{R}^n$ having a compact closure and a finite perimeter. In addition, assume that int Ω^+ and $\mathbb{R}^n \backslash \overline{\Omega^+}$ are connected sets. Then if condition (B) is fulfilled,

1) for every finite charge $\Sigma^+ \in C_V$ with vanishing total mass there exists up to an additive constant exactly one solution of the internal Neumann problem belonging to the class $C(\overline{\Omega^+}) \cap BV(\text{int } \Omega^+)$;

2) for every finite charge $\Sigma^- \in C_V$ there exists precisely one solution of the external Neumann problem belonging to the class $C(\mathbb{R}^n \backslash \text{int } \Omega^+) \cap BV^{(\text{loc})}(\mathbb{R}^n \backslash \overline{\Omega^+})$ and tending to zero at infinity.

Let us in conclusion cite a few more applications of the boundary integral equations method to boundary value problems in domains with irregular boundary. The Dirichlet problem with discontinuous boundary function is considered in Netuka [1975]. A generalization of Theorem 16 to domains with multiply-connected boundary can be found in the book Král [1980]. The third boundary value problem for Laplace's equation is discussed in Sapozhnikova [1966] and Netuka [1972a], [1972b], [1972c]. In the paper Král, Netuka [1977] it is shown that the operator T^2 is a contraction if the region Ω^+ is convex. This offers the possibility of using the method of successive approximation for the solution of the integral equations arising from boundary value problems in convex regions (see also the book Král [1980]).

§3. Boundary Integral Equations in the Space L_p on Lipschitz Surfaces

During the last decade the theory of boundary equations for boundary value problems in Lipschitz domains and with data from $L_p(S)$ has reached a very advanced stage. This development has its beginning in Calderón's paper [1977], where the boundedness of the Cauchy singular integral on L_p over Lipschitz curves with sufficiently small Lipschitz constant was established (see also A.P. Calderón, C.P. Calderón, Fabes, Jodeit, Riviere [1978] and Sec. 1 of Chap. 3 of S. Prössdorf's article in this volume). The restriction to small Lipschitz constants was then removed by Coifman, McIntosh, Meyer [1982]. Fabes, Jodeit, and Riviere [1978] employed Calderón's theorem to develop a potential theory in L_p over curves of the class C^1. Subsequently, in the work of Verchota [1984], Kenig [1984], Fabes [1985], and others, the solvability of boundary integral equations for various boundary value problems in $L_p(S)$ was proved for Lipschitz domains on the basis of the result of Coifman, McIntosh, Meyer [1982].

In the present section we first discuss the integral equations of the Dirichlet and Neumann problems in Lipschitz domains. We also touch the boundary integral equations method for the Lamé and Stokes equations in Lipschitz domains.

3.1. Preliminaries. Throughout, Ω^+ denotes a bounded simply connected domain in $\mathbb{R}^n (n \geq 2)$ and Ω^- is $R^n \backslash \overline{\Omega^+}$.

Definition. The boundary S of $\Omega^\pm$ is said to be *Lipschitzian* (or to belong to the *class $C^{0,1}$*) if for each point $\xi \in S$ there exist a ball $B_\delta(\xi)$, a Cartesian system of coordinates in $\mathbb{R}^n$ with the origin at ξ, and a function $\varphi \colon \mathbb{R}^{n-1} \to \mathbb{R}^1$ with compact support such that $\varphi(0) = 0$, φ satisfies a Lipschitz condition, i.e. $|\varphi(x) - \varphi(y)| \leq c_\xi |x - y|$ for all $x, y \in \mathbb{R}^{n-1}$ and some $c_\xi < \infty$, and, last but not least,

$$\Omega^+ \cap B_\delta(\xi) = \{(x, t) \colon x \in \mathbb{R}^{n-1}, t < \varphi(x)\} \cap B_\delta(\xi).$$

If, in addition, the function φ is continuously differentiable and (which is no loss of generality) $(\partial \varphi / \partial x_i)(0) = 0$ for $i = 1, \ldots, n - 1$, then S is said to belong to the *class C^1*. In the latter case at each point $\xi \in S$ the normal to S exists; the unit vector of the normal directed into Ω^- is denoted by $\mathbf{n}_\xi$. Furthermore, given an arbitrary $\beta \in (0, 1)$ we can find an $\varepsilon > 0$ such that the cones

$$\Gamma_\beta^\pm(\xi) = \{x \in B_\varepsilon(\xi) : \mp (x - \xi, \mathbf{n}_\xi) > \beta |x - \xi|\}$$

and $\Omega^\mp$ are disjoint.

In case S is of the class $C^{0,1}$, a normal vector to S exists almost everywhere on S (with respect to Lebesgue measure). For such surfaces, however, there exist two families $\{\Gamma^\pm(\xi)\}_{\xi \in S}$ of nontangential circular cones contained in $\Omega^\pm$ such that $\Gamma^+(\xi)$ and $\Gamma^-(\xi)$ are symmetric with respect to ξ and such that the values of the angles at the tops of the cones are bounded away from zero throughout S.

Definition. A function u defined in a domain $\Omega^\pm$ of the class $C^{0,1}$ is said to have the *nontangential limit* a at the point $\xi \in S$ if $\lim_{x \in \Gamma^\pm(\xi)} u(x) = a$.

Further, in dependence on where the function u is given, one defines the following three *nontangential maximal functions*:

$$u_*^\pm(\xi) = \sup\{|u(x)| : x \in \Gamma^\pm(\xi)\}, \qquad \xi \in S,$$

$$u_*(\xi) = \sup\{|u(x)| : x \in \Gamma^+(\xi) \cup \Gamma^-(\xi)\}, \qquad \xi \in S.$$

Provided S belongs even to the class C^1, for each $\beta \in (0, 1)$ the nontangential maximal functions $u_\beta^\pm(\xi)$ and $\mu_\beta(\xi)$ are defined analogously but with $\Gamma^\pm(\xi)$ replaced by the cones $\Gamma_\beta^\pm(\xi)$.

Using a partition of unity, the space $W_p^1(S)$ can be defined in a standard way for $p \in (1, \infty)$ and S of the class $C^{0,1}$. If $f \in W_p^1(S)$, then the tangential gradient $\nabla_t f(\xi)$ exists for almost all $\xi \in S$ (see e.g. Nečas [1967]).

3.2. Properties of Simple and Double Layer Potentials.
The harmonic simple and double layer potentials are defined on $\mathbb{R}^n \backslash S$ in the usual manner (recall formulas (1.4) and (1.5)).

The following result was established in Verchota [1984] with the help of the theorem on the boundedness of the Cauchy singular operator on L_p over Lipschitz curves contained in Coifman, McIntosh, Meyer [1982].

Theorem 17. *Let $S \in C^{0,1}$ and consider the family of integrals*

$$\left\{ \int_{S \backslash B_\varepsilon(x)} \chi(\xi) \frac{\partial}{\partial \mathbf{n}_\xi} \mathscr{E}(x, \xi) d_\xi S \right\}_{\varepsilon > 0}. \tag{4.10}$$

Then if $\chi \in L_p(S)$ $(1 < p < \infty)$, the integrals (4.10) converge in $L_p(S)$ and almost everywhere on S as $\varepsilon \to 0$.

The limit so defined is denoted by $(W_0 \chi)(x)$ and referred to as the *direct value of the double layer potential*.

For surfaces S in $\mathbb{R}^n$ $(n \geq 3)$ of the class C^1 the existence of $W_0 \chi$ in the sense of Theorem 17 was first proved in Fabes, Jodeit, Riviere [1978]. In this case it turned out that W_0 is a compact operator on the spaces $L_p(S)$ and $W_p^1(S)$ $(1 < p < \infty)$. However, if S merely belongs to the class $C^{0,1}$, then the operator W_0 is in general not compact on $L_p(S)$ (see Sec. 2).

Theorem 18 (Verchota [1984]). *If S is in the class $C^{0,1}$ and χ belongs to $L_p(S)$ $(1 < p < \infty)$, then the potential $W\chi$ has nontangential limits $(W_{\pm}\chi)(x)$, from the side of $\Omega^{\pm}$, for almost all $x \in S$. Furthermore, we have*

$$W_{\pm}\chi = \pm\tfrac{1}{2}\chi + W_0\chi. \tag{4.11}$$

In the paper Fabes, Jodeit, Riviere [1978] it is shown that if $S \subset \mathbb{R}^n$ $(n \geq 3)$ is a surface of the class C^1, then for each $\beta \in (0, 1)$ the nontangential maximal function $(W\chi)_{\beta}^{+}$ belongs to $L_p(S)$ and admits the estimate

$$\|(W\chi)_{\beta}^{+}\|_{L_p(S)} \leq C_{\beta}\|\chi\|_{L_p(S)}.$$

If, moreover, $\chi \in W_p^1(S)$ then

$$\|\nabla(W\chi)|_{\beta}^{+}\|_{L_p(S)} \leq C_{\beta}\|\chi\|_{W_p^1(S)}$$

(this is proved in the same paper).

We now pass to properties of the simple layer potential.

Theorem 19 (Verchota [1984]). *If $S \in C^{0,1}$ and $\varrho \in L_p(S)$ $(1 < p < \infty)$, then the direct value of the normal derivative of the simple layer potential, i.e. the limit*

$$\frac{\partial(V\varrho)}{\partial\mathbf{n}_0}(x) = \lim_{\varepsilon \to 0} \int_{S \backslash B_{\varepsilon}(x)} \varrho(\xi)\frac{\partial}{\partial\mathbf{n}_x}\mathscr{E}(x, \xi)d_{\xi}S$$

exists. This limit may be understood both in the sense of convergence in $L_p(S)$ or in the sense of pointwise convergence for almost all $x \in S$. The operator $\partial V/\partial\mathbf{n}_0$ is the adjoint of the operator W_0.

The normal derivative $\partial(V\varrho)/\partial\mathbf{n}_x$ possesses nontangential limits $\partial(V\varrho)/\partial\mathbf{n}_{\pm}$ from the side of $\Omega^{\pm}$ for almost all $x \in S$. These limits are tied in with the direct values by the equalities

$$\frac{\partial(V\varrho)}{\partial\mathbf{n}_{\pm}} = \mp\frac{1}{2}\varrho + \frac{\partial(V\varrho)}{\partial\mathbf{n}_0}. \tag{4.12}$$

Furthermore, the nontangential maximal function of the gradient of the potential $V\varrho$ admits the estimate

$$\|\nabla(V\varrho)|_{*}\|_{L_p(S)} \leq C\|\varrho\|_{L_p(S)}. \tag{4.13}$$

Fabes, Jodeit, and Riviere [1978] showed that in the case $n \geq 3$ an estimate of the form (4.13) also holds for the nontangential maximal functions corresponding to an arbitrarily picked $\beta \in (0,1)$.

3.3. Integral Equations for the Dirichlet and Neumann Problems in Ω^{+}. Looking for a solution of the Dirichlet problem in Ω^{+} in the form of the potential $W\chi$, we get, by (4.11), the integral equation

$$\chi + T\chi = 2\varphi \qquad (T = 2W_0) \tag{4.14}$$

for the unknown density χ.

If S is of the class C^1, then the operator W_0 is compact on the space $L_p(S)$ $(1 < p < \infty)$, and so the solvability problem for equation (4.14) and its adjoint can be settled with the help of the classical tools (see Sec. 2 of Chap. 1): the Fredholm theory and the uniqueness theorems for boundary value problems. In this way it was shown in Fabes, Jodeit, Riviere [1978] that equation (4.14) has a unique solution in $L_p(S)$ $(1 < p < \infty)$ and that the potential $W\chi$ is really a solution of the Dirichlet problem in Ω^+. Furthermore, for each $\beta \in (0, 1)$ one has the estimate

$$\|(W\chi)^+_\beta\|_{L_p(S)} \leqq C_\beta \|\varphi\|_{L_p(S)}. \tag{4.15}$$

If all what we know is that S belongs to the class $C^{0,1}$, then in order to prove the closedness and density of the range of the operator $I + T^*$ in $L_2(S)$ (implying the solvability of equation (4.14)), we have to invent some machinery capable of substituting the Fredholm theory. For this purpose Verchota [1984] employed the following inequality.

Lemma. *Let $f \in L_2(S)$ for $n \geq 3$ and suppose $f \in L_{2,0}(S)$ in case $n = 2$, where $L_{p,0}(S) = \{f \in L_p(S) : \int_S f dS = 0\}$. Then*

$$\|f \mp T^*f\|_{L_2(S)} \leqq C\left\{ \|f \pm T^*f\|_{L_2(S)} + \left| \int_S V_0 f dS \right| \right\}, \tag{4.16}$$

with some constant C depending solely on the Lipschitz constant of S.

Here is a sketch of the *proof* of (4.16). The point of departure is *Rellich's identity* (see Nečas [1967]):

$$\int_S (\mathbf{n}, \mathbf{h})|\nabla u|^2 dS = 2 \int_S \frac{\partial u}{\partial \mathbf{n}} (\mathbf{h}, \nabla u) dS + \int_{\Omega^+} [\mathrm{div}(|\nabla u|^2 \mathbf{h}) - 2(\nabla \mathbf{h} \cdot \nabla u, \nabla u)]dx,$$

$\mathbf{h}$ being any smooth vector field on $\mathbb{R}^n$ such that $(\mathbf{n}, \mathbf{h}) \geqq C > 0$ on S.

From Rellich's identity one infers that

$$\left\| \frac{\partial u}{\partial \mathbf{n}} \right\|^2 \leqq C \left\{ \|\nabla_t u\|^2 + \|\nabla_t u\| \left\| \frac{\partial u}{\partial \mathbf{n}} \right\| + \int_{\partial \Omega_j} u \frac{\partial u}{\partial \mathbf{n}} dS \right\},$$

where $\{\Omega_j\}$ is any sequence of smooth domains monotonically approximating Ω^+ from within and $\| \cdot \|$ denotes the norm in $L^2(\partial \Omega_j)$. It then follows that

$$\left\| \frac{\partial u}{\partial \mathbf{n}} \right\|_{L_2(\partial \Omega_j)} \leqq C \|\nabla_t u\|_{L_2(\partial \Omega_j)}$$

and inserting $u = Vf$ into this inequality and passing to the limit $j \to \infty$ gives

$$\|f - T^*f\|_{L_2(S)} \leqq 2C \|\nabla_t V_0 f\|_{L_2(S)} \tag{4.17}$$

(notice that the limit of the tangential derivates Vf exists).

On approximating S by smooth surfaces from outside and arguing in an analogous fashion, one arrives at the inequality

$$\|\nabla_t V_0 f\|_{L_2(S)} \leqq C \left\{ \|f + T^*f\|_{L_2(S)} + \left| \int_S V_0 f dS \right| \right\}.$$

The latter inequality together with (4.17) completes the proof. $\square$

The next result is established in Verchota [1984] as a consequence of the previous lemma.

Theorem 19. (i) *If $S \in C^{0,1}$, then equation (4.14) is uniquely solvable in $L_2(S)$.*
(ii) *If $\varphi \in W_2^1(S)$, then the solution χ of equation (4.14) also belongs to $W_2^1(S)$ and $|V(W\chi)|_*^+$ is in $L_2(S)$.*

To reduce the Dirichlet problem to an integral equation, Calderón [1985] resorts to the potential

$$(W^{(1)}\chi)(x) = \int_S \chi(\xi)\frac{\partial}{\partial \mathbf{l}_\xi}\mathscr{E}(x, \xi)d_\xi S,$$

where $S \in C^{0,1}$, $\chi \in L_p(S)$, and $\mathbf{l}$ is a Lipschitzian vector-function such that $|\mathbf{l}| = 1$ and $(\mathbf{l}, \mathbf{n}) \geq \delta > 0$. This leads to an integral equation for the Dirichlet problem whose adjoint is just the boundary equation for the oblique derivative problem.
    Calderón [1985] shows that the equations so obtained are Fredholm in $L_p(S)$ for all $p \in (p_0, p_0/(p_0 - 1))$, where $p_0 \in (1, 2)$ is a value defined by the local oscillation of the normal vector ($\lim_{\varepsilon \to 0} \sup_{|x-\xi|<\varepsilon} |\mathbf{n}_x - \mathbf{n}_\xi|$). This allows him to establish the unique solvability of the Dirichlet problem with data from $L_p(S)$ · ($p_0 < p \leq 2$) and also the solvability of the oblique derivative problem for functions in $L_p(S)$ ($p_0 < p < p_0/(p_0 - 1)$) satisfying a finite number of orthogonality conditions.
    Seeking the solution of the Neumann problem in Ω^+ as a potential $V\varrho$ yields the integral equation

$$\varrho - T^*\varrho = -2\psi \tag{4.18}$$

for the density ϱ. The following result is proved in Verchota [1984] again with the aid of inequality (4.16).

Theorem 20. *If $S \in C^{0,1}$, then equation (4.18) is uniquely solvable in $L_{2,0}(S)$.*

Remark (Fabes, Jodeit, Riviere [1978]). If S belongs to C^1, then equation (4.18) has a unique solution in $L_{p,0}(S)$ for all $p \in (1, \infty)$ and $|V(V\varrho)|_\beta^+$ satisfies an inequality like (4.13).

We finally want to hint at that the Dirichlet problem in a Lipschitz domain Ω^+ may be reduced to an integral equation of the first kind by means of a simple layer potential (cf. Sec. 2.4 of Chap. 1). In this connection notice the following result of Verchota [1984].

Theorem 21. *If $\varphi \in W_p^1(S)$ ($1 < p \leq 2$) and $\Omega^+ \subset \mathbb{R}^n$ ($n \geq 3$) is a domain whose boundary S belongs to $C^{0,1}$, then the solution of the Dirichlet problem in Ω^+ is representable as a potential $V\varrho$. The density $\varrho \in L_p(S)$ of this potential is the unique solution of the equation*

$$V_0\varrho = \varphi \qquad \text{a.e. on } S.$$

Furthermore, we have the estimate

$$\||V(V\varrho)|_*\|_{L_p(S)} \leq C\|\varphi\|_{W_p^1(S)}.$$

3.4. Integral Equations for the Lamé and Stokes Equations in L^2 on Lipschitz Surfaces. Some of the results on harmonic potentials quoted in 3.2 and 3.3 have been recently carried over to elastic and hydro-dynamic potentials (Fabes [1985], Kenig [1984]). Following the latter two papers, we limit ouselves to the case of a Lipschitz domain of the special form $\Omega^+ = \{(x_1, x_2, x_3) \in \mathbb{R}^3 : x_3 > \varphi(x_1, x_2)\}$, φ being a Lipschitzian function.

Fabes [1985] and Kenig [1984] showed that for the potentials $W^{(\varkappa)}\chi$ and $V\varrho$ (introduced in Sec. 1.3 of Chap. 2) with densities in $L_2(S)$ the usual theorems on the jumps are in force, and they also established the estimates

$$\|(W^{(\varkappa)}\chi)_*^+\|_{L_2(S)} + \|(W^{(\varkappa)}\chi)_*^-\|_{L_2(S)} + \|W^{(\varkappa)}\chi\|_{H^{1/2}(\Omega^+)} \leqq C\|\chi\|_{L_2(S)},$$

$$\|V(V\varrho)|_*^+\|_{L_2(S)} + \|V(V\varrho)|_*^-\|_{L_2(S)} + \|V\varrho\|_{H^{3/2}(\Omega^+)} \leqq C\|\varrho\|_{L_2(S)}.$$
(4.19)

The proof of these inequalities relies upon the theorem by Coifman, McIntosh, Meyer [1982] on the L_p-boundedness of the Cauchy integral over Lipschitz curves.

As in the case of harmonic potentials (for which see the lemma in 3.3), the solvability of the integral equations (2.12) and (2.14) is derived from the inequality

$$\|\mathbf{f} \pm (T^{(\varkappa)})^*\mathbf{f}\|_{L_2(S)} \leqq C\|\mathbf{f} \mp (T^{(\varkappa)})^*\mathbf{f}\|_{L_2(S)},$$
(4.20)

where $\mathbf{f} \in L_2(S)$ and $T^{(\varkappa)} = 2W_0^{(\varkappa)}$.

The estimate (4.20) was set up in Fabes [1985] and Kenig [1984] under the assumption that $\varkappa \neq \mu$, which prevents us from considering the problem with given stresses. The restriction to the case $\varkappa \neq \mu$ has its source in the proof of the auxiliary inequality

$$\sum_{k=1}^{3} \int_S |V_t u_k|^2 dS \leqq C\|\mathcal{T}_\varkappa \mathbf{u}\|_{L_2(S)}^2.$$

Notice that the opposite inequality

$$\|\mathcal{T}_\varkappa \mathbf{u}\|_{L_2(S)}^2 \leqq C \sum_{k=1}^{3} \int_S |V_t u_k|^2 dS,$$

which results from Rellich's indentity for the Lamé equations (Nečas [1967]) and which is also used to derive the estimate (4.20), was shown to hold for all $\varkappa > 0$.

The following theorem is the main result of Fabes [1985] and Kenig [1984] on elastic potentials.

Theorem 22. *On condition that $\varkappa \neq \mu$, the integral equations (2.12) and (2.14) of the first and generalized second problems of elasticity theory in a Lipschitz domain Ω^+ are uniquely solvable.*

The analogue of this theorem for the Stokes equations given in Fabes [1985] and Kenig [1984] is as follows.

Theorem 23. *The integral equation of the Dirichlet problem for the Stokes equations in Ω^+ has a unique solution in $L_2(S)$. The double layer potential generated by the solution of the integral equation and representing the velocity field belongs to the space $H^{1/2}(\Omega^+)$ (see Lions, Magenes [1968]).*

Chapter 5
Boundary Integral Equations on
Piecewise Smooth Surfaces

In the preceding chapter we focussed on some classes of irregular surfaces well adjusted to the elaboration of a theory of boundary integral equations in the spaces C and L_p. On the one hand, we considered surfaces meeting the condition (B) on the local variation of the solid angle, and on the other, we dealt with surfaces that can in local Cartesian coordinates be described by Lipschitz functions. Notwithstanding the great generality involved by such surfaces, they are nevertheless insufficient for many applications. First of all, such simple surfaces as polyhedra or smooth cones are not exhaustively covered by the aforementioned classes of surfaces. Furthermore, the two classes quoted are badly fitted in with the investigation of the smoothness and the local singularities of the solutions.

The difficulties arising when attempting to answer similar questions for arbitrary piecewise smooth surfaces have their origin to a large extent in the narrowness of the existing theories of integral and pseudodifferential operators on non-smooth manifolds.

It proved that this obstacle may be avoided by resorting to the fact that the solutions of boundary integral equations can be expressed in terms of the solutions of certain auxiliary external and internal boundary value problems. Such an approach, permitting the possibility of developing a potential theory without recoursing to the theories of Fredholm and singular integral operators, was proposed in Maz'ya [1981a], [1985], [1986]. In this connection note that the theory of elliptic boundary value problems in domains with piecewise smooth boundary is presently worked out sufficiently well (see, for example, Kondrat'ev [1967], Maz'ya, Plamenevskiĭ [1977a], [1978a], [1978b], [1978c], [1979], [1983], Solonnikov [1983], Zaĭonchkovskiĭ, Solonnikov [1983]). Therefore this theory may be fairly conveniently used to establish theorems on the solvability of boundary integral equations. In this way one can also study differentiability properties of the solutions and their asymptotics near the singularities of the boundary.

The present chapter is devoted to several applications of the approach alluded to in the previous paragraph. In the first section, this approach is exemplified by the two fundamental boundary value problems for Lamé's equations. The boundaries of the domains are allowed to have edges, conical points, and polyhedral corners (without zero cusps). We also consider the system of boundary integro-differential equations corresponding to the mixed boundary value problem of elasticity in the case of a surface with edges on which different boundary conditions collide. In addition, in the same section we provide inversion formulas for boundary integral operators and estimates for the kernels of these operators coming about from the inversion formulas we give. These results bear general character, but we here bound ourselves to harmonic potentials on surfaces with a conical singularity. Section 1 is concluded by a formula for the Fredholm radii of the boundary integral operators of the Dirichlet and Neumann problems for the Laplace equation in spaces of Hölder type on surfaces with edges. The subject of Section 2 is the asymptotic behavior of the solutions of the boundary integral equations of those boundary value problems nearby an angular or conical point of the boundary.

§1. Solvability of Boundary Integral Equations on Piecewise Smooth Surfaces

1.1. Domains and Functional Spaces. Let Ω^+ be a domain in $\mathbb{R}^3$ with compact closure. Denote the boundary of Ω^+ by S and put $\Omega^- = \mathbb{R}^3 \setminus \overline{\Omega^+}$. We assume that S is the union of a finite number of "faces" $\{F\}$, "edges" $\{E\}$, and "vertices" $\{Q\}$. We leave it at this visual description and refer the reader who cares about precise definitions to Maz'ya, Plamenevskiĭ [1983]. Notice that all polyhedra are members of the class under consideration.

Without loss of generality suppose the origin lies in Ω^+. Let $\{U\}$ be a sufficiently fine covering of Ω^+ by open sets which meets the following two requirements: (a) each set U contains at most one vertex Q; (b) if $\overline{U}$ contains no vertex, then there is no more than one edge having points in common with $\overline{U}$. With each vertex Q and each edge E we associate, respectively, a certain real number β_Q and γ_E which will be specified below.

The *spaces* $C_{\beta,\gamma}^{1,\alpha}(\Omega^+)$ $(0 < \alpha < 1, \beta = \{\beta_Q\}, \gamma = \{\gamma_E\})$ are defined with the help of a partition of unity subordinate to $\{U\}$. If U does not contain a singularity of S, the norm in $C_{\beta,\gamma}^{1,\alpha}(U)$ of a function supported in U is equivalent to the norm of the function in the usual Hölder space $C^{1,\alpha}$. In case $U \cap E \neq \varnothing$, $\overline{U} \cap \{Q\} = \varnothing$, supp $u \subset U$ we have

$$\|u\|_{C_{\beta,\gamma}^{1,\alpha}(\Omega^+)} \sim \sup_{x \in \Omega^+} |u(x)| + \sup_{x,y \in \Omega^+} \frac{|r_E(x)^{\gamma_E}\nabla u(x) - r_E(y)^{\gamma_E}\nabla u(y)|}{|x - y|^\alpha},$$

where $r_E(x)$ is the distance between x and E. If U contains the vertex Q and supp $u \subset U$, then

$$\|u\|_{C_{\beta,\gamma}^{1,\alpha}(\Omega^+)} \sim \sup_{x \in \Omega^+} |u(x)|$$

$$+ \sup_{x,y \in \Omega^+} \frac{|\varrho_Q(x)^{\beta_Q} \prod_{\{E:Q \in \overline{E}\}} r_E(x)^{\gamma_E}\nabla u(x) - \varrho_Q(y)^{\beta_Q} \prod_{\{E:Q \in \overline{E}\}} r_E(y)^{\gamma_E}\nabla u(y)|}{|x - y|^\alpha},$$

where $\varrho_Q = |x - Q|$.

Replacing Ω^+ by Ω^- and $\sup_{x \in \Omega^+}|u(x)|$ by $\sup_{x \in \Omega^-}(1 + |x|)|u(x)|$ gives the definition of the *space* $C_{\beta,\gamma}^{1,\alpha}(\Omega^-)$. The *space* $C_{\beta,\gamma}^{1,\alpha}(S)$ is the space of traces on the union $\bigcup F$ of the functions in $C_{\beta,\gamma}^{1,\alpha}(\Omega^+)$ or $C_{\beta,\gamma}^{1,\alpha}(\Omega^-)$.

One more *space* $C_{\beta,\gamma}^{0,\alpha}(S)$ of functions on $\bigcup F$ is needed. It is defined as follows. For $U \cap E \neq \varnothing$, $U \cap \{Q\} = \varnothing$, supp $u \subset U$, we have

$$\|u\|_{C_{\beta,\gamma}^{0,\alpha}(S)} \sim \sup_{x \in \bigcup F} \gamma_E(x)^{\gamma_E - \alpha}|u(x)| + \sup_{x,y \in \bigcup F} \frac{|r_E(x)^{\gamma_E}u(x) - r_E(y)^{\gamma_E}u(y)|}{|x - y|^\alpha}.$$

If U contains the vertex Q and supp $u \subset U$, then

$$\|u\|_{C_{\beta,\gamma}^{0,\alpha}(S)} \sim \sup_{x \in \bigcup F} \varrho_Q(x)^{\beta_Q} \prod_{\{E:Q \in \bar{E}\}} r_E(x)^{\gamma_E} \sum_{\{E:Q \in \bar{E}\}} \frac{1}{r_E^\alpha} |u(x)|$$

$$+ \sup_{x,y \in \bigcup F} \frac{|\varrho_Q(x)^{\beta_Q} \prod_{\{E:Q \in \bar{E}\}} r_E(x)^{\gamma_E}u(x) - \varrho_Q(y)^{\beta_Q} \prod_{\{E:Q \in \bar{E}\}} r_E(y)^{\gamma_E}u(y)|}{|x - y|^\alpha}.$$

Finally, if no singularity of the boundary is located in U and supp $u \subset U$, then the norm of u in $C_{\beta,\gamma}^{0,\alpha}(S)$ is equivalent to the one in $C^{0,\alpha}(S)$.

The same notations are employed in the case of vector-valued functions.

1.2. Boundary Value Problems in Elasticity. We now consider the internal and external Dirichlet problems

$$\mu \Delta \mathbf{u}^+ + (\lambda + \mu)\nabla \operatorname{div} \mathbf{u}^+ = 0 \quad \text{in } \Omega^+, \qquad \mathbf{u}^+ = \boldsymbol{\varphi}^+ \quad \text{on } \bigcup F; \quad (\mathscr{D}^+)$$

$$\mu \Delta \mathbf{u}^- + (\lambda + \mu)\nabla \operatorname{div} \mathbf{u}^- = 0 \quad \text{in } \Omega^-, \qquad \mathbf{u}^- = \boldsymbol{\varphi}^- \quad \text{on } \bigcup F,$$
$$\mathbf{u}^-(x) = o((1 + |x|)^{-1}) \tag{$\mathscr{D}^-$}$$

as well as the internal and external Neumann problems

$$\mu \Delta \mathbf{u}^+ + (\lambda + \mu)\nabla \operatorname{div} \mathbf{u}^+ = 0 \quad \text{in } \Omega^+, \qquad \mathscr{T}\mathbf{u}^+ = \boldsymbol{\psi}^+ \quad \text{on } \bigcup F; \quad (\mathscr{N}^+)$$

$$\mu \Delta \mathbf{u}^- + (\lambda + \mu)\nabla \operatorname{div} \mathbf{u}^- = 0 \quad \text{in } \Omega^-, \qquad \mathscr{T}\mathbf{u}^- = \boldsymbol{\psi}^- \quad \text{on } \bigcup F,$$
$$\mathbf{u}^-(x) = o((1 + |x|)^{-1}), \tag{$\mathscr{N}^-$}$$

where $\mathscr{T} = \mathscr{T}_\mu$ is the stress operator (see Sec. 1.2 of Chap. 2).

The collection $\{\delta_E\}$ of real numbers appearing in the next theorem is defined in the following way. Let $\varphi(z) \in (0, 2\pi)$ be the angle contained by the tangential half-planes at the point z on the side of Ω^+ and put $\omega_E = \inf_{z \in E}(\pi + |\pi - \varphi(z)|)$. Then δ_E denotes the root of the equation $\sin(\omega_E \delta) + \delta \sin \omega_E = 0$ with minimal positive real-part. Note that δ_E is actually a real number belonging to the interval $(1/2, 1)$ and that δ_E decreases as ω_E increases.

Also in the following theorem, $\{\gamma_E\}$ and $\{\beta_Q\}$ are any collections of real numbers such that

$$0 < 1 - \gamma_E + \alpha < \delta_E \qquad \text{for all } E, \tag{5.1}$$

$$\left|\beta_Q + \sum_{\{E:Q \in \bar{E}\}} \gamma_E - \alpha - 3/2\right| < \varepsilon_Q \qquad \text{for all } Q, \tag{5.2}$$

where $\{\varepsilon_Q\}$ is a certain collection of postive numbers in the interval $(0, 1)$

depending on the geometric properties of the tangential cone to S whose vertex is Q. We remark that the numbers ε_Q may be determined by taking refuge in certain spectral boundary value problems in domains on the unit sphere (see Kondrat'ev [1967]). In Maz'ya, Plamenevskiĭ [1981] it was shown that $\varepsilon_Q > 1/2$ for the problems $(\mathscr{D}^+)$ and $(\mathscr{D}^-)$. Under some extra assumption, the same inequality was established for the problems $(\mathscr{N}^+)$ and $(\mathscr{N}^-)$ in Kozlov, Maz'ya [1988].

The numbers γ_E and β_Q prevailing in the previous paragraph are the same as those entering into the collections $\{\gamma_E\}$ and $\{\beta_E\}$ introduced and employed in 1.1.

Now we are in a position to state a theorem on the solvability of the boundary value problems $(\mathscr{D}^\pm)$ and $(\mathscr{N}^\pm)$ which will suffice for our purposes.

Theorem 1. (i) *The problems $(\mathscr{D}^+)$ and $(\mathscr{D}^-)$ are uniquely solvable in $C^{1,\alpha}_{\beta,\gamma}(\Omega^+)$ and $C^{1,\alpha}_{\beta,\gamma}(\Omega^-)$, respectively, for every $\varphi^\pm \in C^{1,\alpha}_{\beta,\gamma}(S)$.*
(ii) *Problem $(\mathscr{N}^-)$ has a unique solution in $C^{1,\alpha}_{\beta,\gamma}(\Omega^-)$ for every $\psi^- \in C^{0,\alpha}_{\beta,\gamma}(S)$.*
(iii) *Problem $(\mathscr{N}^+)$ is solvable in $C^{1,\alpha}_{\beta,\gamma}(\Omega^+)$ for every $\psi^+ \in C^{0,\alpha}_{\beta,\gamma}(S)$ provided the principal vector and the principal moment equal zero. In that case the solution is unique up to a rigid displacement.*

For the problems $(\mathscr{D}^+)$ and $(\mathscr{D}^-)$ this result is a special case of Theorem 11.5 in Maz'ya, Plamenevskiĭ [1983]. The proof given there applies with only minor modifications to the problems $(\mathscr{N}^+)$ and $(\mathscr{N}^-)$ as well.

1.3. Solution of the Problems $(\mathscr{D}^+)$ and $(\mathscr{D}^-)$ by Means of the Simple Layer Potential. Let $V\varrho$ be the elastic simple layer potential with density ϱ (see formula (2.8)).

Theorem 2. *If $\{\gamma_E\}$ and $\{\beta_Q\}$ satisfy the inequalities (5.1) and (5.2), then the operators*

$$C^{0,\alpha}_{\beta,\gamma}(S) \to C^{1,\alpha}_{\beta,\gamma}(\Omega^-), \qquad \varrho \mapsto V\varrho|_{\Omega^-},$$

$$C^{0,\alpha}_{\beta,\gamma}(S) \to C^{1,\alpha}_{\beta,\gamma}(\Omega^+), \qquad \varrho \mapsto V\varrho|_{\Omega^+},$$

are bounded, and there exists a bounded inverse operator $V^{-1}: C^{1,\alpha}_{\beta,\gamma}(S) \to C^{0,\alpha}_{\beta,\gamma}(S)$.

The first part of the theorem can ascertained straightforwardly. The second half follows from Theorem 1 in conjunction with the equalities $\varrho = \mathscr{T}\mathbf{u}^+ - \mathscr{T}\mathbf{u}^-$, where $\mathbf{u}^+$ and $\mathbf{u}^-$ are the restrictions of $V\varrho$ to Ω^+ and Ω^-, respectively.

1.4. Solution of the Problems $(\mathscr{D}^+)$, $(\mathscr{D}^-)$, $(\mathscr{N}^+)$, $(\mathscr{N}^-)$ by Means of the Double Layer Potential. We let $W\chi$ denote the double layer potential with density χ defined by the formulas (2.5) and (2.6).

If $\mathbf{u}^+ = W\chi^+$, then χ^+ fulfils the system of singular integral equations

$$\chi^+ + 2W_0\chi^+ = 2\varphi^+ \quad \text{on} \quad \bigcup F. \tag{5.3}$$

The solution of problem $(\mathscr{D}^-)$ may be represented as a sum of the form $(W\chi)(x) + \Gamma(x,0)\mathbf{a} + \text{rot } \Gamma(x,0)\mathbf{b}$, where $\mathbf{a}$ and $\mathbf{b}$ are unknown (constant) vectors. The triple $(\chi, \mathbf{a}, \mathbf{b})$ then satisfies the system of equations

$$-\chi^- + 2W_0\chi^- + 2\Gamma(\cdot, 0)\mathbf{a} + 2 \text{ rot } \Gamma(\cdot, 0)\mathbf{b} = -2\boldsymbol{\psi}^-. \qquad (5.4)$$

Representing the solution of the problems $(\mathcal{N}^+)$ and $(\mathcal{N}^-)$ in the form $V\chi^\pm$, we arrive at the systems

$$-\chi^+ + 2W_0^*\chi^+ = 2\boldsymbol{\psi}^+, \qquad (5.5)$$

$$\chi^- + 2W_0^*\chi^- = 2\boldsymbol{\psi}^-. \qquad (5.6)$$

Theorem 3. *Let $\{\gamma_E\}$ and $\{\beta_Q\}$ be subject to the conditions (5.1) and (5.2). Then:* (i) *the operators W_0 and W_0^* are bounded on $C_{\beta,\gamma}^{0;\alpha}(S)$ and $C_{\beta,\gamma}^{1;\alpha}(S)$;* (ii) *for $\boldsymbol{\varphi}^+ \in C_{\beta,\gamma}^{1;\alpha}(S)$, the systems (5.3) and (5.4) are uniquely solvable in $C_{\beta,\gamma}^{1;\alpha}(S)$ and $C_{\beta,\gamma}^{1;\alpha}(S) \times \mathbb{R}^3 \times \mathbb{R}^3$, respectively.*

Here now is an outline of the *proof* of the portion of the theorem concerning the solvability of (5.3). Let $\mathbf{u}^+$ and $\mathbf{u}^-$ be the solutions of the problems $(\mathcal{D}^+)$ and $(\mathcal{D}^-)$ for $\boldsymbol{\varphi}^+ = \boldsymbol{\varphi}^-$. Then $-\boldsymbol{\varphi}^+ = V(\mathcal{T}\mathbf{u}^+ - \mathcal{T}\mathbf{u}^-)$. Denote the solution of problem $(\mathcal{N}^-)$ with $\boldsymbol{\psi}^- = -(\mathcal{T}\mathbf{u}^+ - \mathcal{T}\mathbf{u}^-)$ by $\mathbf{v}^-$. Since $\mathbf{v} = -(W\mathbf{v}^- - V\mathcal{T}\mathbf{v}^-)$ in Ω^-, we have $(1/2)\mathbf{v}^- + W_0\mathbf{v}^- = V\mathcal{T}\mathbf{v}^- - \boldsymbol{\varphi}^+$ on $\bigcup F$. Hence the vector-valued function $\chi^+ = \mathbf{v}^-|_{\bigcup F}$ is a solution of (5.3) and so Theorem 1 immediately implies that $\chi^+ \in C_{\beta,\gamma}^{1;\alpha}(S)$.

To show the uniqueness of the solution of system (5.3) it suffices to verify that the formally adjoint system (5.6) has a solution in $C_{\beta,\gamma}^{0;\alpha}(S)$ whenever $\boldsymbol{\psi}^- \in C_{\beta,\gamma}^{0;\alpha}(S)$. But this is easy: denote by $\mathbf{v}^-$ the solution of problem $(\mathcal{N}^-)$, let ϱ be the density for which the simple layer potential $V\varrho$ coincides with $\mathbf{v}^-$ on Ω^- (see Theorem 2), and finally notice that the density ϱ satisfies the system (5.6). $\square$

The argument just used involves all the essential issues of the proof of the following theorem on the solvability of the systems (5.5) and (5.6).

Theorem 4. *Let γ_E and β_E satisfy the inequalities (5.1) and (5.2). Then there exists a continuous inverse of the operator $I + 2W_0^*$ on the space $C_{\beta,\gamma}^{0;\alpha}(S)$. The system (5.5) has a solution in $C_{\beta,\gamma}^{0;\alpha}(S)$ for all $\boldsymbol{\psi}^+$ with vanishing principal vector and principal moment.*

Theorems on the rise of smoothness and on the change of the collections $\{\gamma_E\}$ and $\{\beta_Q\}$ akin to the corresponding theorems for boundary value problems established in Maz'ya, Plamenevskiĭ [1983] also hold for the solutions of the integral equations considered here. Before stating a result in this direction we need a further class of functional spaces.

Let $l \geq 1$ be an integer and let $0 < \alpha < 1$. Replacing in the definition of the space $C_{\beta,\gamma}^{1;\alpha}(\Omega^\pm)$ the gradient by $V_l \sim \{\partial^l/\partial x_1^{\alpha_1} \partial x_2^{\alpha_2} \partial x_3^{\alpha_3}\}$ gives the definition of the space $C_{\beta,\gamma}^{l;\alpha}(\Omega^\pm)$. Denote by $C_{\beta,\gamma}^{l;\alpha}(S)$ the space of traces on S of functions from $C_{\beta,\gamma}^{l;\alpha}(\Omega^\pm)$.

The following result supplements Theorem 3.

Theorem 5. *Let $\{\gamma_E\}, \{\beta_E\}$ be collections for which (5.1), (5.2) are fulfilled and let $\{\gamma_E^* - l + 1\}, \{\beta_Q^* - l + 1\}$ be collections subject to the same conditions with γ_E, β_E replaced by $\gamma_E^* - l + 1$, $\beta_E^* - l + 1$. If $\boldsymbol{\varphi}^\pm$ belong to $C_{\beta,\gamma}^{1;\alpha}(S) \cap C_{\beta^*,\gamma^*}^{l,\alpha}(S)$ and $\chi^\pm \in C_{\beta,\gamma}^{1;\alpha}(S)$ are the solutions of the systems (5.3) and (5.4), then $\chi^\pm \in C_{\beta^*,\gamma^*}^{l,\alpha}(S)$.*

Similar additions can also be made to the Theorems 2 and 4.

The potential theory of boundary value problems in the plane can be erected proceeding along the same lines as in three dimensions. If S is understood as a piecewise smooth curve without cusps and $\{Q\}$ means a finite number of corner points, then the Theorems 1 by 5 remain in force without any change.

For harmonic and hydrodynamic potentials theorems analogous to the five ones quoted above are also valid and can be proved by the method exploited here, too.

1.5. The Equations of the Potential Theory for the Mixed Problem. The material of this subsection is taken from Maz'ya [1986].

Assume that S does not have vertices and that the components of the set $S\setminus(\bigcup E)$ are divided into two groups. The unions of the components of the first and second groups are denoted by S_1 and S_2. Also assume that $\bigcup E$ is the common boundary of the sets S_1 and S_2.

We here consider Lamé's equations in Ω^+ with the boundary conditions

$$\mathbf{w}^+|_{S_1} = \boldsymbol{\varphi}^+, \qquad \mathscr{T}(\partial_x, \mathbf{n})\mathbf{w}^+|_{S_2} = \boldsymbol{\psi}^+. \qquad (\mathscr{X}^+)$$

Problem $(\mathscr{X}^-)$ is obtained from the latter problem by replacing Ω^+ with Ω^- and adding the requirement that $\mathbf{w}(x) = O((1 + |x|)^{-1})$ as $|x| \to \infty$.

Put $\varkappa = (\lambda + 3\mu)/(\lambda + \mu)$ and let $\zeta(z)$ denote the solution with minimal positive real-part of the equation

$$\varkappa \sin^2[\zeta(\pi \pm \pi - \varphi(z))] = (1 + \varkappa)^2 - \zeta^2 \sin^2 \varphi(z).$$

Throughout this subsection we shall suppose that

$$0 < 1 - \beta + \alpha < \min\{1, \operatorname{Re}\zeta(z)\} \quad \text{for all } z \in \bigcup E.$$

Let $\alpha \in (0, 1)$, $\beta \in \mathbb{R}^1$, and denote by $d(x)$ the distance between x and $\bigcup E$. We define $C_\beta^{1,\alpha}(\Omega^+)$ to be the space of all functions on Ω^+ for which the norm $\|u\|_{C_\beta^{1,\alpha}(\Omega^+)}$ given by

$$\sup_{x,y \in \Omega^+} \frac{|d(x)^\beta \nabla u(x) - d(y)^\beta \nabla u(y)|}{|x - y|^\alpha} + \sup_{x \in \Omega^+} |u(x)|$$

is finite. Analogously, $C_\beta^{1,\alpha}(\Omega^-)$ is the space of all functions on Ω^- such that the norm $\|u\|_{C_\beta^{1,\alpha}(\Omega^-)}$ defined by

$$\sup_{x,y \in \Omega^-} \frac{|d(x)^\beta \nabla u(x) - d(y)^\beta \nabla u(y)|}{|x - y|^\alpha} + \sup_{x \in \Omega^-} (1 + |x|)|u(x)|$$

takes on finite values. Further, $C_\beta^{1,\alpha}(S_k)$ refers to the space of limit values on S_k of functions in either of the spaces $C_\beta^{1,\alpha}(\Omega^+)$ and $C_\beta^{1,\alpha}(\Omega^-)$.

One more space is $C_\beta^{0,\alpha}(S_k)$, the norm $\|u\|_{C_\beta^{0,\alpha}(S_k)}$ of which is defined as

$$\sup_{x,y \in S_k} \frac{|d(x)^\beta u(x) - d(y)^\beta u(y)|}{|x - y|^\alpha} + \sup_{x \in S_k} |d(x)^{\beta - \alpha} u(x)|.$$

And finally, we let $\overset{\circ}{C}{}_\beta^{1,\alpha}(S_k)$ refer to the subspace of $C_\beta^{1,\alpha}(S_k)$ consisting of all the functions vanishing identically on $\bigcup E$.

Theorem 6. *There exist continuous inverses of the operators generated by the problems* $(\mathscr{L}^{+})$ *and* $(\mathscr{L}^{-})$ *which map* $C_{\beta}^{1,\alpha}(S_1) \times C_{\beta,\gamma}^{0,\alpha}(S_2)$ *into* $C_{\beta}^{1,\alpha}(\Omega^{+})$ *and* $C_{\beta}^{1,\alpha}(\Omega^{-})$, *respectively.*

For the sake of definiteness, we content ourselves to the problem $(\mathscr{L}^{-})$ (all the following arguments apply with obvious modifications to the internal problem likewise). Let $V_k\varrho$ and $W_k\varrho$ be the potentials resulting from the definition of $V\varrho$ and $W\varrho$ by replacing the surface S with the sets S_k $(k = 1, 2)$.

The solution of problem $(\mathscr{L}^{-})$ is sought in the form $W_1\chi + V_2\varrho$. Then the vector-valued functions ϱ and χ, defined on S_2 and S_1, respectively, must satisfy the system

$$
\begin{aligned}
-\tfrac{1}{2}\chi + W_{10}\chi + V_{20}\varrho &= \varphi^{-} && \text{on } S_1, \\
\tfrac{1}{2}\varrho + \mathscr{T}W_1\chi + \mathscr{T}V_2\varrho &= \psi^{-} && \text{on } S_2.
\end{aligned}
\tag{5.7}
$$

Notice that this system does not comprise integral equations alone: the second equation involves the pseudodifferential operator of the first order $\mathscr{T}W_1$. We also draw the reader's attention to the circumstance that the function χ appearing in the next theorem is required to vanish on $\bigcup E$.

Theorem 7. *The operator*

$$
\mathring{C}_{\beta}^{1,\alpha}(S_1) \times C_{\beta}^{0,\alpha}(S_2) \to C_{\beta}^{1,\alpha}(S_1) \times C_{\beta}^{0,\alpha}(S_2), \qquad (\chi, \varrho) \mapsto (\varphi^{-}, \psi^{-}) \tag{5.8}
$$

induced by the equations (5.7) *is continuous and has an inverse defined on* $C_{\beta}^{1,\alpha}(S_1)$ $\times C_{\beta}^{0,\alpha}(S_2)$.

The *proof* relies on Theorem 6 and the following argument. Let $\mathbf{w}^{+}$ be the solution of problem $(\mathscr{L}^{+})$, and let $\mathbf{q}$ denote the solution of Lamé's equations in Ω^{+} satisfying the conditions

$$
\mathbf{q} = -(\mathbf{w}^{+} - \mathbf{w}^{-}) \quad \text{on } S_2, \qquad \mathscr{T}\mathbf{q} = -(\mathscr{T}\mathbf{w}^{-} - \mathscr{T}\mathbf{w}^{+}) \quad \text{on } S_1.
$$

Using Betti's formula one may ascertain that the vector functions $\chi = \mathbf{q}|_{S_1}$ and $\varrho = (\mathscr{T}\mathbf{q})|_{S_2}$ fulfil the equations (5.7). From the solvability of the system adjoint to problem $(\mathscr{L}^{+})$, in which S_1 and S_2 interchange their roles, we deduce the uniqueness of the solution of (5.7). □

1.6. Representations and Estimates for the Inverses of the Operators Generated by Integral Equations.

We here consider the integral equations of the theory of harmonic potentials on surfaces and curves which are everywhere smooth except for a finite number of conical points. We follow the paper Grachev, Maz'ya [1989], in which estimates of the kernels of the inverse integral operators are provided. These estimates yield, in particular, the solvability of the integral equation for the Dirichlet problem in the space of continuous functions without any extra assumption on the nature of the boundary of the domain. Earlier the solvability in C was established under the hypothesis that the essential norm of the integral operator is less than one (see Sec. 1 of Chap. 4).

Let Ω^+ be a simply connected region in $\mathbb{R}^n$ ($n \geq 2$) with compact closure, put $S = \partial(\Omega^+)$ and assume $0 \in S$. Furthermore, suppose that $S \setminus \{0\}$ is a smooth surface and that Ω^+ in a vicinity of the point 0 coincides with a cone excising the open set G^+ on the unit sphere $\mathbb{S}^{n-1}$. Let $\Omega^- = \mathbb{R}^n \setminus \overline{\Omega^+}$ and $G^- = \mathbb{S}^{n-1} \setminus \overline{G^+}$.

We denote by T the doubled direct value of the double layer potential and by T^* the operator adjoint to T. Finally, $C^\infty(S)$ is the space of restrictions to S of the functions from $C^\infty(\mathbb{R}^n)$, $(\cdot, \cdot)$ refers to the inner product in $L_2(S)$, and $\mathbf{n}$ denotes the outer normal to S.

Theorem 8. *Let D^+ and D^- (resp. N^+ and N^-) be the inverse operators of the Dirichlet problems in Ω^+ and Ω^-(resp. the Neumann problems in Ω^+ and Ω^-), the operator N^+ being defined on functions orthogonal to the constants on S and satisfying the equality $(N^+\psi^+, 1) = 0$.*

(i) *If $\varphi \in C^\infty(S)$, then*

$$(I + T)^{-1}\varphi = \frac{1}{2}\left(I - N^- \frac{\partial}{\partial \mathbf{n}} D^+\right)\varphi, (I + T^*)^{-1}\varphi = \frac{1}{2}\left(I - \frac{\partial}{\partial \mathbf{n}} D^+ N^-\right)\varphi.$$

(ii) *If $\varphi, \psi \in C^\infty(S)$, then the solutions of the equations*

$$(I - T)\chi = \varphi, (\varphi, \sigma_0) = 0 \text{ and } (I - T^*)\varrho = \psi, (\psi, 1) = 0$$

(where $\sigma_0 = \partial(D^-1)/\partial\mathbf{n}$) can be obtained by means of the formulas

$$\chi = \frac{1}{2}\left(I - N^+ \frac{\partial}{\partial \mathbf{n}} D^-\right)\varphi + c \text{ and } \varrho = \frac{1}{2}\left(1 - \frac{\partial}{\partial \mathbf{n}} D^- N^+\right)\psi + c\sigma_0.$$

In what follows $\delta(\delta + n - 2)$ and $\gamma(\gamma + n - 2)$ are the first eigenvalues of the Dirichlet problem in G^+ and the Neumann problem in G^- for the Beltrami operator ($\delta, \gamma > 0$).

Lemma 1 (see Verzhbinskiĭ, Maz'ya [1974] and Maz'ya, Plamenevskiĭ [1979]). *The kernel $P^+(x, \xi)$ of the operator D^+ admits the estimates*

$$|D_x^\alpha D_\xi^\sigma P^+(x, \xi)| \leq c|x|^{\delta - |\alpha|}/|\xi|^{n-1+\delta+|\sigma|} \text{ for } |x| < |\xi|/2,$$

$$|D_x^\alpha D_\xi^\sigma P^+(x, \xi)| \leq c|x - \xi|^{1-n-|\alpha|-|\sigma|} \text{ for } |\xi|/2 < |x| < 2|\xi|,$$

$$|D_x^\alpha D_\xi^\sigma P^+(x, \xi)| \leq c|\xi|^{\delta - 1 - |\sigma|}/|x|^{n-2+\delta+|\alpha|} \text{ for } |x| > 2|\xi|.$$

The following lemma from Maz'ya, Plamenevskiĭ [1979] provides analogous information about the kernel $N^-(x, \xi)$ of the operator N^-.

Lemma 2. *Let B_R be a ball containing the set $\overline{\Omega^+}$. Then in $B_R \setminus \overline{\Omega^+}$ for the kernel $N^-(x, \xi)$ the following inequalities hold:*

$$|D_x^\alpha D_\xi^\sigma(N^-(x, \xi) - N^-(0, \xi))| \leq c|x|^{\gamma - |\alpha|}/|\xi|^{n-2+\gamma+|\sigma|} \text{ for } |x| < |\xi|/2,$$

$$|D_x^\alpha D_\xi^\sigma N^-(x, \xi)| \leq c|x - \xi|^{2-n-|\alpha|-|\sigma|} \text{ for } |\xi|/2 < |x| < 2|\xi|,$$

$$|D_x^\alpha D_\xi^\sigma(N^-(x, \xi) - N^-(x, 0))| \leq c|\xi|^{\gamma - |\sigma|}/|x|^{n-2+\gamma+|\alpha|} \text{ for } |x| > 2|\xi|.$$

Here $N^-(0, \eta) = N^-(\eta, 0) = c_0|\eta|^{2-n} + c_1 + R(\eta)$, where $|D_\eta^\alpha R(\eta)| \leq c|\eta|^{\gamma - |\alpha|}$.

Combining Theorem 8 with the two preceding lemmas produces sharp estimates for the kernels of the inverses occuring in the theorem. We confine ourselves to stating the result for the operator $(I + T)^{-1}$.

Theorem 9. *Let* $\mu = \min\{\delta, \gamma, 1\}$. *Then* $(I + T)^{-1}$ *decomposes into a sum* $I + M_1 + M_2$ *with certain integral operators* M_1 *and* M_2 *on S whose kernels in the case* $\delta \neq \gamma$ *admit the following estimates:*

$$|M_1(x, y)| \leqq c(1 + |y|^{\mu-1}),\tag{5.9}$$

$$|M_2(x, y)| \leqslant \begin{cases} c|y|^{1-n}(|x|/|y|)^{\mu} & for\ |x| < |y|/2, \\ c|y|^{-1}|x - y|^{2-n} & for\ |y|/2 < |x| < 2|y|, \\ c|y|^{-1}|x|^{2-n}(|y|/|x|)^{\mu} & for\ |x| > 2|y|. \end{cases}\tag{5.10}$$

If $\delta = \gamma$, *these estimates become true after adding* $c|y|^{\mu-1}|\log|y||$ *to the right of* (5.9) *and multiplying the right of* (5.10) *by* $(1 + |\log(|x|/|y|)|)$.

Here is a consequence of the previous theorem.

Theorem 10. (i) *Assume that either* $1 \leqq p < \infty$, $(1 - n)/p < \beta < \mu + n - 2 + (1 - n)/p$ *or* $p = \infty, 0 \leqq \beta < \mu + n - 2$. *Then the operator* $(I + T)^{-1}$ *is continuous on the space* $L_{p,\beta}(S)$ *of all functions on S for which the norm* $\|u\|_{L_{p,\beta}(S)} = \||x|^{\beta}u\|_{L_p(S)}$ *is finite.*

(ii) *If* $0 < \alpha < \mu$, *then the operator* $(I + T)^{-1}$ *is continuous on the Hölder space* $C^{0,\alpha}(S)$.

All what was said in this subsection also applies to the systems of integral equations (2.12) and (2.36) of elasticity and hydrodynamics. The roles of δ and γ are then played by the real parts δ^*, γ^* of the eigenvalues of certain operator bundles of boundary value problems in the sets G^+ and G^- on the unit sphere. That δ^* is positive was proved in Maz'ya, Plamenevskiĭ [1981] and the positivity of γ^* was under the additional assumption that S is explicitly defined in some Cartesian coordinate system near the vertex of a cone established in Kozlov, Maz'ya [1988]. The second part of the theorem, which concerns the solvability in $C^{0,\alpha}(S)$ of the integral equations for the Lamé an Stokes equations, has therefore a proof only under the extra assumption just pointed to.

Up to obvious modifications, the representations and estimates quoted above are also valid for $n = 2$ in case the curve is piecewise smooth and has no cusps.

1.7. The Fredholm Radius of Operators of Double Layer Potential Type on Piecewise Smooth Surfaces. In this section we provide a formula for the Fredholm radius of the integral operators of the theory of *n*-dimensional harmonic potentials in domains with edges on the boundary. The integral operators are considered on certain weighted space of Hölder type (see Grachev, Maz'ya [1986]). The formula mentioned is, in essence, derived from some theorems on the operators associated with certain auxiliary boundary value problems. A straightforward approach to the computation of the Fredholm radius of the operators under consideration was in the $n = 2$ case developed by Radon [1919b] for the spaces C and C^* and by Shelepov [1969] for the spaces L_p (see Sec. 1 of Chap. 4).

Let $\{E\}$ denote a closed subset of S which is a C^∞-submanifold of the dimension $n - 2$ of $\mathbb{R}^n$. Suppose furthermore that $S\backslash\{E\}$ is also a C^∞-submanifold of $\mathbb{R}^n$. Finally assume that in some neighborhood of each point of $\{E\}$ the set $\overline{\Omega^+}$ is diffeomorphic to some n-dimensional dihedral angle. Thus, we may assign two well-defined $(n - 1)$-dimensional tangential half-spaces $\Gamma_1(z)$ and $\Gamma_2(z)$ to each point $z \in \{E\}$. Let $\varphi(z) \in (0, 2\pi)$ denote the angle between $\Gamma_1(z)$ and $\Gamma_2(z)$ on the side of Ω^+.

We now define the functional spaces used in the following. Let $0 < \alpha < 1$, let l be an integer, and let $d(x)$ refer to the distance between x and $\{E\}$.

For $l \geq 1$ and $0 < \beta < l + \alpha$, the space $C_\beta^{l,\alpha}(\Omega^+)$ is the space of all functions in Ω^+ for which the norm $\|u\|_{C_\beta^{l,\alpha}(\Omega^+)}$ given by

$$\sup_{x,y \in \Omega^+} \frac{|d(x)^\beta \nabla_l u(x) - d(y)^\beta \nabla_l u(y)|}{|x - y|^\alpha} + \|u\|_{C^{l+\alpha-\beta}(\Omega^+)}, \qquad (5.11)$$

is finite. Here $\nabla_l = \{\partial^l/\partial x_1^{\alpha_1} \ldots \partial x_n^{\alpha_n}\}$, and $C^\varrho(\Omega^+)$ is the ordinary Hölder space. The space $C_\beta^{l,\alpha}(\Omega^-)$ consists of all functions in Ω^- such that the norm obtained by replacing Ω^+ with Ω^- in (5.11) and adding $\sup_{x \in \Omega^-} |x|^{n-2}|u(x)|$ to (5.11) takes on finite values. We let $C_\beta^{l,\alpha}(S)$ denote the space of the limit values on $S\backslash\{E\}$ of the functions from either of the spaces $C_\beta^{l,\alpha}(\Omega^+)$ and $C_\beta^{l,\alpha}(\Omega^-)$.

For $l \geq 0$ and $\beta \geq l + d$, we define the space $C_\beta^{l,\alpha}(S)$ as the space of all functions u on $S\backslash\{E\}$ such that the norm $\|u\|_{C_\beta^{l,\alpha}(S)}$ given by

$$\sup_{x,y \in S\backslash\{E\}} \frac{|d(x)^\beta \nabla_l u(x) - d(y)^\beta \nabla_l u(y)|}{|x - y|^\alpha} + \sup_{x,y \in S\backslash\{E\}} d(x)^{\beta-l-\alpha}|u(x)|$$

is finite.

Theorem 11. *Suppose* $0 < \alpha < 1$, $0 < 1 + \alpha - \beta < 1$, *let* l *be a non-negative integer, and set*

$$\varkappa = \min_{z \in \{E\}} |\sin(\pi(1 + \alpha - \beta))/\sin((\pi - \varphi(z))(1 + \alpha - \beta))|$$

(notice that $\varkappa > 1$ *if* $1 + \alpha - \beta < \pi/(\pi + |\pi - \varphi(z)|)$*, whereas* $\varkappa \leq 1$ *otherwise).*

Then the Fredholm radius of the operators T^* *and* T *on the spaces* $C_{\beta+l}^{l,\alpha}(S)$ *and* $C_{\beta+l}^{l+1,\alpha}(S)$*, respectively, equals* $\varkappa$*.*

§2. Asymptotics of the Solutions of Boundary Integral Equations Near Corner Points

In this section we consider the integral equations of the theory of harmonic potentials. Our first subject is the derivation of the asymptotics of the solutions near the corner points of the contour, including the computation of the coefficients appearing in the asymptotic formulas. As in Section 1, results on the solutions of the integral equations are deduced from well-known results on the asymptotic behavior of the solutions of the internal and external Dirichlet and Neumann problems (for the latter ones see Kondrat'ev [1967] and Maz'ya, Plamenevskiĭ [1974], [1975b], [1979]). Our exposition is adopted from the paper Zargaryan, Maz'ya [1984]. We conclude this section by some results on the asymptotics of the solution of the boundary system corresponding to the mixed problem for the Laplace equation (Zargaryan, Maz'ya [1984]).

Let S be a simple piecewise-smooth curve with a finite number of corner points making angles distinct from 0 and 2π, and let Ω^+ and Ω^- denote the domain inside and outside of S, respectively. We first seek the asymptotics of the solution χ^+ of the integral equation of the internal Dirichlet problem; at the points of S which are no corner points this equation coincides with (1.13). Let $u^\pm$ be the solutions of the Dirichlet problems for the Laplace equation in the domains $\Omega^\pm$ with the same function $\varphi^+ = \varphi^- = \varphi$ prescribed on S (see Secs. 2.1 by 2.4 of Chap. 1). We know that then

$$\varphi(x) = \int_S \left(\frac{\partial u^-}{\partial \mathbf{n}_\xi} - \frac{\partial u^+}{\partial \mathbf{n}_\xi} \right) \mathscr{E}(x, \xi) d_\xi S + u^-(\infty) \tag{5.12}$$

if only $x \in S$ is not a corner point.

Let v^- denote the solution of the external Neumann problem

$$\Delta v^- = 0 \quad \text{in } \Omega^-, \qquad \frac{\partial v^-}{\partial \mathbf{n}} = \frac{\partial u^-}{\partial \mathbf{n}} - \frac{\partial u^+}{\partial \mathbf{n}} \quad \text{on } S,$$

$$v^-(x) = o(1) \qquad \text{as } |x| \to \infty. \tag{5.13}$$

The Ω^- versions of (1.3) and (1.8) imply that

$$\frac{1}{2} v^- + W_0 v^- = \int_S \left(\frac{\partial u^-}{\partial \mathbf{n}} - \frac{\partial u^+}{\partial \mathbf{n}} \right) \mathscr{E}(x, \xi) d_\xi S.$$

Combining this and (5.12) we see that the function $v^- + u^-(\infty)$ satisfies equation (1.13) and thus coincides with χ^+.

For simplicity, assume that Ω^+ in a vicinity of any one of the corner points of S is equal to the sector $\{x_1 + ix_2 = re^{i\vartheta}: 0 < r < \delta, 0 < \vartheta < \alpha\}$. First let $0 < \alpha < \pi$. It was shown by Kondrat'ev [1967] that then as $r \to 0$,

$$u^+(x) = \varphi(0) + \frac{\partial \varphi}{\partial x_1}(0)x_1 + \frac{\partial \varphi}{\partial x_2}(0)x_2 + O(r^{1+\varepsilon}), \qquad \varepsilon > 0, x \in \Omega^+;$$

$$u^-(x) = \varphi(0) + C_1 r^{\pi/(2\pi-\alpha)} \sin \frac{\pi\vartheta}{2\pi - \alpha}$$

$$+ \frac{\partial \varphi}{\partial x_1}(0)x_1 + \frac{\partial \varphi}{\partial x_2}(0)x_2 + O(r^{1+\varepsilon}), \qquad \varepsilon > 0, x \in \Omega^-.$$

Since these equalities may be differentiated, we have

$$\frac{\partial u^-}{\partial \mathbf{n}} - \frac{\partial u^+}{\partial \mathbf{n}} = \frac{\pi c_1}{2\pi - \alpha} r^{(\alpha - \pi)/(2\pi - \alpha)} + O(r^\varepsilon), \qquad x \in S \setminus \{0\},$$

and hence (5.13) gives that, as $r \to 0$,

$$
\begin{aligned}
v^-(x) - v^-(0) &\sim c_0 r^{\pi/(2\pi - \alpha)} \cos\frac{\pi\vartheta}{2\pi - \alpha} \\
&\quad + c_1 r^{\pi/(2\pi - \alpha)} \sin\frac{\pi\vartheta}{2\pi - \alpha}, \qquad x \in \Omega^-.
\end{aligned}
\tag{5.14}
$$

What results is that if $0 < \alpha < \pi$, then

$$\chi^+(x) - \chi^+(0) \sim \pm c_0 r^{\pi/(2\pi - \alpha)},$$

the plus and minus corresponding to the rays $\vartheta = 0$ and $\vartheta = 2\pi - \alpha$, respectively.

The constant c_0 can be determined by the method of Maz'ya, Plamenevskiĭ [1974], [1977]. Let η be any function in $C^\infty(0, \infty)$ which is identically 1 on $(0, \delta)$ and vanishes identically on $(2\delta, \infty)$. Put

$$w(x) = v(x) - v(0) - c_1 \eta(r) r^{\pi/(2\pi - \alpha)} \sin\frac{\pi\vartheta}{2\pi - \alpha}.$$

By virtue of (5.13), the latter function is the solution of the problem

$$\Delta w = c_1 \Delta \left(\eta(r) r^{\pi/(2\pi - \alpha)} \sin\frac{\pi\vartheta}{2\pi - \alpha} \right) \qquad \text{in } \Omega^-,$$

$$\frac{\partial w}{\partial \mathbf{n}} = \frac{\partial u^-}{\partial \mathbf{n}} - \frac{\partial u^+}{\partial \mathbf{n}} - c_1 \frac{\partial}{\partial \mathbf{n}} \left(\eta(r) r^{\pi/(2\pi - \alpha)} \sin\frac{\pi\vartheta}{2\pi - \alpha} \right) \qquad \text{on } S.$$

By (5.14),

$$w(x) \sim c_0 r^{\pi/(2\pi - \alpha)} \sin\frac{\pi\vartheta}{2\pi - \alpha} \qquad \text{as } r \to 0, x \in \Omega^-.$$

Hence (see Maz'ya, Plamenevskiĭ [1977]),

$$
\begin{aligned}
c_0 = {} & \frac{1}{\pi} \int_{\Omega^-} \zeta^- c_1 \Delta \left(\eta(r) r^{\pi/(2\pi - \alpha)} \sin\frac{\pi\vartheta}{2\pi - \alpha} \right) dx \\
& + \frac{1}{\pi} \int_S \zeta^- \left\{ \frac{\partial u^-}{\partial \mathbf{n}} - \frac{\partial u^+}{\partial \mathbf{n}} - c_1 \frac{\partial}{\partial \mathbf{n}} \left(\eta(r) r^{\pi/(2\pi - \alpha)} \sin\frac{\pi\vartheta}{2\pi - \alpha} \right) \right\} dS,
\end{aligned}
$$

where ζ^- is a function having the following properties: it is continuous outside every neighborhood of the origin, it is harmonic in Ω^-, its normal derivative vanishes at every point on S which is not a corner point, and its asymptotics is given by

$$\zeta^-(x) \sim r^{-\pi/(2\pi - \alpha)} \cos\frac{\pi\vartheta}{2\pi - \alpha} \qquad \text{as } r \to 0.$$

From here we arrive after some simplifications at the formula

$$c_0 = -\frac{1}{\pi}\int_S \zeta^- \frac{\partial u^+}{\partial \mathbf{n}}\,dS. \tag{5.15}$$

If we let Z^+ denote any harmonic extension of ζ^- into Ω^+ which is continuous on $S\backslash\{0\}$, then

$$c_0 = -\frac{1}{\pi}\int_S [\varphi - \varphi(0)]\frac{\partial Z^+}{\partial \mathbf{n}}\,dS.$$

Analogously one may show that if $\pi < \alpha < 2\pi$, then

$$\chi^+(x) - \chi^+(0) \sim \tan\frac{\pi^2}{\alpha}D_1 r^{\pi/\alpha} \qquad \text{as } r \to 0,$$

where

$$D_1 = \frac{1}{\pi}\int_S [\varphi - \varphi(0)]\frac{\partial \zeta^+}{\partial \mathbf{n}}\,dS \tag{5.16}$$

and ζ^+ is a function harmonic in Ω^+ and equal to zero on $S\backslash\{0\}$ such that $\zeta^+ \sim r^{-\pi/\alpha}\sin(\pi\vartheta/\alpha)$.

We now pass to the asymptotics of the solution ϱ^- of the integral equation of the external Neumann problem. This equation coincides with (1.16) away from the corner points on S and is the adjoint of the equation for χ^+. The asymptotics of ϱ^- is derived from the equality

$$\varrho^-(x) = \frac{\partial v^-}{\partial \mathbf{n}} - \frac{\partial v^+}{\partial \mathbf{n}},$$

v^- being the solution of the original Neumann problem in Ω^- and v^+ being the solution of the Dirichlet problem

$$\Delta v^+ = 0 \quad \text{in } \Omega^+, \qquad v^+ = v^- \quad \text{on } S.$$

If $0 < \alpha < \pi$, the asymptotic behavior is of the form

$$\varrho^- \sim \pm\frac{\pi}{2\pi - \alpha}c_2 r^{(\alpha-\pi)/(2\pi-\alpha)}\cot\frac{\pi\alpha}{2(2\pi-\alpha)},$$

plus and minus corresponding to the rays $\vartheta = \alpha$ and $\vartheta = 0$, respectively. By Maz'ya, Plamenevskiĭ [1977], the coefficient c_2 equals

$$c_2 = -\frac{1}{\pi}\int_S \zeta^- \psi^-\,dS,$$

where ζ^- is the same function as in (5.15).

In case $\pi < \alpha < 2\pi$, we have

$$\varrho^- \sim \frac{\pi}{\alpha}D_2 r^{(\pi-\alpha)/\alpha}, \qquad D_2 = \frac{1}{\pi}\int_S [v^- - v^-(0)]\frac{\partial \zeta^+}{\partial \mathbf{n}}\,dS,$$

where ζ^+ is as in (5.16). Another expression for the coefficient D_2 is given in Zargaryan, Maz'ya [1984]:

$$D_2 = \frac{1}{\pi} \int_S Z^- \psi^- dS,$$

Z^- being the solution of the boundary value problem

$$\Delta Z^- = 0 \quad \text{on } \Omega^-, \qquad \frac{\partial Z^-}{\partial \mathbf{n}} = \frac{\partial \zeta^+}{\partial \mathbf{n}} \quad \text{on } S \setminus \{0\}$$

which vanishes at infinity and has the asymptotics

$$Z^-(x) = r^{-\pi/\alpha} \frac{\sin \alpha^{-1} \pi(\theta - \pi)}{\cos \alpha^{-2} \pi^2} + O(1)$$

at the origin.

The asymptotics of the solutions of boundary integral equations for the Dirichlet and Neumann problems were also found in the case of threedimensional regions with conical points on the boundary (A.V. Levin and the author). In particular, for the straight circular cone with an angle α at the vertex we have the following results.

Let (r, φ, ϱ) be spherical coordinates such that the vertex and the rotational axis of the cone are described by $r = 0$ and $\varphi = \varrho = 0$, respectively. Then if $0 < \alpha < \pi$, we have as $r \to 0$

$$\chi^+(r, \varphi) = c_0 + r^\delta(c_1 \cos \varphi + c_2 \sin \varphi) + O(r^{1+\varepsilon}),$$

$$\varrho^-(r, \varphi) = r^{\delta-1}(D_1 \cos \varphi + D_2 \sin \varphi) + O(1).$$

Here, as above, χ^+ and ϱ^- are the solutions of the boundary integral equations for the internal Dirichlet and external Neumann problems, and δ is the least positive root of the equation $P_\delta^0(-\cos(\alpha/2)) = 0$, where P_s^m is the generalized Legendre function of the first kind.

In the case where $\pi < \alpha < 2\pi$, the asymptotics assume the form

$$\chi^+(r, \varphi) = c_3 + c_4 r^\lambda + O(r^{1+\varepsilon}),$$

$$\varrho^-(r, \varphi) = D_3 r^{\lambda-1} + O(1),$$

λ being the least positive root of the equation $(P_\lambda^1)'(\cos(\alpha/2)) = 0$.

Notice that in either case the gradient of the solution χ^+ and the solution ϱ^- itself become infinite at the corner or conical point. That in the case $\alpha < \pi$ the solution of the corresponding boundary value problem owns no singularity can be revealed by the form of the asymptotics of the functions χ^+ and ϱ^-.

The approach to finding the asymptotics of the solutions of the boundary integral equations for the Laplace equation outlined here applies to other types of equations as well. For instance, Zargaryan [1983a], [1983b] provided asymptotic formulas for the solutions of the systems of singular integral equations for the fundamental problems of plane elasticity theory, and Zargaryan, Maz'ya [1984] determined the asymptotics of the solutions of the boundary equations for the plane Zaremba problem (mixed problem) of the theory of harmonic potentials.

Consider the *external Zaremba problem*

$$\Delta u^- = 0 \quad \text{in } \Omega^-, \qquad u^- = \varphi^- \quad \text{on } S_1, \qquad \partial u^-/\partial \mathbf{n} = \psi^- \quad \text{on } S_2,$$

where S_1 is the union of a finite number of open arcs $p_{2j-1}p_{2j}$ $(1 \leq j \leq m)$ and $S_2 := S \backslash \bar{S}_1$. A bounded solution of this boundary value problem is sought in the form of a sum of two potentials and an unknown constant, i.e.

$$u^-(x) = \int_{S_1} \chi(\xi) \frac{\partial}{\partial \mathbf{n}_\xi} \mathscr{E}(x, \xi) d_\xi S + \int_{S_2} \varrho(\xi) \mathscr{E}(x, \xi) d_\xi S + A.$$

This ansatz leads to the following system of equations for χ, ϱ, A:

$$\chi(x) - 2 \int_{S_1} \chi(\xi) \frac{\partial}{\partial \mathbf{n}_\xi} \mathscr{E}(x, \xi) d_\xi S - 2 \int_{S_2} \varrho(\xi) \mathscr{E}(x, \xi) d_\xi S - 2A$$

$$= - 2\varphi^-(x) \qquad \text{on } S_1,$$

$$\varrho(x) + 2 \int_{S_1} \chi(\xi) \frac{\partial}{\partial \mathbf{n}_\xi} \mathscr{E}(x, \xi) d_\xi S + 2 \int_{S_2} \varrho(\xi) \frac{\partial}{\partial \mathbf{n}_x} \mathscr{E}(x, \xi) d_\xi S$$

$$= 2\psi^-(x) \qquad \text{on } S_2.$$

As in the case of Lame's equations (cf. Sec. 1.5), the unknown vector (χ, ϱ, A) may be expressed by the solutions of certain auxiliary boundary value problems. The asymptotics of the solutions of those problems near the point p_j then imply that

$$\chi \sim C_r^{\pi/[2(\pi+|\pi-\alpha_j|)]}, \qquad \varrho \sim D r^{\pi/[2(\pi+|\pi-\alpha_j|)]-1}.$$

Another approach to the computation of the asymptotics of the solutions of the integral equations of potential theory over piecewise smooth contours without cusps was worked out by Costabel [1983] and Costabel, Stephan [1983]. This method is based on solving the model equation on the legs of the angle by means of the Mellin transform (cf. Lopatinskiĭ [1953] and Eskin [1973]).

Notes on the Bibliography

The classical theory of simple and double layer potentials for harmonic functions in domains with smooth boundaries is contained in the books Kellog [1929], Günter [1953], Vladimirov [1967], Mikhlin [1977], Steklov [1983]. Second order elliptic equations with variable coefficients are studied with the help of the method of potentials in the monograph Miranda [1955].

The books Kupradze et al. [1976], Parton, Perlin [1977], Ugodchikov, Khutoryanskiĭ [1979], Burchuladze, Gegelia [1985] are devoted to the integral equations consistently used in the mechanics of deformable solid bodies and, in particular, in elasticity theory. The theory of hydrodynamic potentials and boundary equations for the Stokes equations is reflected in the books Ladyzhenskaya [1970], Belonosov, Chernous [1985].

The investigation of boundary integral equations on non-smooth boundaries has its beginnings in the works Zaremba [1904], Carleman [1916], Radon [1919b]. For the basic results on the solvability of the boundary equations of potential theory in the space C on irregular curves and surfaces we refer to the books Burago, Maz'ya [1967], Danilyuk [1975], Král [1980]. The monograph Danilyuk [1975] also deals with these questions (in the plane case) for the spaces L_p.

In the case of Lipschitz surfaces, the L_p theory of boundary integral equations was worked out in the papers Verchota [1984], Calderón [1985] for harmonic potentials and in the works Kenig [1984], Fabes [1985] for the Lamé and Stokes equations. Boundary equations in the space L_p on surfaces of the class C^1 are thoroughly studied in Carleman [1916], Fabes, Jodeit, Riviere [1978]. To become acquainted with other approaches to boundary value problems in Lipschitz domains consult, for example, the papers Dahlberg [1979], Costabel [1985].

The works Maz'ya [1981a], [1985], [1986], Zargaryan [1983a], [1983b], Zargaryan, Maz'ya [1983], [1984], Grachev, Maz'ya [1986], [1989], Maz'ya, Solov'ev [1988a], [1988b] deal with several aspects of the method described in Chap. 4, which bases the investigation of integral equations over piecewise smooth manifolds upon the theory of boundary value problems.

There is extensive literature on the application of the method of potentials to a great amount of different boundary value problems. The boundary integral equations for biharmonic potentials are considered in Mikhlin [1949], Lopatinskiĭ [1953], Panich [1960], [1961], [1966], Muskhelishvili [1968], Parton, Perlin [1977], Cohen, Gosselin [1983]. The works Kleinman, Wendland [1977], Romanov [1980], Angell, Kleinman [1982], Costabel, Stephan [1983], Ahner [1986] focus on the Helmholtz equation and the papers John [1950], Vainberg, Maz'ya [1973], Maz'ya [1977], Kuznetsov, Maz'ya, [1988] concentrate on the integral equations of the theory of surface waves. The boundary equations for the heat potentials are covered by Müntz [1934], Tikhonov [1938], Mikhlin [1949], Kamynin [1965], [1966], Dont [1975a], [1975b], [1976], [1981], [1983], Veselý [1973], [1975], Fabes, Jodeit, Lewis [1977] and their generalizations to parabolic equations (and systems of such equations) are the subject of Mikhaĭlov [1960], Èĭdel'man [1964], Solonnikov [1965]. A potential method for the wave equation is elaborated in the papers Mikhlin, Sapozhnikova [1977], Bamberger, Ha Duong [1986].

The classical reference to numerical methods for boundary integral equations is the monograph Kantorovich, Krylov [1962]. Modern accounts of this topic are, e.g. Brebbia, Walker [1980], Banerjee, Butterfield [1981], Crouch, Starfield [1983]. The latter books also embark on applications to engineering sciences. A large number of works is concerned with questions pertaining to the convergence of various algorithms for the numerical solution of boundary equations (see Hsiao, Wendland [1981], Wendland [1981], [1982], for example).

Fairly exhaustive discussions of the history of the boundary integral equations method can be found in Sologub [1975], Lonseth [1977], Banerjee, Butterfield [1981], Crouch, Starfield [1983].

Bibliography*

Achenbach, J.D., Chao, C.C. [1962]: A three-parameter viscoelastic model particularly suited for dynamic problems. J. Mech. Phys. Solids. *10* (3), 245–252. Zbl.109,172

Agmon, S. [1957]: Multiple layer potentials and the Dirichlet problem for higher order elliptic equations in the plane. Commun. Pure Appl. Math. *10*, 179–239. Zbl.81,98

Ahner, J.F. [1986]: A scattering trinity: the reproducing kernel, null-field equations and modified Green's function. Q.J. Mech. Appl. Math. *39* (1), 153–162. Zbl.576.35112

Angell, T.S., Hsiao, G.C., Kleinman, R.E. [1986]: An integral equation for the floating-body problem. J. Fluid Mech. *166*, 161–171. Zbl.601.76011

Angell, T.S., Kleinman, R.E. [1982]: Boundary integral equations for the Helmholtz equation: the third boundary value problem. Math. Methods Appl. Sci. *4*, 164–193. Zbl.505.35026

Angell, T.S., Kleinman, R.E., Král, J. [1986]: Double layer potentials on boundaries with corners and edges. Commentat. Math. Univ. Carol. *27*, 419

Atkinson, F.V. [1951]. Normal solvability of linear equations in normed spaces. Mat. Sb., Nov. Ser. *28* (70), 3–14. Zbl.42,120

Bamberger, A., Ha Duong, T. [1986]: Formulation variationelle espace-temps pour le calcul par potentiel retarde de la diffraction d'une onde acoustique. I. Math. Methods Appl. Sci. *8* (3), 405–435. Zbl.618.35069

Banerjee, P.K., Butterfield, R. [1981]: Boundary Element Methods in Engineering Science. McGraw-Hill: London. Zbl.499.73070

Bazaliĭ, B.V., Shelepov, V.Yu. [1980]: On the spectrum of the double layer potential on a curve of bounded rotation. In: Boundary Value Problems for Differential Equations, Collect. Sci. Works, Naukova Dumka: Kiev 13–30. Zbl.462.47034

Belonosov, S.M., Chernous, K.A. [1985]: Boundary Value Problems for the Navier-Stokes Equations. Nauka: Moscow. Zbl.589.76002

Brebbia, C.A., Walker, C. [1980]: Boundary Element Techniques in Engineering. Butterworth: London. Zbl.444.73065

Burago, Yu.D., Maz'ya, V.G. [1967]: Multidimensional potential theory and the solution of boundary value problems in domains with non-regular boundaries. Zap. Nauchn. Semin. Leningr. Otd. Mat. Inst. Steklova *3*, 1–152. Zbl.172,149. Engl. Translation: Sem. Math. V.A. Steklov Math. Inst. Leningr. 3, 1–68 (1969)

Burago, Yu.D., Maz'ya, V.G., Sapozhnikova, V.D. [1962]: On the double layer potential for non-regular domains. Dokl. Akad. Nauk SSSR *147* (3), 523–525. English transl.: Sov. Math., Dokl. *3* (1962), 1640–1642 (1963). Zbl.142,383

Burago, Yu.D., Maz'ya, V.G., Sapozhnikova, V.D. [1966]: On the theory of simple and double layer potentials for domains with non-regular boundaries. Probl. Mat. Analiza, Kraevye Zadachi Integral. Uravn. Leningrad, 3–34. Zbl.183,112. English transl.: Probl. Math. Analysis *1*, 1–30 (1968)

Burchuladze, T.V., Gegelia, T.G. [1985]: The Development of the Method of Potentials in Elasticity Theory. Metsniereba: Tbilisi. Zbl.594.73026

Calderón, A.P. [1963]: Boundary value problems for elliptic equations. In: Soviet-American Symp. on Partial Differential Equations, Novosibirsk, 303–304, Akad. Nauk SSSR: Moscow

Calderón, A.P. [1977]: Cauchy integrals on Lipschitz curves and related operators. Proc. Natl. Acad. Sci. USA *74*, 4, 1324–1327. Zbl.373.44003

Calderón, A.P. [1985]: Boundary value problems for the Laplace equation in Lipschitzian domains. In: Recent Progress in Fourier Analysis, Sci. Publ.: Amsterdam 33–48. Zbl.608.31001

---

* For the convenience of the reader, references to reviews in Zentralblatt für Mathematik (Zbl.), compiled using the MATH database, and Jahrbuch über die Fortschritte der Mathematik (Jrb.) have, as far as possible, been included in this bibliography.

Calderón, A.P., Calderón, C.P., Fabes, E.B., Jodeit, M., Riviere, N.M. [1978]: Applications of the Cauchy integral along Lipschitz curves. Bull. Am. Math. Soc. *84*, 287–290. Zbl.389.30025

Carleman, T. [1916]: Über das Neumann-Poincarésche Problem für ein Gebiet mit Ecken. Ålmquist and Wiksell: Diss. Uppsala. (193p) Jbr.46,732

Cohen, J., Gosselin, J. [1983]: The Dirichlet problem for the biharmonic equation in a C^1 domain in the plane. Indiana Univ. Math. J. *32* (5), 635–685. Zbl.534.31003

Coifman, R.R., McIntosh, A., Meyer, Y. [1982]: L'intégrale de Cauchy définit un opérateur borné sur L_2 pour les courbes Lipschitziennes. Ann. Math. II. Ser. *116*, 361–387. Zbl.497.42012

Costabel, M. [1983]: Boundary integral operators on curved polygons. Ann. Mat. Pura Appl., IV. Ser. *133*, 305–326. Zbl.533.45009

Costabel, M. [1985]: Boundary integral operators on Lipschitz domains: elementary results. Preprint Nr. 898, Technische Hochschule Darmstadt

Costabel, M., Stephan, E. [1983a]: Curvature terms in the asymptotic expansions for solutions of boundary integral equations on curved polygons. J. Integral Equations *5*, 353–371. Zbl.538.35022

Costabel, M., Stephan, E. [1983b]: A direct boundary equation method for transmission problems. Preprint Nr. 753, Technische Hochschule Darmstadt and J. Math. Anal. Appl. *106*, 367–413 (1985). Zbl.597.35021

Crouch, S.L., Starfield, A.M. [1983]: Boundary Element Methods in Solid Mechanics. G. Allen and Unwin: London. Zbl.528.73083

Dahlberg, B.E.J. [1979]: On the Poisson integral for Lipschitz and C^1 domains. Stud. Math. *66* (1), 13–24. Zbl.422.31008

Danilyuk, I.I. [1975]: Nonregular Boundary Value Problems in the Plane. Nauka: Moscow. Zbl.302.45007

Danilyuk, I.I., Shelepov, V.Yu. [1967]: On the boundedness in L_p of the singular operator with a Cauchy kernel along a curve of bounded rotation. Dokl. Akad. Nauk SSSR *174* (3), 514–517. English transl.: Sov. Math., Dokl. *8*/A, 654–657 (1967). Zbl.155,179

De Hoop, A.T. [1958]: Representation Theorems for the Displacement in an Elastic Solid and Their Application to Elastodynamic Diffraction Theory. Dr. Sci. Thesis, Tech. Hogeschool: Delft

Dont, M. [1972]: Non-tangential limits of the double layer potential. Čas. Pěstovani Mat. *97*, 231–258. Zbl.237.31012

Dont, M. [1975a]: On a heat potential. Czech. Math. J. *25*, 84–109. Zbl.304.35051

Dont, M. [1975b]: On a boundary value problem for the heat equation. Czech. Math. J. *25*, 110–133. Zbl.304.35052

Dont, M. [1976]: A note on a heat potential and the parabolic variation. Čas. Pěstovani Mat. *101*, 28–44. Zbl.325.35043

Dont, M. [1981, 1982]: Third boundary value problem for the heat equation. I, II. Čas. Pěstovani Mat. *106*, 376–394 (1981) and *107*, 7–22 (1982). Zbl.483.35039, Zbl.488.35037

Dont, M. [1983]: Flows of heat and time moving boundary. Čas. Pěstovani Mat. *108*, 146–182. Zbl.547.35053

Duduchava R. [1984]: On general singular integral operators of the plane theory of elasticity, Rendiconti del Seminario Matematico Universita e Politecnico di Torino 42, 3, 15–41

Duduchava R. [1986]: General singular integral equations and basic problems of planar elasticity theory, Trudy Tbilisskogo Matematicheskogo Instituta 82, 45–89

Duvaut, G., Lions, J.-L. [1972]: Les inéquations en mécanique et en physique. Dunod: Paris. Zbl.298.73001

Egorov, Yu.V. [1984]: Linear Differential Equations of the Principal Type. Nauka: Moscow. Zbl.574.35001

Egorov, Yu.V. Kondrat'ev, V.A.: The oblique derivative problem. Mat. Sb. *78* (1969), 148–176. Zbl.165,122. English transl.: Math. USSR, Sb. 7, 139–169 (1969)

Eidel'man, S.D. [1964]: Parabolic Systems. Nauka: Moscow. Zbl.121,319

Eskin, G.I. [1973]: Boundary Value Problems for Elliptic Pseudodifferential Equations. Nauka: Moscow. Zbl.262.35001, Zbl.458.35002. English transl.: Transl. Math. Monogr. *52* (1981).

Fabes, E.B. [1985]: Boundary value problems of linear elastostatics and hydrostatics on Lipschitz domains. Proc. Cent. Math. Anal. Aust. Natl. Univ. *9*, 27–45. Zbl.608.35014

Fabes, E.B., Jodeit, M., Lewis, J.E. [1977]: Double layer potentials for domains with corners and edges. Indiana Univ. Math. J. 26 (1), 95–114. Zbl.363.35010

Fabes, E.B., Jodeit, M., Riviere, N.M. [1978]: Potential techniques for boundary value problems in C^1 domains. Acta Math. 141 (3–4), 165–186. Zbl.402.31009

Federer, H. [1958]: A note on the Gauss-Green theorem. Proc. Am. Math. Soc. 9, 447–451. Zbl.87,273

Federer, H. [1969]: Geometric Measure Theory. Springer-Verlag: Berlin, Heidelberg, New York. Zbl.176,8

Gelfand, I.M., Shilov, G.E. [1958]: Generalized Functions. Fizmatgiz: Moscow. Zbl.91,111. English transl.: Academic Press: New York (1964)

Giraud, G. [1934]: Equations à intégrales principales. Ann. Sci. Ec. Norm. Supér., III. Ser. 51 (3–4), 251–372. Zbl.11,216

Giroire, J., Nedelec, J.C. [1978]: Numerical solution of an exterior Neumann problem using a double layer potential. Math. Comput. 32 (144), 973–990. Zbl.405.65060

Gohberg, I., Krein, M.G. [1957]: The fundamentals on defect numbers, root numbers, and indices of linear operators. Usp. Mat. Nauk 12, Nr. 2 (74), 43–118. Zbl.88,321. English transl.: Transl., II. Ser., Am. Math. Soc. 13, 185–264 (1960)

Grachev, N.V., Maz'ya, V.G. [1986]: On the Fredholm radius of operators of the double layer potential type on piecewise smooth boundaries. Vestn. Leningr. Univ., Ser. I 1986, No. 4, 60–64. Zbl.639.31003. English transl.: Vestn. Leningr. Univ., Math. 19, No. 4, 20–25 (1986)

Grachev, N.V., Maz'ya, V.G. [1988]: Representations and estimates for inverse operators of the integral equations of potential theory on surfaces with conical points. Soobshch. Akad. Nauk Gruz. SSR 132, No. 1, 21–24

Günter, N.M. [1953]: Potential Theory and Its Application to Problems of Mathematical Physics. Gostekhizdat: Moscow. Zbl.52,105. French original: Paris 1934. Zbl.9,113

Hadamard, J. [1932]: Le problème de Cauchy et les équations aux dérivées partielles linéaires hyperboliques. Hermann: Paris. Zbl.6,205

Hörmander, L. [1963]: Linear Partial Differential Operators. Springer-Verlag: Berlin, Heidelberg, New York. Zbl.108,93

Hörmander, L. [1966]: Pseudo-differential operators and non-elliptic boundary problems. Ann. Math. II. Ser. 83 (1), 129–209. Zbl.132,74

Hsiao, G.C., Wendland, W.L. [1981]: The Aubin-Nitsche lemma for integral equations. J. Integral Equations 3, 299–315. Zbl.478.45004

Igumnov, L.A., Khutoryanskiĭ, N.M. [1983]: Numerical investigation of the displacement and stress fields in a viscoelastic medium induced by concentrated impulse sources. In: Applied Problems of Strength and Plasticity. Algorithmization and Automatization of the Solution of Problems of Elasticity and Plasticity, 42–51, Gor'k. State Univ.: Gor'kiĭ

John, F. [1950]: On the motion of floating bodies. II. Commun. Pure Appl. Math. 3 (1), 45–101

Kamynin, L.I. [1965, 1966]: On the smoothness of heat potentials. I, II. Differ. Uravn. 1 (6), 799–839 and 2 (5), 647–687. Zbl.145,140 and Zbl.162,415. English transl.: Differ. Equations 1 (1965), 613–647 (1967) and 2 (1966), 337–356 (1969)

Kantorovich, L.V., Krylov, V.I. [1962]: Approximate Methods of Higher Analysis. Fizmatgiz: Moscow (last Russian edition). Zbl.83,352. English transl.: Interscience: New York (1958)

Kellogg, O.D. [1929]: Foundations of Potential Theory. Springer-Verlag: Berlin. Jrb.55,282

Kenig, C.E. [1984]: Boundary value problems of linear elastostatics and hydrostatics on Lipschitz domains. Semin. Goulaouic-Meyer-Schwartz Equation Deriv. Partielles 1983–1984, Exp. N 21, 1–12. Zbl.547.73007

Khutoryanskiĭ, N.M. [1978]: On the method of generalized delaying potentials and integral equations for non-stationary dynamic problems of elasticity theory. In: Applied Problems of Strength and Plasticity, Vol. 9, 8–18, Gor'k. State Univ.: Gor'kiĭ

Khutoryanskiĭ, N.M. [1979]: The method of boundary time-dependent integral equations for non-stationary dynamic problems of viscoelasticity. In: Applied Problems of Strength and Plasticity. Mechanics of deformable Systems, 11–17, Gor'k. State Univ.: Gor'kiĭ

Khutoryanskiĭ, N.M. [1981]: Boundary integral and integro-differential equations of the second kind for the fundamental mixed problem of elasticity theory. In: Applied Problems of Strength and Plasticity. Statics and Dynamics of Deformable Systems, 3–13, Gor'k. State Univ.: Gor'kiĭ

Khutoryanskiĭ, N.M. [1983]: Boundary time-dependent integral equations of quasistatic viscoelasticity and projection-iteration methods for their solution. In: Applied Problems of Strength and Plasticity. Methods for Solving Problems of Elasticity and Plasticity, 18–30, Gor'k. State Univ.: Gor'kiĭ

Khutoryanskiĭ, N.M. [1985]: The Green tensor of non-stationary dynamic elasticity for the anisotropic homogeneous unbounded medium. In: Applied Problems of Strength and Plasticity. Statics and Dynamics of Deformable Systems, 23–31, Gor'k. State Univ.: Gor'kiĭ

Khutoryanskiĭ, N.M. [1986a]: Discrete equations with positive-definite and symmetric matrices in the boundary element method for the fundamental problems of elasticity theory. In: Applied Problems of Strength and Plasticity. Automatization of Investigations in Strength, 35–40, Gor'k. State Univ.: Gor'kiĭ

Khutoryanskiĭ, N.M. [1986b]: Discrete equations with positive-definite matrices in the boundary element method for non-stationary dynamic problems of elasticity theory. In: Applied Problems of Strength and Plasticity. Algorithmization and Software for Problems of Strength, 20–25, Gor'k. State Univ.: Gor'kiĭ

Khutoryanskiĭ, N.M., Igumnov, L.A. [1982]: Construction of fundamental solutions of the non-stationary dynamic theory of viscoelasticity for certain differential models of a stable isotropic homogeneous medium. In: Applied Problems of Strength and Plasticity. Statics and Dynamics of Deformable Systems, 12–20, Gor'k. State Univ.: Gor'kiĭ

Kleinman, R.E. [1978]: Low frequency electromagnetic scattering. In: Electromagnetic Scattering, 1–28, Academic Press: New York

Kleinman, R.E., Wendland, W. [1977]: On Neumann's method for the exterior Neumann problem for the Helmholtz equation. J. Math. Anal. Appl. 57 (1), 170–202. Zbl.351.35022

Kondrat'ev, V.A. [1967]: Boundary value problems for elliptic equations in regions with conical or angular points. Tr. Mosk. Mat. O. -va. 16, 209–292. Zbl.162.163. English transl.: Trans. Mosc. Math. Soc. 16, 227–313 (1967)

Korn, A. [1901–1902]: Abhandlungen zur Potentialtheorie in 5 Heften. Dummler: Berlin. Jrb.32,770

Kozlov, V.A., Maz'ya, V.G. [1988]: Spectral properties of operator bundles generated by elliptic boundary value problems in a cone. Funkts. Anal. Prilozh. 22, No. 2, 38–46. English transl.: Funct. Anal. Appl. 22, No. 2, 114–121 (1988)

Král, J. [1964]: On the double layer potential in the multidimensional space. Dokl. Akad. Nauk SSSR 159 (6), 1218–1220. English transl.: Sov. Math., Dokl. 5/B, 1677–1680 (1965). Zbl.139,287

Král, J. [1965]: The Fredholm radius of an operator in potential theory. Czech. Math. J. 15 (3–4), 454–473 and 565–588. Zbl.145,370

Král, J. [1966]: The Fredholm method in potential theory. Trans. Am. Math. Soc. 125 (3), 511–547. Zbl.149,79

Král, J. [1970]: Limits of double layer potentials. Atti. Accad. Naz. Lincei, Rend., Cl. Sci. Fis. Mat. Nat., VIII. Ser. 48, 39–42. Zbl.195,116

Král, J. [1976]: On the boundary behavior of double layer potentials. Tr. Semin. S.L. Soboleva 2, 14–34

Král, J. [1980]: Integral Operators in Potential Theory. Lect. Notes Math., 823, Springer-Verlag: Berlin, Heidelberg, New York. Zbl.431.31001

Král, J. [1985]: Boundary regularity and normal derivatives of logarithmic potentials. Report No. 199-A, Univ. of Delaware: Newark. Published in Proc. R. Soc. Edinb., Sect. A 106, 241–258 (1987). Zbl.644.31002

Král, J., Netuka, I. [1977]: Contractivity of C. Neumann's operator in potential theory. J. Math. Anal. Appl. 61, 607–619. Zbl.372.31004

Král, J., Wendland, W. [1986]: Some examples concerning applicability of the Fredholm-Radon method in potential theory. Apl. Mat. 31, 293–308. Zbl.615.31005

Kresin, G.I. [1981]: On the Fredholm radius and the spectral radius. Cand. Dissert., Inst. of Appl. Math. and Mech.: Donezk.

Kresin, G.I., Maz'ya, V.G. [1979]: On the essential norm of an operator of the double layer potential type in the space C_m. Dokl. Akad. Nauk SSSR *246* (2), 272–275. English transl.: Sov. Math., Dokl. *20*, 459–462 (1979). Zbl.447.47035

Kresin, G.I., Maz'ya, V.G. [1981]: On the essential norm of an operator of the double layer potential type in the space C_m. In: Funct. Anal. and Numer. Math., Collect. Artic., Alma-Ata, 131–165. Zbl.494.47030

Kupradze, V.D. [1963]: Potential Methods in Elasticity Theory. Fizmatgiz: Moscow. Zbl.115,187. English transl.: New York (1965)

Kupradze, V.D., Gegelia, T.G., Basheleĭshvili, M.O., Burchuladze, T.V. [1976]: Three-Dimensional Problems of the Mathematical Theory of Elasticity and Thermoelasticity. Nauka: Moscow. Zbl.406.73001

Kuznetsov, N.G., Maz'ya, V.G. [1988]: On the unique solvability of the two-dimensional Neumann-Kelvin problem. Mat. Sb., Nov. Ser. *135*, No. 4, 440–462. English transl: Math. USSR, Sb. *63*, No. 2, 425–446 (1989)

Ladyzhenskaya, O.A. [1961]: Mathematical Questions of the Dynamics of Viscous Incompressible Liquids. Nauka: Moscow. Zbl.106,394. English transl.: New York/London (1963)

Ladyzhenskaya, O.A., Solonnikov, V.A., Ural'tseva, N.N. [1967]: Linear and Quasilinear Equations of Parabolic Type. Nauka: Moscow. Zbl.164.123. English transl.: Transl. Math. Monogr. 23, Am. Maths. Soc. (1968)

Levin, A.V., Maz'ya, V.G. [1989]: Asymptotics of densities of harmonic potentials near the vertex of a cone. Z. Anal. Anwend. *8* (6), 501–514

Lichtenstein, L. [1928]: Über einige Existenzprobleme der Hydrodynamik III. Math. Z. *28*, 387–415. Jrb.54,905

Lions, J.-L., Magenes, E. [1968]: Problemes aux limites non homogenes et applications. Dunod: Paris. Zbl.165,108 (I & II), Zbl.197,67 (III)

Lonseth, A.T. [1977]: Sources and applications of integral equations. SIAM Rev. *19* (2), 241–278. Zbl.363.45001

Lopatinskiĭ, Ya.B. [1953]: On a method of reducing boundary problems for systems of differential equations of elliptic type to regular integral equations. Ukr. Mat. Zh. *5* (2), 123–151. Zbl.52,102

Lopatinskiĭ, Ya.B. [1963]: On a type of singular integral equations. Teor. Prikl. Mat., Vyp. *2*, 53–57

Malyutov, M.B. [1969]: On the boundary value problem of Poincaré. Tr. Mosk. Mat. O. -va *20*, 173–204. Zbl.226.35029

Martin, P.A. [1984]: On the null-field equations for water-wave scattering problems. IMA J. Appl. Math. *33* (1), 55–69. Zbl.547.76022

Maz'ya, V.G. [1972]: On the degenerate oblique derivative problem. Mat. Sb., Nov. Ser. *87*, 417–454. English transl.: Math. USSR, Sb. *16*, 429–469 (1972). Zbl.262.35024

Maz'ya, V.G. [1977]: On the stationary problem of small oscillations of a fluid in the presence of an imbedded solid. Tr. Semin. S.L. Soboleva *2*, 57–79. Zbl.456.35084

Maz'ya, V.G. [1981a]: The integral equations of potential theory in domains with piecewise smooth boundary. Usp. Mat. Nauk *36* (4), 229–230

Maz'ya, V.G. [1981b]: On the solvability of the integral equations of classical elasticity theory in domains with piecewise smooth boundary. In: Proc. Conf. on General Mechanics of Elasticity Theory, Telavi (1981), 55–56, Metsniereba: Tbilisi

Maz'ya, V.G. [1985]: Boundary integral equations of elasticity in domains with piecewise smooth boundaries. Equadiff 6, Proc. Int. Conf., Brno/Czech., Lect. Notes Math. 1192, 235–242. Zbl.614.73088

Maz'ya, V.G. [1986]: Potential theory for the Lamé equations in domains with piecewise smooth boundary. In: Proc. All-Union Symp., Tbilisi, April 21–23 (1982), 123–129, Metsniereba: Tbilisi

Maz'ya, V.G., Paneyakh, B.P. [1974]: Degenerate elliptic pseudodifferential operators and the oblique derivative problem. Tr. Mosk. Mat. O. -va. *31*, 237–295. Zbl.331.35056. English transl.: Trans. Mosc. Math. Soc. *31*, 247–305 (1976).

Maz'ya, V.G., Plamenevskiĭ, B.A. [1974]: On the coefficients in the asymptotics of the solutions of elliptic boundary value problems in a cone. Dokl. Akad. Nauk SSSR *219* (2), 286–289. Zbl.318.35005. English transl.: Sov. Math., Dokl. *15*, 1570–1575 (1975)

Maz'ya, V.G., Plamenevskiĭ, B.A. [1975a]: On boundary value problems for the second order elliptic equation in a domain with edges. Vestn. Leningr. Univ., Mat. Mekh. 1/1975, 102–108. Zbl.296.35029. English transl.: Vestn. Leningr. Univ., Math. 8, 99–106 (1980)

Maz'ya, V.G., Plamenevskiĭ, B.A. [1975b]: On the coefficients in the asymptotics of the solutions of elliptic boundary value problems in a cone. Zap. Nauchn. Semin. Leningr. Otd. Mat. Inst. Steklova 52, 110–127. Zbl.351.35010. English transl.: J. Sov. Math. 9, 750–764 (1978)

Maz'ya, V.G., Plamenevskiĭ, B.A. [1977a]: Elliptic boundary value problems on manifolds with singularities. In: Probl. Mat. Anal. 6, 85–142. Zbl.453.58022

Maz'ya, V.G., Plamenevskiĭ, B.A. [1977b]: On the coefficients in the asymptotics of the solutions of elliptic boundary value problems in domains with conical points. Math. Nachr. 76, 29–60. Zbl.359.35024. English transl.: Transl., II. Ser., Am. Math. Soc. 123, 57–88 (1984)

Maz'ya, V.G., Plamenevskiĭ, B.A. [1978a]: L_p-estimates of the solutions of elliptic boundary value problems in domains with edges. Tr. Mosk. Mat. O. -va. 37, 49–93. Zbl.441.35028. English transl.: Trans. Mosc. Math. Soc. 1, 49–97 (1980)

Maz'ya, V.G., Plamenevskiĭ, B.A. [1978b]: Estimates of the Green function and Schauder estimates of the solutions of elliptic boundary value problems in a dihedral angle. Sib. Mat. Zh. 19 (5), 1064–1082. English transl.: Sib. Math. J. 19, 752–764 (1979). Zbl.408.35014

Maz'ya, V.G., Plamenevskiĭ, B.A. [1978c]: Schauder estimates of the solutions of elliptic boundary value problems in domains with edges on the boundary. Tr. Semin. S.L. Soboleva 2, 69–102. English transl.: Transl., II. Ser., Am. Math. Soc. 123, 141–169 (1984). Zbl.423.35021

Maz'ya, V.G., Plamenevskiĭ, B.A. [1979]: On the asymptotics of fundamental solutions of elliptic boundary value problems in domains with conical points. In: Probl. Math. Anal. 7, 100–145. English transl.: Sel. Math. Sov. 4, 363–397 (1985). Zbl.417.35014

Maz'ya, V.G., Plamenevskiĭ, B.A. [1981]: On the properties of solutions of three-dimensional problems of elasticity and hydrodynamics in domains with isolated singular points. Din. Sploshnoj Sredy 50, 99–120. English transl.: Transl., II. Ser., Am. Math. Soc. 123, 109–123 (1984). Zbl.561.73020

Maz'ya, V.G., Plamenevskiĭ, B.A. [1983]: The first boundary value problem for the classical equations of mathematical physics in domains with piecewise smooth boundary. I, II. Z. Anal. Anwend. 2 (4), 335–359 and 2 (6), 523–551. Zbl.532.35065 and Zbl.554.35099

Maz'ya, V.G., Sapozhnikova, V.D. [1964]: A remark on the regularization of the singular system of isotropic elasticity. Vestn. Leningr. Univ. 19, No 7, 165–167; Zbl.127,148. Correction: Vestn. Leningr. Univ. 19/1977, p. 160

Maz'ya, V.G., Solov'ev, A.A. [1988a]: Asymptotics of the solution of the integral equation of the Neumann problem in a place domain with cusps on the boundary. Soobshch. Akad. Nauk Gruz. SSR 130 (1), 17–20. Zbl.647.45003

Maz'ya, V.G., Solov'ev, A.A. [1988b]: Solvability of the integral equation of the Dirichlet problem in a plane domain with cusps on the boundary. Dokl. Akad. Nauk SSSR 298 (6), 1312–1315

Mikhaĭlov, V.P. [1960]: Solution of the mixed problem for parabolic systems by the potential method. Dokl. Akad. Nauk SSSR 132 (2), 291–294. English transl.: Sov. Mat., Dokl. 1/A, 556–560 (1960). Zbl.144,350

Mikhlin, S.G. [1949]: Integral Equations and Their Applications. Gostekhizdat: Moscow. Zbl.39,113

Mikhlin, S.G. [1962]: Vorlesungen über lineare Integralgleichungen. VEB Deutscher Verlag der Wissenschaften: Berlin. (Russian Original: Moscow 1959). Zbl.87,100

Mikhlin, S.G. [1962]: Multidimensional Singular Integrals and Integral Equations. Fizmatgiz: Moscow. Zbl.105,303. English tansl.: Pergamon Press: New York (1965)

Mikhlin, S.G. [1977]: Linear Partial Differential Equations. Vysshaya Shkola: Moscow

Mikhlin, S.G. Morozov, N.F., Paukshto, M.V. [1986]: Boundary Integral Equations and Problems of Elasticity Theory. Leningr. State Univ.: Leningrad

Mikhlin, S.G. Sapozhnikova, V.D. [1977]: Potentials of the wave equation. I, II. Izv. Vyssh. Uchebn. Zaved., Mat. 9, 48–64 and 10, 100–108. Zbl.379.35038 and Zbl.384.35037

Miranda, C. [1955]: Equazioni alle derivate parziali di tipo ellittico. Springer-Verlag: Berlin. Zbl.65,85

Müntz, Ch.H. [1932]: Sur la résolution du problème dynamique de l'élasticité. C. R. Acad. Sci., Paris 194 (17), 1456–1459. Zbl.4,315

Müntz, G.M. [1934]: Integral Equations. GTTI: Leningrad, Moscow

Muskhelishvili, N.I. [1966]: Some Basic Problems of the Mathematical Theory of Elasticity. 5th edition, Nauka: Moscow. English transl.: Noordhoff: Groningen (1953). Zbl.52,414

Muskhelishvili, N.I. [1968]: Singular Integral Equations. 3rd edition, Nauka: Moscow. English transl.: Noordhoff: Groningen (1963). Zbl.51,332, Zbl.174,162

Natroshvili, D.G. [1979]: On fundamental matrices of the equations of steady-state oscillations and pseudo-oscillations of the anisotropic elasticity theory. Soobshch. Akad. Nauk Gruz. SSR 96 (1), 49–52. Zbl.416.35018

Natroshvili, D.G. [1981]: On an integral equation of the first kind. Soobshch. Akad. Nauk Gruz. SSR 102 (3), 565–568. Zbl.532,73020

Nečas, J. [1967]: Les méthods directes en théorie des équations élliptiques. Academia: Prague

Netuka, I. [1971]: Smooth surfaces with infinite cyclic variation. Čas. Pěstovani Mat. 96, 86–101. Zbl.204,80

Netuka, I. [1972a]: Generalized Robin problem in potential theory. Czech. Math. J. 22, 312–324. Zbl.241.31008

Netuka, I. [1972b]: An operator connected with the third boundary value problem in potential theory. Czech. Math. J. 22, 462–489. Zbl.241.31009

Netuka, I. [1972c]: The third boundary value problem in potential theory. Czech. Math. J. 22, 554–580. Zbl.242.31007

Netuka, I. [1974]: Double layer potentials and the Dirichlet problem. Czech. Math. J. 24, 59–73. Zbl.308.31008

Netuka, I. [1975]: Fredholm radius of a potential theoretic operator for convex sets. Čas. Pěstovani Mat. 100, 374–383. Zbl.314.31006

Nikol'skiĭ, S.M. [1943]: Linear equations in linear normed spaces. Izv. Akad. Nauk SSSR, Ser. Mat. 7 (3), 147–166. Zbl.61,263

Nowacki, W. [1975]: Elasticity Theory. Mir: Moscow. Zbl.342.73004

Odqvist, F.K.G. [1930]: Über die Randwertaufgaben der Hydrodynamik zäher Flüssigkeiten. Math. Z. 32, 329–375. Jrb.56,713

Panich, O.I. [1960]: On potentials for the polyharmonic equation of fourth order. Mat. Sb., Nov. Ser. 50, 335–368. Zbl.103,70

Panich, O.I. [1961, 1962]: Solution of the fundamental boundary value problem for the poly-harmonic equation of the fourth order by the potential method. Izv. Vyssh. Uchebn. Zaved., Mat. 3, 80–90, 4, 66–77, 6, 89–96 (1961); 1, 118–129 (1962). Zbl.121,80 and Zbl.171,90

Panich, O.I. [1966]: Equivalent regularization and solvability of normally solvable boundary value problems with vanishing index for polyharmonic equations and strongly elliptic systems of the second order in the plane. Sib. Mat. Zh. 7 (3), 591–619. Zbl.163,138

Parton, V.Z., Perlin, P.I. [1977]: Integral Equations in Elasticity. Nauka: Moscow. English transl.: Mir Publishers: Moscow (1982). Zbl.497.73002

Pobedrya, B.E. [1980]: A new formulation of the problem in mechanics of a deformable solid body under stress. Dokl. Akad. Nauk SSSR 253 (2), 295–297. English transl.: Sov. Math., Dokl. 22, 88–91 (1980). Zbl.495.73012

Poincaré, H. [1910]: Leçons de mécanique céleste, 3. Gauthier-Villars: Paris

Radon, J. [1919a]: Über lineare Funktionaltransformationen und Funktionalgleichungen. Sitzungs-ber. Akad. Wiss., Abt. 2a, Wien 128 (7), 1083–1121. Jrb.47,385

Radon, J. [1919b]: Über die Randwertaufgaben beim logarithmischen Potential. Sitzungsber. Akad. Wiss., Abt. 2a, Wien 128 (7), 1123–1167. Jrb.47,457

Riesz, F., Sz.-Nagy, B. [1955]: Functional Analysis. Frederick Ungar Publishing Co.: New York. French original: Akad. Kiado: Budapest (1952). Zbl.51,84

Romanov, A.V. [1980]: The Neumann method in boundary value problems for the Helmholtz equation. Dokl. Akad. Nauk SSSR 251 (2), 288–291. English transl.: Sov. Math., Dokl. 21, 431–435 (1980). Zbl.448.35030

Ryaben'kiĭ, V.S. [1987]: The Method of Difference Potentials for Some Problems in Continuum Mechanics. Nauka: Moscow. Zbl.631.73069

Sapozhnikova, V.D. [1966]: Solution of the third boundary value problem by potential theory methods for domains with non-regular boundaries. In: Probl. Math. Analiza, Boundary Value Problems and Integral Equations, Leningr., 35–44. Zbl.163,351. English transl.: Probl. Math. Analysis 1, 31–40 (1968)

Seeley, R.T. [1966]: Singular integrals and boundary value problems. Am. J. Math. 88, 781–809. Zbl.178,176

Shelepov, V.Yu. [1969]: On the index of an integral operator of the potential type in the space L_p. Dokl. Akad. Nauk SSSR 186 (6), 1266–1268. English transl.: Sov. Math., Dokl. 10/A, 754–757 (1969). Zbl.187,81

Shelepov, V.Yu. [1983]: On investigations by Ya.B. Lopatinskiĭ's method of matrix integral equations in the space of continuous functions. In: General Theory of Boundary Value Problems. Collect. Sci. Works, Naukova Dumka: Kiev 220–226. Zbl.598.45021

Shippy, D.J. [1963]: Application of the boundary-integral equation method to transient phenomena in solids. In: Soviet-American Symp. on Partial Differential Equations, Novosibirsk, 15–30, Akad. Nauk SSSR: Moscow

Sobolev, S.L. [1974]: Introduction to the Theory of Cubature Formulas. Nauka: Moscow. Zbl.294.65013

Sologub, V.S. [1975]: The Development of the Theory of Elliptic Equations in the XVIIIth and XIXth Centuries. Naukova Dumka: Kiev

Solonnikov, V.A. [1965]: On boundary value problems for linear parabolic systems of general form. Tr. Mat. Inst. Steklova 83, 1–162. Zbl.161,84. English transl.: Proc. Steklov Inst. Math. 83 (1965), 1–184 (1967)

Solonnikov, V.A., Zajonczkowski, W. [1983]: Estimates of the solutions of the Neumann problem for the second order elliptic equation in domains with edges on the boundary. Zap. Nauchn. Semin. Leningr. Otd. Mat. Inst. Steklova 127, 7–48. Zbl.518.35030. English transl.: J. Sov. Math. 27, 2561–2586 (1984)

Steklov, V.A. [1983]: The Basic Problems of Mathematical Physics. Nauka: Moscow. Zbl.547.35004

Stoker, J.J. [1957]: Water Waves. Interscience Publishers: New York. Zbl.78,408

Tikhonov, A.N. [1938]: On the equation of heat conduction for several variables. Bull. Moscow State Univ., Sec. A, 1 (9), 1–45. Zbl.24,112

Ugodchikov, A.G., Khutoryanskiĭ, N.M. [1979]: On an approach to solving mixed boundary value problems of elasticity theory by the potential method. In: Proc. All-Union Conf. on Elasticity Theory, 345–347, Akad. Nauk Arm. SSR: Erevan

Ugodchikov, A.G., Khutoryanskiĭ, N.M. [1986]: The Boundary Element Method in Mechanics of the Deformable Solid Body. Kazan State Univ.: Kazan

Vaĭnberg, B.R., Maz'ya, V.G. [1973]: On the problem of the motion of a body immersed in a fluid. Tr. Mosk. Mat. O. -va. 28, 35–56. English transl.: Trans. Mosc. Math. Soc. 28, 33–55 (1975). Zbl.285.35020

Vekua, I.N. [1948]: New Methods of Solving Elliptic Equations. Gostekhizdat: Moscow, Leningrad. Zbl.41,62

Verchota, G. [1984]: Layer potentials and regularity for the Dirichlet problem for Laplace's equation in Lipschitz domains. J. Funct. Anal. 59 (3), 572–611. Zbl.589.31005

Veryuzhskiĭ, Yu.V. [1978]: Numerical Potential Methods in Some Problems of Applied Mechanics. Vishcha Shkola: Kiev. Zbl.413.73077

Verzhbinskiĭ, G.M., Maz'ya, V.G. [1974]: On the closure in L_p of the operator of the Dirichlet problem in a domain with conical points. Izv. Vyssh. Uchebn. Zaved., Mat. 6, 8–19. Zbl.304.35028

Veselý, J. [1973]: On the heat potential of the double distribution. Čas. Pěstovani Mat. 98, 181–198. Zbl.256.31006

Veselý, J. [1975]: On a generalized heat potential. Czech. Math. J. 25, 404–423. Zbl.313.35047

Vladimirov, V.S. [1967]: Equations of Mathematical Physics. Nauka: Moscow. English transl.: Mir Publishers: Moscow (1984). Zbl.207,91

Wehausen, J.V., Laitone, E.V. [1960]: Surface Waves. Encyclopedia of Physics, Vol. 9, Springer-Verlag: Berlin

Wendland, W.L. [1981]: Asymptotic convergence of boundary element methods. Integral equations for mixed boundary value problems. In: The Mathematics of finite elements and applications IV, Proc. Conf., Uxbridge/Middlerex 1981, 281–312. Zbl.561.65086

Wendland, W.L. [1983]: Boundary element methods and their asymptotic convergence. CISM Courses Lect. 277, 135–216. Zbl.618.65109

Zajonczkowski, W., Solonnikov, V.A. [1983]: On the Neumann problem for second order elliptic equations in domains with edges. Zap. Nauchn. Semin. Leningr. Otd. Mat. Inst. Steklova 126, 7–48. Zbl.518.35030. English transl.: J. Sov. Math. 27, 2561–2586 (1984)

Zaremba, S. [1904]: Les fonctions fondamentales de H. Poincaré et méthode de Neumann pour une frontiére composée de polygones curvilignes. J. Math. Pures Appl., V. Ser., 10 (4), 395–444. Jrb.35,355

Zargaryan, S.S. [1983a]: On the asymptotics of the solutions of the system of singular integral equations generated by Lamé's equations in a neighborhood of angular points of the contour. Dokl. Akad. Nauk Arm. SSR 77 (1), 30–35. Zbl.567.73029

Zargaryan, S.S. [1983b]: On the singularities of the system of singular integral equations of plane elasticity in the case of prescribed stresses on the boundary. Dokl. Akad. Nauk Arm. SSR 77 (4), 167–172. Zbl.567.73030

Zargaryan, S.S., Maz'ya, V.G. [1983]: On the singularities of the solutions of the system of integral equations of potential theory for the Zaremba problem. Vestn. Leningr. Univ., Mat. Mekh. 1/1983, 43–48. Zbl.512.31003. English transl.: Vestn. Leningr. Univ., Math. 16, 49–55 (1984)

Zargaryan, S.S., Maz'ya, V.G. [1984]: On the asymptotics of the solutions of the integral equations of potential theory in a neighborhood of angular points of the contour. Prikl. Mat. Mekh. 48 (1), 169–174. English transl.: J. Appl. Math. Mech. 48, 120–124 (1985). Zbl.573.45002

Author Index

Subject Index

Encyclopaedia of Mathematical Sciences
Editor-in-chief: R. V. Gamkrelidze

Dynamical Systems

Volume 1: **D. V. Anosov, V. I. Arnol'd** (Eds.)
Dynamical Systems I
Ordinary Differential Equations and Smooth Dynamical Systems
1988. IX, 233 pp. ISBN 3-540-17000-6

Volume 2: **Ya. G. Sinai** (Ed.)
Dynamical Systems II
Ergodic Theory with Applications to Dynamical Systems and Statistical Mechanics
1989. IX, 281 pp. 25 figs.
ISBN 3-540-17001-4

Volume 3: **V. I. Arnol'd** (Ed.)
Dynamical Systems III
1988. XIV, 291 pp. 81 figs.
ISBN 3-540-17002-2

Volume 4: **V. I. Arnol'd, S. P. Novikov** (Eds.)
Dynamical Systems IV
Symplectic Geometry and its Applications
1989. VII, 283 pp. 62 figs.
ISBN 3-540-17003-0

Volume 5: **V. I. Arnol'd** (Ed.)
Dynamical Systems V
Theory of Bifurcations and Catastrophes
1991. Approx. 280 pp. ISBN 3-540-18173-3

Volume 6: **V. I. Arnol'd** (Ed.)
Dynamical Systems VI
Singularity Theory I
1991. Approx. 250 pp. ISBN 3-540-50583-0

Volume 16: **V. I. Arnol'd, S. P. Novikov** (Eds.)
Dynamical Systems VII
Nonholonomic Dynamical Systems. Integrable Hamiltonian Systems
1991. Approx. 290 pp. ISBN 3-540-18176-8

Several Complex Variables

Volume 7: **A. G. Vitushkin** (Ed.)
Several Complex Variables I
Introduction to Complex Analysis
1989. VII, 248 pp. ISBN 3-540-17004-9

Volume 8: **A. G. Vitushkin, G. M. Khenkin** (Eds.)
Several Complex Variables II
Function Theory in Classical Domains. Complex Potential Theory
1992. Approx. 260 pp. ISBN 3-540-18175-X

Volume 9: **G. M. Khenkin** (Ed.)
Several Complex Variables III
Geometric Function Theory
1989. VII, 261 pp. ISBN 3-540-17005-7

Volume 10: **S. G. Gindikin, G. M. Khenkin** (Eds.)
Several Complex Variables IV
Algebraic Aspects of Complex Analysis
1990. VII, 251 pp. ISBN 3-540-18174-1

Springer-Verlag
Berlin
Heidelberg
New York
London
Paris
Tokyo
Hong Kong
Barcelona

Encyclopaedia of Mathematical Sciences
Editor-in-chief: R. V. Gamkrelidze

Analysis

Volume 13: **R. V. Gamkrelidze** (Ed.)
Analysis I
Integral Representations and Asymptotic Methods
1989. VII, 238 pp. ISBN 3-540-17008-1

Volume 14: **R. V. Gamkrelidze** (Ed.)
Analysis II
Convex Analysis and Approximation Theory
1990. VII, 255 pp. 21 figs.
ISBN 3-540-18179-2

Volume 26: **S. M. Nikol'skij** (Ed.)
Analysis III
Spaces of Differentiable Functions
1991. VII, 221 pp. 22 figs.
ISBN 3-540-51866-5

Volume 15: **V. P. Khavin, N. K. Nikol'skij** (Eds.)
Commutative Harmonic Analysis I
General Survey. Classical Aspects
1991. Approx. 290 pp. ISBN 3-540-18180-6

Volume 19: **N. K. Nikol'skij** (Ed.)
Functional Analysis I
Linear Functional Analysis
1992. Approx. 300 pp. ISBN 3-540-50584-9

Volume 20: **A. L. Onishchik** (Ed.)
Lie Groups and Lie Algebras I
Foundations of Lie Theory. Lie Transformation Groups
1992. Approx. 235 pp. ISBN 3-540-18697-2

Algebra

Volume 11: **A. I. Kostrikin, I. R. Shafarevich** (Eds.)
Algebra I
Basic Notions of Algebra
1989. V, 258 pp. 45 figs. ISBN 3-540-17006-5

Volume 18: **A. I. Kostrikin, I. R. Shafarevich** (Eds.)
Algebra II
Noncommutative Rings. Identities
1991. Approx. 240 pp. ISBN 3-540-18177-6

Topology

Volume 12: **D. B. Fuks, S. P. Novikov** (Eds.)
Topology I
General Survey. Classical Manifolds
1991. Approx. 310 pp. ISBN 3-540-17007-3

Volume 24: **S. P. Novikov, V. A. Rokhlin** (Eds.)
Topology II
Homotopies and Homologies
1992. Approx. 235 pp. ISBN 3-540-51996-3

Volume 17: **A. V. Arkhangel'skij, L. S. Pontryagin** (Eds.)
General Topology I
Basic Concepts and Constructions. Dimension Theory
1990. VII, 202 pp. 15 figs. ISBN 3-540-18178-4

Springer-Verlag
Berlin
Heidelberg
New York
London
Paris
Tokyo
Hong Kong
Barcelona